WHAT EVERY CHEMICAL TECHNOLOGIST WANTS TO KNOW ABOUT...

Volume IV

CONDITIONERS, EMOLLIENTS AND LUBRICANTS

Compiled by

Michael and Irene Ash

Chemical Publishing Co., Inc.
New York, N.Y.

Conditioners, Emollients and Lubricants Volume 4

ISBN: 978-0-8206-0053-6

Chemical Publishing Company:
www.chemical-publishing.com
www.chemicalpublishing.net

First Edition:
© **Chemical Publishing Company, Inc.** - New York 1990
Second Impression:
Chemical Publishing Company, Inc. - 2011

Printed in the United States of America

PREFACE

This reference book is the fourth volume in the set of books entitled WHAT EVERY CHEMICAL TECHNOLOGIST WANTS TO KNOW . . . SERIES. This compendium serves a unique function for those involved in the chemical industry—it provides the necessary information for making the decision as to which trademark chemical product is most suitable for a particular application.

The chemicals included in this fourth book of the series have their major function as conditioners, emollients, and lubricants, however, complete cross-referencing is provided for the multiple functions of all the chemicals.

The first section which is the major portion of each volume contains the most common generic name of the chemicals as the main entry. All these generic entries are in alphabetical order. Synonyms for these chemicals are then listed. The CTFA name appears alongside the appropriate generic name. The structural and/or molecular formula of the chemical is listed whenever possible. The generic chemical is sold under various tradenames and these are listed here in alphabetical order for ease of reference along with their manufacturer in parentheses. The *Category* subheading lists all the possible functions that the chemical can serve. Because of differences in form, activity, etc., individual tradenames of the generic chemical are used in particular applications more frequently. These are delineated in the *Applications* section. The differences in properties, toxicity/handling, storage/handling, and standard packaging are specified in the subsequent sections wherever distinguishing characteristics are known.

The second section of the volume TRADENAME PRODUCTS AND GENERIC EQUIVALENTS helps the user who only knows a chemical by one tradename to locate its main entry in section 1. The user can look up this tradename in this section of the book and be referred to the appropriate, main-entry, generic chemical name.

The third section GENERIC CHEMICAL SYNONYMS AND CROSS REFERENCES provides a way of locating the main entries by knowing only one of the synonyms. If the generic chemical is not in the volume, it will refer you to the volume in which it is contained.

The fourth section TRADENAME PRODUCT MANUFACTURERS lists the full addresses of the companies that manufacture or distribute the tradename products found in the first section.

The following is a list of the six volumes that comprise this series:

This series has been made possible through long hours of research and compilation and the dedication and tireless efforts of Roberta Dakan who helped make this distinctive series possible. Our appreciation is extended to all the chemical manufacturers and distributors who supplied the technical information.

M. and I. Ash

NOTE

The information contained in this series is accurate to the best of our knowledge; however, no liability will be assumed by the publisher for the correctness or comprehensiveness of such information. The determination of the suitability of any of the products for prospective use is the responsibility of the user. It is herewith recommended that those who plan to use any of the products referenced seek the manufacturer's instructions for the handling of that particular chemical.

OTHER BOOKS BY MICHAEL AND IRENE ASH

ABBREVIATIONS

@	at
anhyd.	anhydrous
APHA	American Public Health Association
approx.	approximately
aq.	aqueous
ASTM	American Society for Testing and Materials
avg.	average
B.P.	boiling point
Btu	British thermal unit
C	degrees Centigrade
CAS	Chemical Abstracts Service
cc	cubic centimeter(s)
CC	closed cup
cm	centimeter(s)
cm³	cubic centimeter(s)
COC	Cleveland Open Cup
compd.	compound, compounded
conc.	concentrated, concentration
cP, cps	centipoise
cs, cSt	centistokes
CTFA	Cosmetic, Toiletry and Fragrance Association
DEA	diethanolamine
disp	dispersible, dispersion
dist	distilled
DOT	Department of Transportation
DW	distilled water
EO	ethylene oxide
equiv.	equivalent
F	degrees Fahrenheit
F.P.	freezing point
FDA	Food and Drug Administration
ft³	cubic foot, cubic feet
g	gram(s)
gal	gallon(s)
HLB	hydrophile-lipophile balance
insol.	insoluble
IPA	isopropyl alcohol
kg	kilogram(s)
l, L	liter(s)
lb	pound(s)
M.P.	melting point
M.W.	molecular weight
max	maximum
MEA	monoethanolamine
MEK	methyl ethyl ketone
mfg.	manufacture
MIBK	methyl isobutyl ketone
min	minute(s)
min.	mineral, minimum
MIPA	monoisopropanolamine

misc.	miscible
ml	milliliter(s)
mm	millimeter(s)
NF	National Formulary
no.	number
o/w	oil-in-water
OC	open crucible
PEG	polyethylene glycol
pH	hydrogen-ion concentration
pkgs	packages
PMCC	Pensky Marten closed cup
POE	polyoxyethylene, polyoxyethylated
POP	polyoxypropylene
PPG	polypropylene glycol
pt.	point
R&B	Ring & Ball
RD	Recognized Disclosure
ref.	refractive
rpm	revolutions per minute
R.T.	room temperature
s	second(s)
sol.	soluble, solubility
sol'n.	solution
sp.gr.	specific gravity
SS	stainless steel
std.	standard
SUS	Saybolt Universal seconds
TCC	Taggart closed cup
TEA	triethanolamine
tech.	technical
temp.	temperature
theoret.	theoretical
TLV	threshold limit value
TOC	Taggart open cup
UL	Underwriter's Laboratory
USP	United States Pharmacopoeia
uv, UV	ultraviolet
veg	vegetable
visc.	viscosity, viscous
w/o	water-in-oil
wt	weight
$\approx$	approximately equal to
$<$	less than
$>$	greater than
$\leq$	less than or equal to
$\geq$	greater than or equal to

TABLE OF CONTENTS

Acetylated lanolin (CTFA)

SYNONYMS:
Lanolin, acetates
CAS No.:
61788-48-5
TRADENAME EQUIVALENTS:
Acylan [Croda]
Lanacet 1705 [Henkel/Emery]
Ritacetyl [RITA]
CATEGORY:
Emollient, conditioner
APPLICATIONS:
Cosmetic industry preparations: (Ritacetyl); baby cosmetics (Acylan); hair preparations (Lanacet 1705); hypoallergenic cosmetics (Acylan); skin preparations (Lanacet 1705)
PROPERTIES:
Form:
Paste (Acylan)
Solid (Lanacet 1705; Ritacetyl)
Color:
Yellow (Acylan)
Composition:
100% active (Acylan; Lanacet 1705; Ritacetyl)
Ionic Nature:
Nonionic (Acylan; Lanacet 1705; Ritacetyl)

Acetylated lard glyceride (CTFA)

SYNONYMS:
Glycerides, lard mono-, acetates
CAS No.:
8029-92-3
TRADENAME EQUIVALENTS:
Grindtek AMOS 90 [Grindsted]
Tegin E 66 [Goldschmidt AG]
CATEGORY:
Lubricant, emollient, plasticizer, cosolvent, emulsifier
APPLICATIONS:
Cosmetic industry preparations: (Tegin E 66)
Food applications: food additives (Tegin E 66)
Industrial applications: (Tegin E 66); coatings (Grindtek AMOS 90); plastics (Grindtek AMOS 90)
Industrial cleaners: metal processing surfactants (Grindtek AMOS 90)
PROPERTIES:
Form:
Liquid (Grindtek AMOS 90; Tegin E 66)
Color:
Yellow (Grindtek AMOS 90)
Composition:
100% conc. (Tegin E 66)
Solubility:
Sol. in ethanol (Grindtek AMOS 90)
Sol. in paraffin oil (Grindtek AMOS 90)
Sol. in peanut oil (Grindtek AMOS 90)
Partly sol. warm in propylene glycol (Grindtek AMOS 90)
Sol. in white spirit (Grindtek AMOS 90)
Ionic Nature:
Nonionic (Tegin E 66)
HLB:
1.8 (Grindtek AMOS 90)
2–3 (Tegin E 66)

Acetyl monoethanolamide

SYNONYMS:
Acetamide MEA (CTFA)
Acetamide, N-(2-hydroxyethyl)-
N-Acetyl ethanolamine
N-(2-hydroxyethyl) acetamide

EMPIRICAL FORMULA:

$C_4H_9NO_2$

STRUCTURE:

CAS No.:

142-26-7

TRADENAME EQUIVALENTS:

Incromectant AMEA-70 [Croda]

Lipamide MEAA [Lipo]

Schercomid AME, AME-70 [Scher]

CATEGORY:

Conditioner, solvent, humectant, clarifying agent, intermediate, coupling agent, dispersant

APPLICATIONS:

Cosmetic industry preparations: (Incromectant AMEA-70); beauty masks (Lipamide MEAA); conditioners (Incromectant AMEA-70; Schercomid AME, AME-70); creams and lotions (Incromectant AMEA-70; Lipamide MEAA); hair preparations (Schercomid AME, AME-70); shampoos (Incromectant AMEA-70); skin preparations (Schercomid AME, AME-70); wave sets (Lipamide MEAA)

Industrial applications: pigments (Schercomid AME, AME-70)

Pharmaceutical applications: antiperspirant/deodorant (Lipamide MEAA); depilatories (Lipamide MEAA)

PROPERTIES:

Form:

Liquid (Incromectant AMEA-70)

Clear liquid (Schercomid AME, AME-70)

Color:

Gardner 2.0 max. (Schercomid AME, AME-70)

Odor:

Mild organoleptic (Schercomid AME, AME-70)

Composition:

70% active in water (Incromectant AMEA-70; Schercomid AME-70)

95% active min. (Schercomid AME)

Solubility:

Sol. in alcohols (Schercomid AME, AME-70)

Sol. in diols (Schercomid AME, AME-70)

Sol. in glycol ethers (Schercomid AME, AME-70)

Sol. in glycols (Schercomid AME, AME-70)

Sol. in ketones (Schercomid AME)

Sol. in polyols (Schercomid AME, AME-70)

Acetyl monoethanolamide (cont'd.)

Sol. in triols (Schercomid AME-70)
Sol. in water (Schercomid AME, AME-70)

Ionic Nature:
Nonionic (Schercomid AME, AME-70)

Sp.gr.:
1.07–1.17 (Incromectant AMEA-70)
1.1 ± 0.01 (Schercomid AME-70)
1.120 (Schercomid AME)

Density:
9.2 lb/gal (Schercomid AME-70)
9.3 lb/gal (Incromectant AMEA-70; Schercomid AME)

Flash Pt.:
> 180 C (OC) (Schercomid AME); (anhyd.) (Schercomid AME-70)

Acid No.:
10.0 max. (Schercomid AME, AME-70)

Alkali No.:
15.0 max. (Schercomid AME, AME-70)

Storage Stability:
1 yr min. shelf life in closed containers (Schercomid AME, AME-70)

Ref. Index:
1.4395 (Schercomid AME-70)
1.4700 (Schercomid AME)

pH:
6.0–8.5 (Incromectant AMEA-70; Schercomid AME-70); (50% aq. sol'n.) (Schercomid AME)

TOXICITY/HANDLING:
Conc. surface active agent—avoid prolonged contact with skin (Incromectant AMEA-70)

STORAGE/HANDLING:
Store in a cool, dry place (Incromectant AMEA-70)

STD. PKGS.:
55-gal (450 lb net) Blo-lined drums (Incromectant AMEA-70)

AMP isostearic hydrolyzed animal protein (CTFA)

SYNONYMS:
Isostearic hydrolyzed animal protein, aminomethyl propanol salt

CAS No.:
977010-74-4

AMP isostearic hydrolyzed animal protein *(cont'd.)*

TRADENAME EQUIVALENTS:
Crotein AD, AD Anhyd., ADX [Croda]
CATEGORY:
Conditioner, emulsifier, emollient
APPLICATIONS:
Cosmetic industry preparations: hair preparations (Crotein AD, AD Anhyd., ADX);
skin preparations (Crotein AD, AD Anhyd., ADX)
PROPERTIES:
Form:
Clear liquid (Crotein AD, AD Anhyd., ADX)
Color:
Yellow (Crotein AD, AD Anhyd., ADX)
Composition:
24–26% solids in ethanol/water (Crotein AD)
27–33% solids in ethanol (Crotein AD Anhyd.)
36–40% solids in ethanol (Crotein ADX)
Solubility:
Sol. in alcohols (Crotein AD, ADX)
Sol. in oil (Crotein AD)
Sp.gr.:
0.830–0.850 (Crotein AD Anhyd.)
0.855–0.870 (Crotein ADX)
0.885–0.900 (Crotein AD)
Acid No.:
29–39 (Crotein AD)
35–50 (Crotein AD Anhyd.)
55–70 (Crotein ADX)
pH:
8.0–9.0 (Crotein AD Anhyd.)
8.4–9.2 (Crotein AD, ADX)

Behenamidopropyl dimethylamine (CTFA)

SYNONYMS:
Dimethylaminopropyl behenamide
N-[3-(Dimethylamino) propyl] docosanamide
Docosanamide, N-[3-(dimethylamino) propyl]-

EMPIRICAL FORMULA:
$C_{27}H_{56}N_2O$

STRUCTURE:

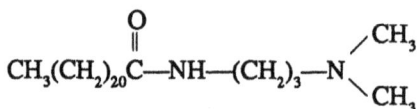

CAS No.:
RD No.: 977063-18-5

TRADENAME EQUIVALENTS:
Incromine BB [Croda]
Lexamine B-13 [Inolex]
Mackine 601 [McIntyre]
Schercodine B [Scher]

CATEGORY:
Conditioner, emulsifier, emollient, lubricant, moisturizer

APPLICATIONS:
Cosmetic industry preparations: conditioners (Lexamine B-13; Mackine 601); hair preparations (Incromine BB; Lexamine B-13; Mackine 601; Schercodine B); shampoos (Lexamine B-13); skin preparations (Incromine BB; Schercodine B)

PROPERTIES:

Form:
Solid (Mackine 601)
Flake (Incromine BB; Lexamine B-13)
Hard wax (Schercodine B)

Color:
Yellow (Incromine BB)
Tan (Schercodine B)

Composition:
98% amide min. (Schercodine B)
100% active (Incromine BB; Lexamine B-13; Mackine 601)

Ionic Nature:
Nonionic (Mackine 601)

6

Behenamidopropyl dimethylamine *(cont'd.)*

Cationic (Lexamine B-13; Schercodine B)
Nonionic/cationic (Incromine BB)
M.W.:
394 (Schercodine B)
420–450 (Incromine BB)
M.P.:
63–68 C (Schercodine B)
70–72 C (Incromine BB)
Acid No.:
5.0 max. (Incromine BB)
Alkali No.:
135–145 (Schercodine B)
TOXICITY/HANDLING:
Avoid prolonged contact with skin (Incromine BB)
STORAGE/HANDLING:
Store in a cool, dry place (Incromine BB)

Behenoyl-PG-trimonium chloride *(CTFA)*

SYNONYMS:
(3-Behenoyloxy-2-hydroxypropyl) trimethyl ammonium chloride
STRUCTURE:

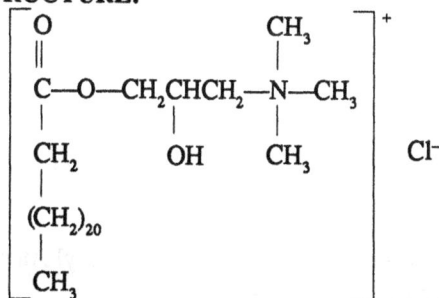

CAS No.:
69537-38-8
TRADENAME EQUIVALENTS:
Akypoquat 131, 132 [Chem-Y GmbH]
CATEGORY:
Conditioner
APPLICATIONS:
Cosmetic industry preparations: hair preparations (Akypoquat 131, 132)

Behenoyl-PG-trimonium chloride *(cont'd.)*

PROPERTIES:
Form:
    Liquid (Akypoquat 131, 132)
Composition:
    70% conc. (Akypoquat 131, 132)
Ionic Nature:
    Cationic (Akypoquat 131, 132)
Biodegradable: (Akypoquat 131)

Behenyl alcohol *(CTFA)*

SYNONYMS:
    1-Docosanol
EMPIRICAL FORMULA:
    $C_{22}H_{46}O$
STRUCTURE:
    $CH_3(CH_2)_{20}CH_2OH$
CAS No.:
    661-19-8
TRADENAME EQUIVALENTS:
    Dehydag Wax 22 (Lanette) [Henkel KGaA]
    Lanette 22 [Henkel Canada]
    Nikkol Behenyl Alcohol 65, 80 [Nikko]
CATEGORY:
    Emollient, consistency agent
APPLICATIONS:
    Cosmetic industry preparations: (Dehydag Wax 22 (Lanette); Nikkol Behenyl Alcohol 65, 80); creams and lotions (Dehydag Wax 22 (Lanette); Lanette 22); emulsions (Dehydag Wax 22 (Lanette)); skin preparations (Lanette 22)
    Pharmaceutical applications: (Dehydag Wax 22 (Lanette); Nikkol Behenyl Alcohol 65, 80); ointments (Dehydag Wax 22 (Lanette))
PROPERTIES:
Form:
    Flakes (Lanette 22)
    Fused flakes (Dehydag Wax 22 (Lanette))
Color:
    White (Lanette 22)
Solubility:
    Sol. in warm ethanol (Nikkol Behenyl Alcohol 65, 80)

Sol. in 2-hexyldecanol (Nikkol Behenyl Alcohol 65, 80)
Sol. in isopropyl myristate (Nikkol Behenyl Alcohol 65, 80)
Sol. in min. oil (Nikkol Behenyl Alcohol 65, 80)
Insol. in water (Nikkol Behenyl Alcohol 65, 80)

Behenyl dimethyl benzyl ammonium chloride

SYNONYMS:
Ammonium, benzyldocosyldimethyl-, chloride
Behenalkonium chloride (CTFA)
Benzyldocosyldimethyl ammonium chloride

STRUCTURE:

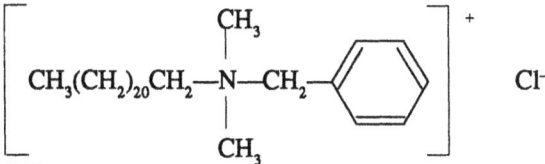

CAS No.:
16841-14-8

TRADENAME EQUIVALENTS:
Incroquat B-85 [Croda]
Incroquat BDQ-25P, Behenyl BDQ/P [Croda Universal]
Kemamine BQ-2802C [Humko/Witco]

CATEGORY:
Conditioner, softener, emulsifier, antisat, dyeing aid, corrosion inhibitor

APPLICATIONS:
Cosmetic industry preparations: conditioners (Kemamine BQ-2802C); creams and lotions (Kemamine BQ-2802C); hair preparations (Incroquat B-85, BDQ-25P; Kemamine BQ-2802C); personal care products (Kemamine BQ-2802C); shampoos (Kemamine BQ-2802C); skin preparations (Incroquat B-85)
Industrial applications: textile/leather processing (Kemamine BQ-2802C)

PROPERTIES:
Form:
Paste (Incroquat BDQ-25P, Behenyl BDQ/P)
Solid (Incroquat B-85)
Color:
Gardner 4 max. (Kemamine BQ-2802C)
Composition:
25% conc. (Incroquat Behenyl BDQ/P)
75% active min. (Kemamine BQ-2802C)
85% conc. (Incroquat B-85)

Behenyl dimethyl benzyl ammonium chloride *(cont'd.)*

Ionic Nature:
Cationic (Kemamine BQ-2802C)
M.W.:
475 (Kemamine BQ-2802C)
pH:
9 max. (5% sol'n.) (Kemamine BQ-2802C)

Behenyl erucate (CTFA)

SYNONYMS:
13-Docosenoic acid, docosyl ester
Docosyl 13-docosenoate
EMPIRICAL FORMULA:
$C_{44}H_{86}O_2$
CAS No.:
18312-32-8
TRADENAME EQUIVALENTS:
Kemester BE [Humko/Witco]
Schercemol BE [Scher]
CATEGORY:
Emollient
APPLICATIONS:
Cleansers: liq. soaps (Kemester BE)
Cosmetic industry preparations: (Kemester BE); creams and lotions (Kemester BE; Schercemol BE); makeup (Kemester BE; Schercemol BE); skin preparations (Schercemol BE)
PROPERTIES:
Form:
Soft solid (Schercemol BE)
Solid (Kemester BE)
Color:
Cream (Schercemol BE)
Gardner 1 (Kemester BE)
Odor:
Slight, typical (Schercemol BE)
Solubility:
Sol. in aliphatic hydrocarbons (Schercemol BE)
Sol. in aromatic hydrocarbons (Schercemol BE)
Sol. in chlorinated hydrocarbons (Schercemol BE)
Sol. in esters (Schercemol BE)
Partly sol. in glycols (Schercemol BE)

Behenyl erucate *(cont'd.)*

Sol. in min. oils (Schercemol BE)
Disp. in polyols (Schercemol BE)
Disp. in triols (Schercemol BE)
Sol. in veg. oils (Schercemol BE)
Insol. in water (Schercemol BE)
M.W.:
361 (theoret.) (Schercemol BE)
Sp.gr.:
0.840 (Schercemol BE)
Density:
7.0 lb/gal (Schercemol BE)
M.P.:
40–44 C (Kemester BE)
44–48 C (Schercemol BE)
Flash Pt.:
> 170 C (OC) (Schercemol BE)
Acid No.:
1.0 (Kemester BE)
2.0 max. (Schercemol BE)
Iodine No.:
50 (Kemester BE)
55 max. (Schercemol BE)
Saponification No.:
80–95 (Schercemol BE)
TOXICITY/HANDLING:
Nontoxic (Schercemol BE)

Behenyl trimethyl ammonium chloride

SYNONYMS:
Behentrimonium chloride (CTFA)
1-Docosanaminium, N,N,N-trimethyl-, chloride
N,N,N,-Trimethyl-1-docosanaminium chloride
EMPIRICAL FORMULA:
$C_{25}H_{54}N \cdot Cl$
STRUCTURE:

$$\left[CH_3(CH_2)_{20}CH_2-\overset{\displaystyle CH_3}{\underset{\displaystyle CH_3}{N}}-CH_3 \right]^+ \quad Cl^-$$

11

Behenyl trimethyl ammonium chloride *(cont'd.)*

CAS No.:
  17301-53-0
TRADENAME EQUIVALENTS:
  Genamin KDM [Hoechst-Celanese]
  Genamin KDM-F [Hoechst-Celanese; Hoechst-Celanese AG]
  Incroquat Behenyl TMC/P [Croda Universal]
CATEGORY:
  Conditioner, antistat, emulsifier
APPLICATIONS:
  Cosmetic industry preparations: conditioners (Genamin KDM); hair preparations
    (Genamin KDM, KDM-F); shampoos (Genamin KDM)
  Pharmaceutical applications: medicated hair preparations (Genamin KDM-F)
PROPERTIES:
Form:
  Paste (Genamin KDM; Incroquat Behenyl TMC/P)
  Waxy flake (Genamin KDM-F)
Color:
  White (Genamin KDM-F)
  Weakly yellowish (Genamin KDM)
Composition:
  25% conc. (Incroquat Behenyl TMC/P)
  80% active (Genamin KDM, KDM-F)
Solubility:
  Sol. cloudy in water (Genamin KDM)
  Sol. in water/alcohol sol'ns. (Genamin KDM, KDM-F)
Ionic Nature:
  Cationic (Genamin KDM-F; Incroquat Behenyl TMC/P)
pH:
  6–7 (1% active in 50% IPA) (Genamin KDM)

Benzalkonium chloride *(CTFA)*

SYNONYMS:
  Alkyl dimethyl benzyl ammonium chloride
STRUCTURE:

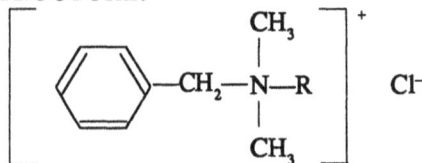

where R represents a mixture of alkyls including all or some of the group beginning with capryl and extending through higher homologs with lauryl, myristyl, and cetyl predominating

CAS No.:

8001-54-5

TRADENAME EQUIVALENTS:

Alkaquat DMB-451, DMB-451 50%, DMB-451 80% [Alkaril]

Arosurf PT50 [Sherex]

Arquad B-50 USP, B-90 USP, B-100, DMCB, DMMCB-50, DMMCB-75 [Akzo Chemie]

Barquat MB-50 (USP), MB-80 (USP), MX-50, MX-80, OJ-50 [Lonza]

Bio-Quat 50-24 (USP), 50-25, 50-28, 50-30, 50-40 (USP), 50-42, 50-60, 50-65, 80-24 (USP), 80-28, 80-40 (USP), 80-42 [Bio-Lab; Lonza]

BTC 50, 50 USP, 65, 65 USP, 100, 835, 2565, 8248, 8249, E-8358 [Millmaster-Onyx]

Carsoquat 621, 621 80% [Lonza]

Catigene CS40, T 50, T 80, T 80 F [Stepan Europe]

Cation G-40 [Sanyo]

Crapol AU-20, AU-21, AU-24 [Pulcra SA]

Cyncal 80% [Hilton-Davis]

Empigen BAC, BAC 90, BCB 50, BCF 80 [Albright & Wilson/Marchon]

FMB 65-15 Quat, 65-28 Quat, 451-5 Quat, 451-8 Quat, 500-15 Quat USP, 4500-5 Quat, 4500-8 Quat [Huntington Lab]

Gardiquat 12H, 1450 (USP), 1480 (USP), SV480 [Albright & Wilson (Australia)]

Hyamine 3500, 3500 NF [Lonza]

JAQ Powdered Quat [Huntington Lab]

Jordaquat 350, 350 80%, 358 [PPG-Mazer]

Lebon GM [Sanyo]

Lutensit K-LC, K-LC80, K-OC [BASF AG]

Maquat LC-12S 50%, 80%, MC-1412 50%, 80%, MC-1416 50%, 80%, MC-6025 50% [Mason]

Protectol KLC 50, 80 [BASF AG]

Quadrilan BC (BP grade) [Harcros UK]

Quatrene CB, CB-50, CB-80, MB-50, MB-80 [Henkel]

Querton 246 [Berol Nobel]

Retarder CA [Hart Chem. Ltd.]

Rewoquat B50 [Rewo GmbH]

Roccal 50% Technical, II 50% (USP), MC-14 (USP) [Hilton-Davis]

Sanisol C, CPR, CR, CR 80%, HTPR, OPR, TPR [Kao]

Swanol CA-101 [Nikko]

Synprolam 35DMBQC [ICI PLC]

Variquat 50AC, 50AE, 50MC, 60LC, 80AC, 80AE, 80MC, 80ME, B345, LC60, LC80 [Sherex]

Benzalkonium chloride (cont'd.)

Vikol RQ [Vikon]
Zoharquat 50, 80 [Zohar]
  Generically sold by:
    [Mason]

CATEGORY:

Conditioner, softener, antimicrobial, bactericide, biocide, sanitizer, germicide, fungicide, mildewcide, algicide, preservative, disinfectant, deodorizing agent, slimicide, detergent, wetting agent, emulsifier, retardant, antistat, flocculant, corrosion inhibitor, surfactant, antistat, retarding agent, leveling agent, raw material, humectant, dispersant

APPLICATIONS:

Cosmetic industry preparations: (Arquad DMMCB-50); hair preparations (Catigene CS40; Cation G-40); medicated cosmetics (Roccal MC-14)

Farm products: (Roccal 50% Technical, II 50%); agricultural oils/sprays (Arosurf PT50); insecticides/pesticides (Variquat 50AC, 50AE, 50MC, 80AC, 80AE, 80MC); soil stabilization (Arquad DMMCB-50)

Food applications: (Alkaquat DMB-451; Querton 246); food/beverage plant cleaners (Alkaquat DMB-451 50%, DMB-451 80%; Arquad B-100; Empigen BAC, BAC 90, BCB 50; Gardiquat 12H; Lutensit K-LC, K-LC 80, K-OC; Rewoquat B50; Roccal 50% Technical, II 50%; Variquat 50AC, 50AE, 50MC)

Household detergents: (Protectol KLC 50, 80; Variquat 50AC, 50AE, 50MC); disinfectant cleaners (Lutensit K-LC, K-LC 80, K-OC); hard surface cleaner (Arquad DMCB; BTC 50, 50 USP, 65, 65 USP, 100, 835, 8248, 8249, E-8358); laundry detergent (Roccal MC-14)

Industrial applications: (Empigen BCF 80; Variquat 50AC, 50AE, 50MC); chemical synthesis (Arosurf PT50; Protectol KLC 50, 80); dyes and pigments (Quadrilan BC; Retarder CA; Rewoquat B50); mining (Arosurf PT50); paper mfg. (Alkaquat DMB-451; Hyamine 3500); petroleum industry (Arquad DMCB; Gardiquat 12H; Quatrene CB, CB-50, CB-80, MB-50, MB-80); preservatives (Alkaquat DMB-451); textile/leather processing (Alkaquat DMB-451 50%, DMB-451 80%; Arquad DMMCB-50; Catigene CS40; Crapol AU-21; Empigen BCB 50; Hyamine 3500; Retarder CA; Rewoquat B50; Roccal II 50%; Synprolam 35DMBQC; Vikol RQ); water treatment (Alkaquat DMB-451, DMB-451 50%, DMB-451 80%; Arquad DMCB; Barquat OJ-50; Bio-Quat 50-24, 50-28, 50-40, 80-24, 80-28, 80-40; BTC 835, 2565, 8248, 8249, E-8358; Crapol AU-20, AU-24; Cyncal 80%; FMB 65-15 Quat, 65-28 Quat, 451-5 Quat, 451-8 Quat, 4500-5 Quat, 4500-8 Quat; JAQ Powdered Quat; Roccal MC-14; Sanisol CPR, CR, CR 80%, HTPR, OPR, TPR; Variquat 60LC, 80AC, 80AE, 80MC, LC60, LC80)

Industrial cleaners: beverage/dairy cleaners (generic—Mason; Alkaquat DMB-451, DMB-451 50%, DMB-451 80%; Empigen BAC, BAC 90, BCB 50; Rewoquat B50); bottle washing (Empigen BAC, BAC 90, BCB 50); institutional cleaners (Alkaquat DMB-451, DMB-451 50%, DMB-451 80%; Arquad B-100; Gardiquat

12H; Lutensit K-LC, K-LC 80, K-OC; Roccal MC-14); sanitizers/germicides (Barquat MB-50, MB-80, MX-50, MX-80; Bio-Quat 50-40, 80-40; BTC 50, 50 USP, 65, 65 USP, 100, 835, 8248, 8249, E-8358; Gardiquat 12H; Hyamine 3500, 3500-NF; Quadrilan BC; Querton 246; Rewoquat B50; Roccal MC-14; Sanisol C, CPR, CR, CR 80%, HTPR, OPR, TPR; Variquat 50AC, 50AE, 50MC, 80AC, 80AE, 80MC, 80ME)

Pharmaceutical applications: (Arosurf PT50; Empigen BAC, BAC 90, BCB 50; FMB 500-15 Quat USP; Hyamine 3500; Querton 246; Sanisol C, CPR, CR, CR 80%, HTPR, OPR, TPR); antidandruff rinses (Alkaquat DMB-451)

PROPERTIES:
Form:

Liquid (generic—Mason; Alkaquat DMB-451, DMB-451 50%; Arosurf PT50; Arquad B-50 USP, B-100, DMCB, DMMCB-50; Barquat MB-50, MB-80, MX-50, MX-80, OJ-50; BTC 50, 50 USP, 65, 65 USP, 835, 2565, 8248, 8249, E-8358; Carsoquat 621; Catigene T50, T80, T80F; Cation G-40; Crapol AU-20, AU-21, AU-24; Empigen BAC, BCB 50, BCF 80; Hyamine 3500, 3500-NF; Jordaquat 350, 350 80%, 358; Lebon GM; Maquat MC-1412 50%, 80%, Maquat MC-1416 50%, 80%, MC-6025 50%; Protectol KLC 50, 80; Quadrilan BC; Quatrene CB, CB-50, CB-80, MB-50, MB-80; Querton 246; Roccal 50% Technical, II 50%; Sanisol C, CPR, CR, OPR, TPR; Swanol CA-101; Synprolam 35DMBQC; Variquat 50AC, 50AE, 60LC, 80AC, 80AE, 80MC, 80ME, B345, LC80; Zoharquat 50, 80)

Clear liquid (Alkaquat DMB-451 80%; Bio-Quat 50-24, 50-25, 50-28, 50-30, 50-40, 50-42, 50-60, 50-65, 80-24, 80-40, 80-42; Carsoquat 621 (80%); Gardiquat 12H, 1450, 1480, SV480; Lutensit K-LC, K-LC 80, K-OC; Retarder CA; Variquat 50MC)

Clear liquid if kept above the gel point (Bio-Quat 80-28)

Low viscosity liquid (Rewoquat B50)

Liq./paste (Catigene CS40; Sanisol HTPR)

Viscous mass (Arquad B-90 USP)

Gel (Empigen BAC 90)

Paste (generic—Mason; Sanisol CR 80%)

Solid (BTC 100)

Powder (Roccal MC-14)

Color:

Water-white (Alkaquat DMB-451 50%; Gardiquat SV480)

Almost water-white (20 C) (Gardiquat 1450, 1480)

White (Roccal MC-14)

Straw (Carsoquat 621 (80%))

Colorless to pale yellow (Gardiquat 12H)

Water-white to pale yellow (Bio-Quat 50-24, 50-25, 50-28, 50-30, 50-40, 50-42, 50-60, 50-65, 80-24, 80-28, 80-40, 80-42; Carsoquat 621)

Pale yellow (Alkaquat DMB-451; Empigen BCB 50; Hyamine 3500-NF; Quadrilan BC; Roccal 50% Technical, II 50%)

Benzalkonium chloride (cont'd.)

Yellowish (Lutensit K-LC 80, K-OC)
APHA 100 max. (Maquat MC-1412 50%, 80%, Maquat MC-1416 50%, 80%)
Gardner 2 max. (Arquad B-50 USP, B-100; Variquat 50MC, 60LC, 80ME, LC80)
Gardner 3 max. (Variquat B345)
Gardner 4 max. (Arquad DMCB)
Gardner 9 max. (Arquad B-90 USP)
VCS 2–5 (Hyamine 3500)

Odor:

Characteristic (Arquad DMCB)
Mild (Roccal 50% Technical)
Mild, characteristic (Quadrilan BC)
Mild pleasant (Roccal II 50%)

Taste:

Bitter (Arquad DMCB)

Composition:

40% active (Cation G-40; Retarder CA)
49.5–51.5% quat (Variquat 60LC)
50% active (Arquad B-50 USP, DMMCB-50; Barquat MB-50, MX-50, OJ-50; Bio-Quat 50-24, 50-25, 50-30, 50-40, 50-42, 50-60, 50-65; Carsoquat 621; Empigen BAC; FMB 65-15 Quat, 451-5 Quat, 500-15 Quat USP, 4500-5 Quat; Gardiquat 1450; Jordaquat 350; Lutensit K-LC, K-OC; Maquat LC-12S 50%, MC-6025 50%; Protectol KLC 50; Quadrilan BC; Quatrene CB-50, MB-50; Rewoquat B50; Roccal 50% Technical, II 50%; Sanisol CR; Synprolam 35DMBQC; Variquat 50MC; Zoharquat 50)
50% active in water (generic—Mason liq.; Hyamine 3500; Maquat MC-1412 50%, MC-1416 50%)
50% active in aq. isopropanol (Arquad B-100)
50% active in ethanol (VBariquat 50AE)
50% active in isopropanol (Bio-Quat 50-28; Variquat 50AC)
50% quat (Arosurf PT50; BTC 50, 50 USP, 65, 65 USP, 835, 2565)
50% active, 10% ethanol, 40% water (Alkaquat DMB-451)
50 ± 1% act. (Crapol AU-21)
50 ± 1% act. in water (Gardiquat 12H)
50 ± 1.5% act. in water (Empigen BCB 50)
50–51% active (Alkaquat DMB-451 50%)
60% quat (Variquat LC60)
75% active (Sanisol CPR, HTPR, OPR, TPR)
75–77% solids (Variquat B345)
79% active in aq. isopropanol (Arquad DMCB)
80% active (Barquat MB-80, MX-80; Bio-Quat 80-24, 80-28, 80-40, 80-42; Emigen BCF 80; FMB 65-28 Quat, 451-8 Quat, 4500-8 Quat; Gardiquat 1480, SV480; Jordaquat 350 80%, 358; Lutensit K-LC 80; Maquat LC-12S 80%; Protectol KLC 80; Quatrene CB, CB-80, MB-80; Sanisol CR 80%; Synprolam 35DMBQC;

Variquat 80MC, LC80; Zoharquat 80)
80% active in IPA (Variquat 80AC)
80% active in ethanol (Variquat 80AE)
80% active in IPA or ethanol (Maquat MC-1412 80%, MC-1416 80%)
80% active, 10% water, 10% alcohol (Carsoquat 621 (80%))
80% quat (BTC 8248, E-8358)
80 ± 2% conc. (Crapol AU-20, AU-24)
80–81% active (Alkaquat DMB-451 80%)
81.5–82.5% quat (Variquat 80ME)
90% active (Arquad B-90 USP; Empigen BAC 90; Roccal MC-14)
90% quat (BTC 8249)
95% active (JAQ Powdered Quat)
95% quat min. (BTC 100)
97–100% (generic—Mason paste)

Solubility:
Sol. in acetone (Alkaquat DMB-451); misc. (Empigen BAC)
Sol. in alcohol (generic—Mason; Lutensit K-LC, K-LC 80, K-OC); sol. in lower alcohols (Cyncal 80%; Roccal 50% Technical, II 50%, MC-14); misc. with alcohol (Empigen BAC); misc. with lower alcohols (Carsoquat 621)
Sol. in many aliphatic solvents (Alkaquat DMB-451)
Sol. in aromatics @ 10% (Alkaquat DMB-451 80%)
Slightly sol. in benzene (generic—Mason)
Sol. in chlorinated hydrocarbons (Lutensit K-LC, K-LC 80)
Sol. in ethanol (Alkaquat DMB-451)
Slightly sol. in ether (generic—Mason)
Sol. in glycols (Cyncal 80%; Roccal 50% Technical, II 50%, MC-14)
Sol. in 5% hydrochloric acid (Lutensit K-LC, K-LC 80)
Sol. in ketones (Cyncal 80%; Roccal 50% Technical, II 50%, MC-14)
Disp. in min. oil @ 1% (Alkaquat DMB-451 80%)
Disp. in min. spirits @ 1% (Alkaquat DMB-451 80%)
Slightly sol. in oil (Lutensit K-LC, K-LC 80, K-OC)
Sol. in organic solvents (Arosurf PT50)
Sol. in perchloroethylene @ 10% (Alkaquat DMB-451 80%)
Sol. in polar solvents (Maquat LC-12S 50%, 80%, MC-1412 50%, 80%, MC-1416 50%, 80%); sol. in common polar organic solvents (Arquad DMCB); misc. with most polar solvents (Carsoquat 621)
Sol. in 5% sodium chloride (Lutensit K-LC, K-LC 80)
Sol. in 5% sodium hydroxide (Lutensit K-LC, K-LC 80, K-OC)
Sol. in water (generic—Mason; Alkaquat DMB-451; Arquad DMCB; Cyncal 80%; Lutensit K-LC, K-LC 80, K-OC; Maquat LC-12S 50%, 80%, MC-1412 50%, 80%, MC-1416 50%, 80%; Roccal 50% Technical, II 50%, MC-14); @ 10% (Alkaquat DMB-451 50%); @ 20 C (Quadrilan BC); disp. in aq. systems (Arosurf PT50); misc. (Carsoquat 621; Empigen BAC; Vikol RQ)

Benzalkonium chloride (cont'd.)

Slightly sol. in white spirit (Lutensit K-LC, K-LC 80, K-OC)
Sol. in xylene (Lutensit K-LC 80)

Ionic Nature:

Cationic (Alkaquat DMB-451; Bio-Quat 50-24, 50-30, 50-40, 50-60, 80-24, 80-28, 80-40; BTC 50, 50 USP, 65, 65 USP, 100, 835, 2565, 8248, 8249, E-8358; Carsoquat 621, 621 (80%); Cation G-40; Crapol AU-20, AU-21, AU-24; Empigen BAC, BAC 90, BCB 50, BCF 80; Hyamine 3500; Lutensit K-LC, K-LC 80, K-OC; Protectol KLC 50, 80; Quadrilan BC; Quatrene CB, CB-50, CB-80, MB-50, MB-80; Retarder CA; Roccal 50% Technical, II 50%, MC-14; Sanisol C, CPR, CR, CR 80%, HTPR, OPR, TPR; Synprolam 35DMBQC; Variquat 50MC, 60LC, 80MC, LC80)

M.W.:

351 (Variquat 60LC)
354 (Arquad DMCB; Variquat LC80)
358 (Arosurf PT50; Maquat MC-1412 50%, 80%; Variquat 50MC, 80MC, 80ME)
359 (Roccal II 50%)
361 (Carsoquat 621 (80%))
380 (Arquad B-100; Maquat MC-1416 50%, 80%)
404 (Roccal MC-14)
417 (Variquat B345)

Sp.gr.:

0.916 (20 C) (Bio-Quat 80-24)
0.922 (20 C) (Bio-Quat 80-28, 80-40)
0.929 (Gardiquat SV480)
0.93 (BTC E-8358
0.935 (Arquad DMCB)
0.94 (BTC 8249; Roccal II 50%)
0.945 (Gardiquat 1480)
0.95 (BTC 100, 8248), (20 C) (Bio-Quat 80-42)
0.96 (Alkaquat DMB-451; BTC 2565; Carsoquat 621), (25/20 C) (BTC 50, 50 USP)
0.961 (20 C) (Bio-Quat 50-60)
0.97 (BTC 65, 65 USP, 835)
0.979 (Bio-Quat 50-28); (20 C) (Bio-Quat 50-40)
0.98 (Arquad B-50 USP; Retarder CA; Roccal 50% Technical); (20 C) (Gardiquat 12H)
0.98–0.99 (20 C) (Bio-Quat 50-42)
0.984 (20 C) (Bio-Quat 50-25)
0.985 (20 C) (Bio-Quat 50-30)
0.987 (20 C) (Bio-Quat 50-65)
0.988 (Bio-Quat 50-24); 20 C (Quadrilan BC)

Density:

0.93 g/ml (Alkaquat DMB-451 80%)
0.96 g/ml (Alkaquat DMB-451 50%)
0.98 g/cm³ (Empigen BCB 50), (23 C) (Lutensit K-LC, K-LC 80, K-OC)

  7.63 lb/gal (Bio-Quat 80-24)
  7.73 lb/gal (Bio-Quat 80-28, 80-40)
  7.8 lb/gal (Roccal II 50%)
  7.8–8.2 lb/gal (Maquat MC-1416 50%, 80%)
  8.0 lb/gal (Carsoquat 621; Hyamine 3500)
  8.0–8.4 lb/gal (Maquat MC-1412 50%, 80%)
  8.01 lb/gal (Bio-Quat 50-60)
  8.16 lb/gal (Bio-Quat 50-28, 50-40)
  8.20 lb/gal (Bio-Quat 50-25; Roccal 50% Technical)
  8.21 lb/gal (Bio-Quat 50-30, 80-42)
  8.23 lb/gal (Bio-Quat 50-24, 50-65)
  8.24 lb/gal (Bio-Quat 50-42)
Visc.:
  42 cps (Hyamine 3500)
  60 cps (Bio-Quat 50-24)
  105 cps (Bio-Quat 50-28, 50-60)
  120 cs (Empigen BCB 50); 20 C (Quadrilan BC)
  350 cps (Bio-Quat 80-24)
  405 cps (Bio-Quat 80-28)
  425 cps (Bio-Quat 50-40)
  106 s (No. 4 cup) (Gardiquat SV480)
  80 mPa•s (23 C) (Lutensit K-LC)
  100 mPa•s (23 C) (Lutensit K-OC)
  500 mPa•s (23 C) (Lutensit K-LC 80)
M.P.:
  60 C min. (Roccal MC-14)
Gel Pt.:
  4–8 C (Bio-Quat 80-28)
Pour Pt.:
  15 F (Hyamine 3500)
  < 0 C (Quadrilan BC)
Flash Pt.:
  None (Bio-Quat 50-25, 50-40)
  75 F (PMCT) (Bio-Quat 80-24, 80-40, 80-42), (PMCC) (Bio-Quat 80-28)
  < 80 F (Arquad B-100, DMMCB-75)
  80 F (PM) (Arquad DMCB)
  81 F (PM) (Variquat B345)
  100 F (BTC E-8358); (PM) (Variquat 80MC, 80ME)
  102 F (PM) (Variquat LC80)
  105 F (TOC) (Hyamine 3500)
  110 F (BTC 2565, 8248)
  116.6 F (Bio-Quat 50-60)
  118 F (BTC 8249)

Benzalkonium chloride (cont'd.)

120 F (PM) (Variquat 50MC)
126 F (BTC 50, 50 USP)
130 F (BTC 835); (PM) (Variquat 60LC)
132 F (BTC 65, 65 USP)
134 F (PMCT) (Bio-Quat 50-30)
135 F (PMCT) (Bio-Quat 50-42, 50-65)
> 200 F (BTC 100); COC (Quadrilan BC)
29 C (PMCC) (Gardiquat SV480)
32 C (PMCC) (Gardiquat 1480)

Cloud Pt.:

–36 F (Arquad DMMCB-75)
–10 C (Lutensit K-LC 80)
–4 C (Lutensit K-LC)
–3 C (Lutensit K-OC)
< 0 C (Empigen BCB 50); (Quick Cool) (Gardiquat SV480)
0 C (Quick Cool) (Gardiquat 12H)
1 C max. (Quick Cool) (Gardiquat 1450, 1480)

Stability:

Good (Alkaquat DMB-451; Carsoquat 621)
Unaffected by light; stable over wide pH range; heat-stable up to b.p. of water (Arquad DMCB)

pH:

6.0 (1% aq. sol'n.) (Lutensit K-LC), (0.1% aq. sol'n.) (Lutensit K-OC)
6.5–9.5 (10% DW) (Alkaquat DMB-451 80%)
6.8–8.0 (10%) (Roccal 50% Technical, II 50%)
7.0 ± 0.5 (1% aq. sol'n.) (Gardiquat 12H), (1% sol'n.) (Gardiquat 1450, 1480)
7.0–8.0 (1% aq.) (Quadrilan BC); (1% sol'n.) (Carsoquat 621 (80%), (10%) (Maquat MC-1412 50%, 80%, Maquat MC-1416 50%, 80%)
7.0–9.0 (Alkaquat DMB-451 50%)
7.0–9.5 (5% aq.) (Empigen BCB 50)
7.2 (1% sol'n.) (Gardiquat SV480)
7.2–8.0 (10%) (Bio-Quat 50-24, 50-25, 50-28, 50-30, 50-40, 50-42, 50-60, 50-65, 80-24, 80-28, 80-40, 80-42)
7.5 ± 0.5 (10%) (Roccal MC-14)

Surface Tension:

33 dynes/cm (1% sol'n.) (Alkaquat DMB-451)
40 dynes/cm (0.01% aq.) (Hyamine 3500)
31 mN/m (23 C, 1 g/l) (Lutensit K-OC)
37 mN/m (23 C, 1 g/l) (Lutensit K-LC, K-LC 80)

Biodegradable: (Arquad DMMCB-75l Quadrilan BC); 80% (Lutensit K-LC, K-LC 80, K-OC)

TOXICITY/HANDLING:

Skin and eye irritant; avoid contact with skin and eyes (Retarder CA)

Causes skin burns, severe eye damage at conc. supplied (Alkaquat DMB-451)

Skin irritant, severe eye irritant (Arquad B-50 USP, B-90 USP)

Irritating to skin and eyes; protective gloves and goggles should be worn (Arquad B-100, DMMCB-75)

Corrosive; causes skin and eye damage; harmful or fatal if swallowed or absorbed through skin; wear protective goggles and gloves; toxic to fish (Arquad DMCB)

May cause irritation on prolonged contact with coc. form; wear protective gloves and goggles; spillages may be slippery (Quadrilan BC)

Avoid contact with eyes, skin, clothing (Carsoquat 621)

Avoid prolonged contact with skin (Gardiquat 12H)

Prolonged contact with skin may cause irritation (Lutensit K-LC, K-LC 80, K-OC)

Avoid contact with skin (Gardiquat 1450, 1480, SV480)

STORAGE/HANDLING:

Avoid contact with strong oxidizing agents, anionics; kills bacteria in biological disposal systems (Arquad B-50 USP, B-90 USP)

Do not mix with strong oxidizing or reducing agents—potentially explosive; store in stainless steel, glass, polythene, or resin-lined equipment (Quadrilan BC)

Flammable (Arquad B-100, DMMCB-50)

Flammable; store away from naked flame (Gardiquat 1450, 1480, SV480)

Flammable; keep away from heat, sparks, open flame; store drums away from heat and cold (Arquad DMCB)

Store in plastic, glass, or coated metal (Lutensit K-LC, K-LC 80, K-OC)

STD. PKGS.:

Drums, T/L (Retarder CA)

55-gal drums (Carsoquat 621)

55-gal epoxy-phenolic lined drums and bulk (Arquad DMCB)

100 kg net open-head polythene drums (Arquad B-50 USP, B-90 USP)

180 kg net lined mild steel closed-head drums (Gardiquat 12H, 1450, 1480, SV480)

N,N-Bis (2-hydroxyethyl) cocamine oxide

SYNONYMS:

Amines, coco alkyl dihydroxyethyl, oxides

Coco di-(hydroxyethyl) amine oxide

Dihydroxyethyl cocamine oxide (CTFA)

Ethanol, 2,2'-iminobis-, N-coco alkyl, N-oxide

2,2'-Iminobisethanol, N-coco alkyl, N-oxide

N,N-Bis (2-hydroxyethyl) cocamine oxide (cont'd.)

STRUCTURE:

CH₂CH₂OH
|
R—N → O
|
CH₂CH₂OH

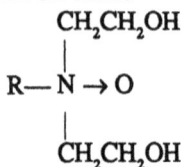

where R represents the coconut radical

CAS No.:

61791-47-7

TRADENAME EQUIVALENTS:

Alkamox C2, C2-0 [Alkaril]

Aromox C/12, C/12-W, CD/12 [Akzo/Armak]

Schercamox CMA [Scher]

CATEGORY:

Emollient, conditioner, lubricant, thickener, wetting agent, emulsifier, foam stabilizer, foaming agent, antistat, corrosion inhibitor, detergent

APPLICATIONS:

Bath products: bubble bath (Schercamox CMA)

Cleansers: body cleansers (Schercamox CMA); soaps (Schercamox CMA)

Cosmetic industry preparations: conditioners (Jorphox KCAO); personal care products (Aromox C/12, C/12-W, CD/12); shampoos (Aromox C/12, C/12-W, CD/12; Schercamox CMA); shaving preparations (Schercamox CMA)

Farm products: agricultural chemicals (Aromox C/12, C/12-W, CD/12)

Industrial applications: chemical specialties (Aromox C/12, C/12-W, CD/12); electroplating (Alkamox C2-0); industrial processing (Aromox C/12, C/12-W, CD/12); lubricating/cutting oils (Aromox C/12, C/12-W, CD/12); metalworking (Aromox C/12, C/12-W, CD/12); paint mfg. (Aromox C/12, C/12-W, CD/12); paper mfg. (Aromox C/12, C/12-W, CD/12); petroleum industry (Aromox C/12, C/12-W, CD/12); plastics (Aromox C/12, C/12-W, CD/12); rubber (Aromox C/12, C/12-W, CD/12); textile/leather processing (Aromox C/12, C/12-W, CD/12)

PROPERTIES:

Form:

Liquid (Alkamox C2)

Clear liquid (Alkamox C2-0; Aromox C/12, C/12-W, CD/12; Schercamox CMA)

Color:

Water-white (Schercamox CMA)

Gardner 1 (Aromox CD/12)

Gardner 2 (Aromox C/12, C/12-W)

Odor:

Mild (Schercamox CMA)

Composition:

29–31% amine oxide (Alkamox C2-0)

30% conc. (Alkamox C2)

40% active (Schercamox CMA); 40% active in water (Aromox C/12-W)

50% active in aq. isopropanol (Aromox C/12, CD/12)

Solubility:

Sol. in alcohols (Schercamox CMA)

Sol. in glycol ethers (Schercamox CMA)

Sol. in glycols (Schercamox CMA)

Disp. in min. oil @ 10% (Alkamox C2-0)

Sol. in polyols (Schercamox CMA)

Sol. in triols (Schercamox CMA)

Sol. in water (Aromox C/12-W; Schercamox CMA); @ 10% (Alkamox C2-0)

Ionic Nature:

Cationic (Aromox C/12, C/12-W, CD/12)

Cationic in nature, nonionic in sol'ns. at pH < 7.0 (Schercamox CMA)

M.W.:

301 avg. (Schercamox CMA)

Sp.gr.:

0.949 (Aromox C/12)

0.949 (Aromox CD/12)

0.99 ± 0.05 (Schercamox CMA)

0.997 (Aromox C/12-W)

Density:

1.0 g/ml (Alkamox C2-0)

Visc.:

52 cp (Aromox C/12, CD/12)

2097 cp (Brookfield #2 spindle) (Aromox C/12-W)

F.P.:

175 F (Aromox C/12, CD/12)

Pour Pt.:

0 F (Aromox C/12, CD/12)

35 F (Aromox C/12-W)

Flash Pt.:

< 80 F (Aromox CD/12)

82 F (Aromox C/12)

> 212 F (Aromox C/12-W)

Cloud Pt.:

18 F (Aromox C/12, CD/12)

Stability:

Good (Schercamox CMA)

pH:

7.0 ± 0.5 (1% sol'n.) (Schercamox CMA)

Surface Tension:

30.8 dynes/cm (0.1%) (Aromox C/12-W)

33 dynes/cm (0.1%) (Aromox C/12)

N,N-Bis (2-hydroxyethyl) cocamine oxide *(cont'd.)*

Biodegradable: (Aromox C/12, C/12-W, CD/12)
TOXICITY/HANDLING:
    Wear protective clothing, gloves, goggles (Aromox C/12, C/12-W, CD/12)
STD. PKGS.:
    55-gal poly-lined drums (Schercamox CMA)
    SS, glass, or fiberglass-reinforced polyester tanks (Aromox C/12, C/12-W, CD/12)

Bis (2-hydroxyethyl) tallow amine oxide

SYNONYMS:
    Amines, tallow alkyl dihydroxyethyl, oxides
    Dihydroxyethyl tallowamine oxide (CTFA)
    Ethanol, 2,2´-iminobis-, N-tallow alkyl, N-oxide
    2,2´-Iminobisethanol, N-tallow alkyl, N-oxide
STRUCTURE:

$$CH_2CH_2OH$$
$$|$$
$$R—N \rightarrow O$$
$$|$$
$$CH_2CH_2OH$$

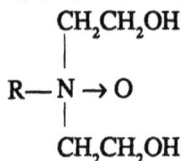

    where R represents the tallow radical
CAS No.:
    61791-46-6
TRADENAME EQUIVALENTS:
    Aromox T/12 [Akzo/Armak]
    Schercamox T-12 [Scher]
CATEGORY:
    Conditioner, softener, wetting agent, emulsifier, foam stabilizer, foaming agent, antistat, corrosion inhibitor
APPLICATIONS:
    Cosmetic industry preparations: hair rinses (Schercamox T-12); personal care products (Aromox T/12); shampoos (Aromox T/12; Schercamox T-12)
    Household detergents: detergent base (Aromoc T/12); dishwashing (Schercamox T-12)
    Industrial applications: dyes and pigments (Aromox T/12); industrial processing (Aromox T/12); metalworking (Aromox T/12); paint mfg. (Aromox T/12); paper mfg. (Aromox T/12); petroleum industry (Aromox T/12); plastics (Aromox T/12); textile/leather processing (Aromox T/12)
PROPERTIES:
Form:
    Clear liquid (Aromox T/12; Schercamox T-12)

Bis (2-hydroxyethyl) tallow amine oxide *(cont'd.)*

Color:
　Gardner 4 (Aromox T/12)
Composition:
　49% amine oxide (Schercamox T-12)
　50% active in isopropyl alcohol (Aromox T/12)
Solubility:
　Sol. in alcohol (Aromox T/12)
Ionic Nature:
　Cationic (Aromox T/12)
Sp.gr.:
　0.94 (Aromox T/12)
Visc.:
　77 cp (Aromox T/12)
Pour Pt.:
　55 F (Aromox T/12)
Flash Pt.:
　90 F (Aromox T/12)
Fire Pt.:
　180 F (Aromox T/12)
Cloud Pt.:
　60 F (Aromox T/12)
Surface Tension:
　33.0 dynes/cm (0.1%) (Aromox T/12)
TOXICITY/HANDLING:
　Wear protective clothing, gloves, and goggles (Aromox T/12)
STD. PKGS.:
　SS, glass, or fiberglass-reinforced polyester tanks (Aromox T/12)

Butyl myristate (CTFA)

SYNONYMS:
　n-Butyl myristate
　Butyl n-tetradecanoate
　Tetradecanoic acid, butyl ester
EMPIRICAL FORMULA:
　$C_{18}H_{36}O_2$
STRUCTURE:

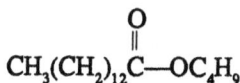

$$CH_3(CH_2)_{12}\overset{\displaystyle O}{\overset{\|}{C}}{-}OC_4H_9$$

Butyl myristate (cont'd.)

CAS No.:
110-36-1
TRADENAME EQUIVALENTS:
Bumyr [Amerchol]
Radia 7070 [Synfina-Oleofin]
Wickenol 141 [CasChem]
CATEGORY:
Chemical intermediate, lubricant, emollient, plasticizer
APPLICATIONS:
Bath products: bath oils (Wickenol 141)
Cosmetic industry preparations: (Bumyr; Wickenol 141); creams and lotions (Wickenol 141); makeup (Wickenol 141)
Industrial applications: chemical synthesis (Radia 7070); lamination (Radia 7070); lubricating/cutting oils (Radia 7070); rust inhibitors (Radia 7070); textile/leather processing (Radia 7070)
PROPERTIES:
Form:
Oily liquid (generic)
Color:
Water-white (generic)
Solubility:
Sol. in acetone (generic)
Sol. in alcohols (Wickenol 141)
Sol. in animal oil (Wickenol 141)
Sol. in castor oil (generic)
Sol. in chloroform (generic)
Insol. in glycerin (Wickenol 141)
Sol. in methanol (generic)
Sol. in min. oil (generic; Radia 7070)
Sol. in oil (Bumyr)
Insol. in propylene glycol (Wickenol 141)
Sol. in most solvents (Radia 7070)
Sol. in toluene (generic)
Sol. in veg. oil (Radia 7070; Wickenol 141)
Insol. in water (generic; Wickenol 141)
Sol. in white oils (Wickenol 141)
M.W.:
282 avg. (Radia 7070)
Sp.gr.:
0.845 (37.8 C) (Radia 7070)
0.850–0.858 (generic)
Visc.:
410 cps (37.8 C) (Radia 7070)

F.P.:
  1–7 C (generic)
B.P.:
  167–197 C (5 mm) (generic)
Flash Pt.:
  172 C (COC) (Radia 7070)
Cloud Pt.:
  0 C (Radia 7070)
Saponification No.:
  193–1203 (generic)
Ref. Index:
  1.4388 (Radia 7070)
Surface Tension:
  30 dynes/cm (Radia 7070)

Butyl stearate (CTFA)

SYNONYMS:
  n-Butyl octadecanoate
  n-Butyl stearate
  Octadecanoic acid, butyl ester
EMPIRICAL FORMULA:
  $C_{22}H_{44}O_2$
STRUCTURE:

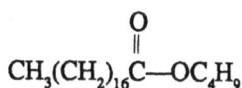

$$CH_3(CH_2)_{16}\overset{\displaystyle O}{\overset{\|}{C}}-OC_4H_9$$

CAS No.:
  123-95-5
TRADENAME EQUIVALENTS:
  Crodamol BS [Croda]
  Emerest 2321, 2325, 2326 [Henkel/Emery]
  Graden Butyl Stearate [Graden]
  Grocor 5410, 5510 [A. Gross]
  Kemester 5510 [Humko]
  Kessco BSC [Stepan]
  Kessco Butyl Stearate Cosmetic, Distilled [Stepan]
  Lipal EB [Aquatec Quimica]
  Nikkol GM-18S [Nikko]
  Radia 7051 [Synfina-Oleofin]
  Rilanit BS [Henkel KGaA]

27

Butyl stearate (cont'd.)

Uniflex BYS CP, Tech. [Union Camp]
Unimate BYS [Union Camp]
  Generically sold by:
    [Amerchol; Croda; Union Camp]
CATEGORY:
  Lubricant, defoamer, plasticizer, wetting agent, emollient, solubilizer, solvent, cosolvent, antidrying agent, chemical intermediate, emulsifier, dispersant
APPLICATIONS:
  Cleansers: cleansing creams (Crodamol BS; Emerest 2326); cleansing lotions (Crodamol BS; Emerest 2326)
  Cosmetic industry preparations: (Kemester 5510; Kessco BSC; Lipal EB); cosmetic base (Emerest 2326; Grocor 5510); creams and lotions (generic—Amerchol, Union Camp; Crodamol BS; Emerest 2325; Kessco Butyl Stearate Cosmetic; Nikkol GM-18S; Unimate BYS); hair preparations (generic—Amerchol, Union Camp; Crodamol BS; Kessco Butyl Stearate Cosmetic, Distilled; Nikkol GM-18S); makeup (generic—Amerchol, Union Camp; Crodamol BS; Emerest 2325, 2326; Kessco Butyl Stearate Cosmetic; Nikkol GM-18S; Unimate BYS); personal care products (Crodamol BS); skin preparations (Kessco BSC)
  Farm products: fertilizers (generic—Croda)
  Industrial applications: ceramics (Rilanit BS); chemical synthesis (Radia 7051); coatings (Uniflex BYS); industrial processing (Grocor 5510); lubricating/cutting oils (generic—Croda; Grocor 5410, 5510; Radia 7051; Rilanit BS); metalworking (generic—Croda; Grocor 5410; Kessco BSC; Emerest 2321, 2326; Rilanit BS; Uniflex BYS Tech.); paper mfg. (Grocor 5410, 5510); plastics (generic—Croda; Kessco Butyl Stearate Cosmetic); rubber (Kessco Butyl Stearate Cosmetic, Distilled); textile/leather processing (generic—Croda; Emerest 2326; Grocor 5410, 5510; Kessco BSC, Butyl Stearate Cosmetic; Radia 7051; Rilanit BS; Uniflex BYS Tech.)
PROPERTIES:
Form:
  Liquid (Crodamol BS; Emerest 2321, 2325, 2326; Grocor 5410, 5510; Kemester 5510; Kessco BSC, Butyl Stearate Cosmetic, Distilled; Lipal EB; Rilanit BS)
  Oily liquid (Graden Butyl Stearate)
  Liquid/semisolid (generic—Croda)
Color:
  Water-white (Crodamol BS; Grocor 5410, 5510)
  APHA 25 (Uniflex BYS CP)
  APHA 30 max. (Kessco Butyl Stearate Cosmetic)
  APHA 35 max. (Graden Butyl Stearate)
  APHA 60 max. (Kessco Butyl Stearate Distilled)
  APHA 80 (Uniflex BYS, tech.)
  Gardner 1 max. (Emerest 2326)
  Gardner 4 (Emerest 2325)

Butyl stearate (cont'd.)

Odor:

Practically none (Graden Butyl Stearate)

Mild (Grocor 5410, 5510)

Low (Uniflex BYS Tech.)

Composition:

98% active min. (Grocor 5410, 5510)

100% active (generic—Croda; Emerest 2321, 2326; Kessco BSC; Lipal EB)

Solubility:

Sol. in acetone (Graden Butyl Stearate)

Sol. in benzol (Graden Butyl Stearate)

Sol. in carbon tetrachloride (Graden Butyl Stearate)

Sol. in castor oil (generic—Amerchol, Union Camp; Graden Butyl Stearate); sol. warm (Nikkol GM-18S)

Sol. in ethanol (generic—Amerchol, Union Camp); sol. warm (Nikkol GM-18S)

Sol. in ethyl acetate (Graden Butyl Stearate)

Insol. in glycerin (Unimate BYS)

Insol. in glycerol (generic—Amerchol, Union Camp)

Sol. in heptane (Graden Butyl Stearate)

Sol. in IPA (generic—Amerchol, Union Camp; Emerest 2325, 2326; Graden Butyl Stearate; Grocor 5410, 5510)

Sol. in IPM (Emerest 2325)

Partly sol. in methanol (Graden Butyl Stearate)

Sol. in min. oil (Graden Butyl Stearate; Radia 7051; Unimate BYS); @ 5% (Emerest 2325, 2326)

Sol. in oil (Grocor 5410)

Sol. warm in oleyl alcohol (Nikkol GM-18S)

Sol. warm in olive oil (Nikkol GM-18S)

Sol. in peanut oil (Graden Butyl Stearate)

Insol. in propylene glycol (generic—Amerchol, Union Camp; Unimate BYS)

Sol. in most solvents (Radia 7051)

Sol. in toluene @ 5% (Emerest 2326)

Sol. in toluol (Grocor 5410, 5510)

Sol. in triolein @ 5% (Emerest 2325)

Sol. in veg. oil (Radia 7051)

Insol. in water (generic—Amerchol, Union Camp; Emerest 2325, 2326; Nikkol GM-18S; Unimate BYS)

Sol. in xylene @ 5% (Emerest 2326)

Ionic Nature:

Nonionic (Emerest 2321, 2326; Grocor 5410, 5510)

M.W.:

325 avg. (Radia 7051)

Sp.gr.:

0.850–0.860 (20/25 C) (Kessco Butyl Stearate Cosmetic)

Butyl stearate (cont'd.)

0.853–0.859 (20/25 C) (Kessco Butyl Stearate Distilled; Graden Butyl Stearate)
0.855 (Grocor 5410, 5510)
0.856 (Uniflex BYS CP)
0.857 (Uniflex BYS tech.)
0.884 (37.8 C) (Radia 7051)

Density:
7.1 lb/gal (Graden Butyl Stearate; Kessco Butyl Stearate Cosmetic, Distilled)
7.2 lb/gal (Emerest 2326)

Visc.:
7 cSt (100 F) (Emerest 2326; Kessco Butyl Stearate)
10 cs (Grocor 5410, 5510)
10.5 cSt (Uniflex BYS CP, BYS tech.)
6.10 cps (37.8 C) (Radia 7051)
7 cps (Graden Butyl Stearate)

F.P.:
18 C (Kessco Butyl Stearate Distilled)
19 C (Graden Butyl Stearate; Kessco Butyl Stearate Cosmetic)
20 C (Grocor 5410, 5510)

B.P.:
196.5 C (Graden Butyl Stearate)
200 C (4 mm Hg) (Kessco Butyl Stearate Cosmetic, Distilled)

M.P.:
20 C (Uniflex BYS tech.)
23 C (Uniflex BYS CP)

Pour Pt.:
20 C (Emerest 2326)

Flash Pt.:
369 F (COC) (Graden Butyl Stearate)
370 F (COC) (Kessco Butyl Stearate Cosmetic, Distilled)
375 F (Emerest 2326)
200 C (COC) (Radia 7051)
202 C (COC) (Uniflex BYS CP, tech.)

Fire Pt.:
221 C (COC) (Uniflex BYS CP, tech.)

Cloud Pt.:
22.5 C (Radia 7051)

Acid No.:
0.4 (Uniflex BYS CP)
0.8 (Uniflex BYS tech.)
1 (Emerest 2326)

Iodine No.:
0.5 max. (Kessco Butyl Stearate Cosmetic)
1.0 (Uniflex BYS CP)

5 (Uniflex BYS tech.)
6.0 max. (Graden Butyl Stearate; Kessco Butyl Stearate Distilled)
Saponification No.:
170 (Uniflex BYS CP)
172 (Emerest 2326)
177 (Uniflex BYS tech.)
Ref. Index:
1.4328 (Grocor 5410, 5510)
1.442 (Graden Butyl Stearate; Kessco Butyl Stearate Cosmetic, Distilled)
1.4432 (Radia 7051)
Surface Tension:
32 dynes/cm (Radia 7051)
STD. PKGS.:
55 gal (380 lb net) steel drum (bung type) (Graden Butyl Stearate)
Bulk or drums (Grocor 5410, 5510)

Calcium stearate (CTFA)

SYNONYMS:
Calcium octadecanoate
Octadecanoic acid, calcium salt

EMPIRICAL FORMULA:
$C_{18}H_{36}O_2 \cdot \frac{1}{2} Ca$

STRUCTURE:
$[CH_3(CH_2)_{16}COO]_2Ca$

CAS No.:
1592-23-0

TRADENAME EQUIVALENTS:
Akrochem Calcium Stearate [Akron]
Flowco [Mallinckrodt]
Harwick Calcium Stearate [Harwick]
Hy Dense Calcium Stearate HP Gran., RSN Powd. [Mallinckrodt]
Interstab CA-18-1 [Akzo/Interstab]
Lubracal 48, 53, 60 [Witco]
Nopcote C-104 [Henkel/Process, Henkel-Nopco]
Norfox CS [Norman, Fox]
Petrac CP-11, CP-11LS, CP-11 LSG, CP-12, CP-22G [Synthetic Products]
Radiacid 1060 [Synfina-Oleofin]
Radiastar 1060 [Synfina-Oleofin]
Witco Calcium Stearate A, EA, F, G, Kosher, LM, M Impalpable, Polymer Grade, Polymer Grade S, Regular, Superfine [Witco]

CATEGORY:
Release agent, lubricant, plasticizer, water repellent, anticaking agent, binder, processing aid, emulsifier, suspension aid, flow control agent, dusting agent, stabilizer

APPLICATIONS:
Cosmetic industry preparations: (Interstab CA-18-1; Witco Calcium Stearate Regular)
Food applications: (Radiastar 1060; Witco Calcium Stearate EA, Kosher)
Industrial applications: cement, clay, concrete (Petrac CP-11, CP-11 LS, CP-11 LSG, CP-12; Witco Calcium Stearate F, Superfine); ceramics (Witco Calcium Stearate Regular); dyes and pigments (Petrac CP-11, CP-11 LS, CP-11 LSG, CP-12; Witco Calcium Stearate Regular); metallurgical applications (Interstab CA-18-1); paint mfg. (Petrac CP-11, CP-11 LS, CP-11 LSG, CP-12; Witco Calcium Stearate Regular); paper mfg. (Flowco; Lubracal 48, 53, 60; Nopcote C-104; Petrac CP-11, CP-11 LS, CP-11 LSG, CP-12; Witco Calcium Stearate Regular); plastics (Hy Dense Calcium Stearate HP Gran., RSN Powd.; Interstab CA-18-1; Petrac CP-11,

CP-11 LS, CP-11 LSG, CP-12; Radiacid 1060; Witco Calcium Stearate A, F, G, LM, M Impalpable, Polymer Grade, Polymer Grade S, Regular, Superfine); printing inks (Petrac CP-11, CP-11 LS, CP-11 LSG, CP-12); rubber (Harwick Calcium Stearate; Interstab CA-18-1)

PROPERTIES:

Form:
Dispersion (Flowco)
Granular (Hy Dense Calcium Stearate HP Gran.; Petrac CP-22G)
Fine powder (Akrochem Calcium Stearate; Interstab CA-18-1)
Powder (Hy Dense Calcium Stearate RSN Powd.; Lubracal 48, 53, 60; Petrac CP-11, CP-11 LS, CP-11 LSG, CP-12; Witco Calcium Stearate A, EA, F, G, Kosher, LM, M Impalpable, Polymer Grade, Polymer Grade S, Regular, Superfine)
Fluffy powder (Radiacid 1060; Radiastar 1060)
Impalpable powder (Norfox CS)
Free-flowing powder (Harwick Calcium Stearate)

Fineness:
97% through 100 mesh (Petrac CP-11, CP-12)
90% through 200 mesh (Petrac CP-11 LS)
98.5% through 325 mesh (Akrochem Calcium Stearate)
99% through 325 mesh (Witco Calcium Stearate LM)
99.5% through 325 mesh (Petrac CP-11 LSG)
99.9% through 325 mesh (Lubracal 48, 53, 60; Witco Calcium Stearate A, EA, F, G, Kosher, M Impalpable, Polymer Grade, Polymer Grade S, Regular, Superfine)

Color:
White (Akrochem Calcium Stearate; Harwick Calcium Stearate; Lubracal 48, 53, 60; Petrac CP-11; Radiacid 1060; Radiastar 1060; Witco Calcium Stearate A, EA, F, G, Kosher, LM, M Impalpable, Polymer Grade, Polymer Grade S, Regular, Superfine)

Composition:
40–55% solids in water (Flowco)
50.1% solids (Lubracal 48)
56.0% solids (Lubracal 53)
60.0% solids (Lubracal 60)
100% conc. (Petrac CP-11, CP-22G

Solubility:
Sol. in benzene (Witco Calcium Stearate A, EA, F, G, Kosher, LM, M Impalpable, Polymer Grade, Polymer Grade S, Regular, Superfine)
Sol. in carbon tetrachloride (Witco Calcium Stearate A, EA, F, G, Kosher, LM, M Impalpable, Polymer Grade, Polymer Grade S, Regular, Superfine)
Sol. in min. oil (Witco Calcium Stearate A, EA, F, G, Kosher, LM, M Impalpable, Polymer Grade, Polymer Grade S, Regular, Superfine)
Sol. in oleic acid (Witco Calcium Stearate A, EA, F, G, Kosher, LM, M Impalpable, Polymer Grade, Polymer Grade S, Regular, Superfine)
Sol. in pyridine (Witco Calcium Stearate A, EA, F, G, Kosher, LM, M Impalpable,

Calcium stearate *(cont'd.)*

  Polymer Grade, Polymer Grade S, Regular, Superfine)
  Sol. in toluene (Witco Calcium Stearate A, EA, F, G, Kosher, LM, M Impalpable, Polymer Grade, Polymer Grade S, Regular, Superfine)
  Sol. hot in turpentine (Witco Calcium Stearate A, EA, F, G, Kosher, LM, M Impalpable, Polymer Grade, Polymer Grade S, Regular, Superfine)
  Sol. in veg. oil (Witco Calcium Stearate A, EA, F, G, Kosher, LM, M Impalpable, Polymer Grade, Polymer Grade S, Superfine)
  Disp. in water (Nopcote C-104); insol. or sparingly sol. in water (Witco Calcium Stearate A, EA, F, G, Kosher, LM, M Impalpable, Polymer Grade, Polymer Grade S, Regular, Superfine)
  Sol. in waxes (Witco Calcium Stearate A, EA, F, G, Kosher, LM, M Impalpable, Polymer Grade, Polymer Grade S, Regular, Superfine)
  Sol. in xylene (Witco Calcium Stearate A, EA, F, G, Kosher, LM, M Impalpable, Polymer Grade, Polymer Grade S, Regular, Superfine)
Sp.gr.:
  1.00 (Lubracal 48, 53)
  1.02 (Lubracal 60)
  1.03 (Petrac CP-11, CP-11 LS, CP-11 LSG, CP-12; Witco Calcium Stearate A, EA, F, G, Kosher, LM, M Impalpable, Polymer Grade, Polymer Grade S, Regular, Superfine)
  1.04 (Harwick Calcium Stearate)
Density:
  1.9 lb/gal (apparent) (Akrochem Calcium Stearate)
  25 lb/ft^3 (bulk, tamped) (Petrac CP-12)
  26 lb/ft^3 (Petrac CP-11, CP-11 LS)
Visc.:
  110 cps (Lubracal 48)
  350 cps (Lubracal 60)
  700 cps (Lubracal 53)
M.P.:
  145–160 C (Akrochem Calcium Stearate)
  151 C (Petrac CP-11)
  154 C (Petrac CP-11 LS, CP-11 LSG)
  158 C (Harwick Calcium Stearate)
Softening Pt.:
  150 C (Petrac CP-11 LS, CP-11 LSG)
  151 C (Petrac CP-12)
  160 C (Witco Calcium Stearate A, EA, F, G, Kosher, LM, M Impalpable, Polymer Grade, Polymer Grade S, Regular, Superfine)
Acid No.:
  205 (Petrac CP-11)
STD. PKGS.:
  Bags, cartons (Harwick Calcium Stearate)

Cetearyl alcohol (CTFA)

SYNONYMS:
  Cetostearyl alcohol
  Cetyl/stearyl alcohol
CAS No.:
  8005-44-5
TRADENAME EQUIVALENTS:
  Adol 63, 630, 640 [Sherex]
  Amphocerin K [Henkel; Henkel KGaA]
  Cetax 50 [Aquatec Quimica]
  Cetostearyl Alcohol BP, NF [Croda]
  Cetyl-Stearyl Alcohol, NF [Alcolac]
  Crodacol CS50 [Croda]
  Cyclogol Cetyl-Stearyl Alcohol [Alcolac]
  Dehydag Wax O [Henkel; Henkel KGaA]
  Hydrenol DD [Henkel Canada]
  Lanette O [Henkel Canada; Henkel KGaA]
  Lanol CS [Seppic]
  Laurex CS, CS/D [Albright & Wilson; Albright & Wilson/Marchon]
  Niox EO-26 [Pulcra SA]
  Tewax TC1, TC2 [Tessilchimica]
CATEGORY:
  Emollient, consistency agent, superfatting agent, lubricant, raw material, base
APPLICATIONS:
  Cosmetic industry preparations: (Cyclogol Cetyl-Stearyl Alcohol; Lanol CS; Tewax
    TC1, TC2); conditioners (Cetyl-Stearyl Alcohol, NF); cosmetic base (Amphocerin
    K); creams and lotions (Cetax 50; Cetyl-Stearyl Alcohol, NF; Cyclogol Cetyl-
    Stearyl Alcohol; Hydrenol DD; Lanette O; Niox EO-26); hair dyeing (Tewax TC1,
    TC2); hair preparations (Cetax 50; Crodacol CS50); makeup (Cetyl-Stearyl Alco-
    hol, NF); personal care products (Dehydag Wax O; Laurex CS/D); shampoos
    (Laurex CS); skin preparations (Crodacol CS50; Hydrenol DD; Lanette O)
  Industrial applications: (Cetyl-Stearyl Alcohol, NF)
  Pharmaceutical applications: (Dehydag Wax O; Laurex CS/D); ointments (Am-
    phocerin K; Lanette O)
PROPERTIES:
Form:
  Soft solid (Amphocerin K)
  Solid (Laurex CS)
  Flake (Cetax 50; Cetostearyl Alcohol BP, NF; Cetyl-Stearyl Alcohol, NF; Cyclogol
    Cetyl-Stearyl Alcohol; Hydrenol DD; Laurex CS; Tewax TC1, TC2)
  Granular (Dehydag Wax O; Lanette O)
  Wax (Lanol CS)
Color:
  Translucent (Amphocerin K)

Cetearyl alcohol (cont'd.)

White (Cetostearyl Alcohol BP, NF; Dehydag Wax O; Hydrenol DD; Lanette O; Laurex CS)
Off-white (Cyclogol Cetyl-Stearyl Alcohol)
APHA 40 (Adol 640)
APHA 100 (Adol 630)
Lovibond 5Y/0.5R (Adol 63)

Composition:
100% conc. (Dehydag Wax O; Lanette O; Lanol CS)

Solubility:
Sol. in min. oil (Cyclogol Cetyl-Stearyl Alcohol)
Sol. in oil (Crodacol CS50)
Insol. in water (Cyclogol Cetyl-Stearyl Alcohol; Lanette O; Niox EO-26)

Ionic Nature:
Nonionic (Cyclogol Cetyl-Stearyl Alcohol; Dehydag Wax O; Lanette O; Lanol CS)

M.W.:
253 (Adol 640)
264 (Adol 63)
265 (Adol 630)

Sp.gr.:
0.816 (60/25 C) (Adol 63, 630)
0.820 (50/50 C) (Adol 640)

Density:
0.4 g/cm^3 (Laurex CS/D)
0.816 g/ml (60 C) (Dehydag Wax O)

M.P.:
43–46 C (Adol 640)
43–53 C (Cetyl-Stearyl Alcohol NF)
48–53 C (Cyclogol Cetyl-Stearyl Alcohol; Laurex CS/D)
50–54 C (Laurex CS)
52–56 C (Cetyl-Stearyl Alcohol)

Solidification Pt.:
48–52 C (Dehydag Wax O)

Flash Pt.:
150 C (Laurex CS/D)

Acid No.:
0.3 (Laurex CS)
1.0 max. (Adol 63, 640)
2.0 max. (Adol 630)

Iodine No.:
1.5 (Laurex CS)
1.5 max. (Adol 640)
2.0 max. (Adol 63, 630)

Saponification No.:
1.0 max. (Dehydag Wax O; Adol 640)
1.5 (Laurex CS)
2.0 max. (Laurex CS/D; Adol 630)
3.0 max. (Adol 63)
Hydroxyl No.:
204–216 (Laurex CS)
205–218 (Adol 630)
206–218 (Adol 63)
210–223 (Adol 640)

Cetearyl octanoate (CTFA)

SYNONYMS:
Cetyl/stearyl 2-ethylhexanoate
2-Ethylhexanoic acid, cetyl/stearyl ester
CAS No.:
RD No. 977063-73-2
TRADENAME EQUIVALENTS:
Crodamol CAP [Croda Ltd.]
Luvitol EHO [BASF AG]
Schercemol 1688 [Scher]
CATEGORY:
Emollient, water repellent, moisturizer
APPLICATIONS:
Bath products: (Schercemol 1688)
Cosmetic industry preparations: (Crodamol CAP; Luvitol EHO); skin preparations
(Crodamol CAP; Schercemol 1688)
Pharmaceutical applications: (Crodamol CAP; Luvitol EHO)
PROPERTIES:
Form:
Liquid (Crodamol CAP; Luvitol EHO)
Clear liquid (Schercemol 1688)
Color:
Water-white (Crodamol CAP)
APHA 100 max. (Schercemol 1688)
Odor:
Low (Crodamol CAP)
Slight, characteristic (Schercemol 1688)
Composition:
100% conc. (Luvitol EHO)

Cetearyl octanoate *(cont'd.)*

Solubility:
 Sol. in alcohols (Schercemol 1688)
 Sol. in aliphatic hydrocarbons (Schercemol 1688)
 Sol. in aromatic hydrocarbons (Schercemol 1688)
 Sol. in chlorinated hydrocarbons (Schercemol 1688)
 Sol. in esters (Schercemol 1688)
 Sol. in natural fats (Schercemol 1688)
 Sol. in glycol ethers (Schercemol 1688)
 Sol. in higher glycols (Schercemol 1688)
 Sol. in ketones (Schercemol 1688)
 Sol. in min. oil (Schercemol 1688)
 Insol. in water (Schercemol 1688)
Ionic Nature:
 Nonionic (Luvitol EHO)
M.W.:
 388 (theoret.) (Schercemol 1688)
Sp.gr.:
 0.852 (Schercemol 1688)
Density:
 7.1 lb/gal (Schercemol 1688)
F.P.:
 3 C (Schercemol 1688)
Flash Pt.:
 170 C (OC) (Schercemol 1688)
Cloud Pt.:
 4 C (Schercemol 1688)
Acid No.:
 1.0 max. (Schercemol 1688)
Iodine No.:
 1.0 max. (Schercemol 1688)
Saponification No.:
 140–150 (Schercemol 1688)
Ref. Index:
 1.4448 (Schercemol 1688)

Cetyl alcohol *(CTFA)*

SYNONYMS:
 Alcohol C-16
 Cetylic alcohol
 1-Hexadecanol

n-Hexadecyl alcohol
Palmityl alcohol
EMPIRICAL FORMULA:
  $C_{16}H_{34}O$
STRUCTURE:
  $CH_3(CH_2)_{14}CH_2OH$
CAS No.:
  36653-82-4
TRADENAME EQUIVALENTS:
  Adol 52, 52 NF, 54, 520, 520 NF [Sherex]
  Alfol 16 [Vista] (NF)
  Cachalot C-50, C-51 [M. Michel]
  Cetal [Amerchol] (NF)
  Cetax 16 [Aquatec Quimica]
  CO-1695 [Procter & Gamble] (NF)
  Crodacol C [Croda]
  Cycetyl Alcohol NF [Alcolac]
  Cyclogol Cetyl Alcohol NF [Alcolac]
  Dehydag Wax 16 [Henkel KGaA]
  Emery 1787 [Henkel/Emery] (NF)
  Epal 16NF [Ethyl]
  Kalcohl 60 [Kao]
  Lanette 16 [Henkel Canada]
  Lanol C [Seppic]
  Lipocol C [Lipo]
  Michel XO-144, XO-144B [M. Michel]
  Niox EO-10 [Pulcra SA]
  Rilanit G16 [Henkel KGaA]
      Generically sold by:
        [Henkel Canada; Vista]

CATEGORY:
  Emollient, thickener, viscosity builder, viscosity control agent, consistency agent,
    bodying agent, superfatting agent, aux. emulsifier, coemulsifier, intermediate, raw
    material, lubricant, stabilizer, plasticizer, foam control agent, cosolvent, carrier,
    opacifier

APPLICATIONS:
  Cosmetic industry preparations: (Adol 52 NF, 54, 520 NF; Alfol 16; Cachalot C-50,
    C-51; Cetal; Cycetyl Alcohol NF; Cyclogol Cetyl Alcohol NF; Dehydag Wax 16;
    Lanol C); cosmetic base (Niox EO-10); creams and lotions (Cetax 16; Cycetyl
    Alcohol NF; Cyclogol Cetyl Alcohol NF; Dehydag Wax 16; Lanette 16; Lipocol C;
    Niox EO-10); hair preparations (Cetax 16); makeup (Cetal; Lipocol C); skin
    preparations (Lanette 16)

Cetyl alcohol (cont'd.)

Industrial applications: ceramics (Rilanit G16); industrial processing (Adol 52 NF, 54, 520 NF); lubricating/cutting oils (Adol 52 NF, 54, 520 NF; Rilanit G16); metalworking (Rilanit G16); paper mfg. (Adol 52 NF, 54, 520 NF); plastics (Alfol 16); printing inks (Adol 52 NF, 54, 520 NF); textile/leather processing (Adol 52 NF, 54, 520 NF; Rilanit G16)

Pharmaceutical applications: (Dehydag Wax 16); ointments (Dehydag Wax 16)

PROPERTIES:

Form:

Liquid (Michel XO-144; Rilanit G16)

Solid (Kalcohl 60)

Wax (Lanol C)

Waxy solid (Alfol 16; Cetal; Epal 16NF)

Beads (Kalcohl 60)

Flakes (generic—Henkel; Cetax 16; Cyclogol Cetyl Alcohol NF; Dehydag Wax 16; Emery 1787; Lanette 16)

Waxy flakes (Adol 52, 520)

Color:

White (Alfol 16; Cetal; Lanette 16)

Off-white (Cyclogol Cetyl Alcohol NF)

APHA 50 max. (Adol 520 NF)

Lovibond 5Y/0.5R ($5^1/_4$ in. max.) (Adol 52 NF, 54)

Odor:

Mild, characteristic (Cetal)

Typical fatty alcohol (Alfol 16)

Composition:

98.9% active (Alfol 16)

100% conc. (Emery 1787; Epal 16NF; Kalcohl 60; Lanol C)

Solubility:

Sol. in acetone (generic—Vista; Adol 52 NF, 54, 520 NF; Alfol 16; Cachalot C-50, C-51)

Sol. in alcohols (generic—Vista; Alfol 16; Cachalot C-50, C-51)

Sol. in aromatic hydrocarbons (Cachalot C-50, C-51)

Sol. in benzene (Adol 52 NF, 54, 520 NF)

Sol. in carbon disulfide (Cachalot C-50, C-51)

Sol. in chloroform (Cachalot C-50, C-51)

Sol. in diglycol ethers (Cachalot C-50, C-51)

Misc. in ethanol/water mixtures (Michel XO-144B)

Sol. in ether (generic—Vista; Alfol 16)

Sol. in ethyl ether (Adol 52 NF, 54, 520 NF)

Sol. in fats (Crodacol C); misc. (Cachalot C-50, C-51)

Sol. in glycol ethers (Cachalot C-50, C-51)

Sol. in isopropanol (Adol 52 NF, 54, 520 NF)

Sol. in kerosene (Adol 52 NF, 54, 520 NF)

Sol. in min. oil (Cycetyl Alcohol NF; Cyclogol Cetyl Alcohol NF); sol. in light min. oil (Adol 52 NF, 54, 520 NF); misc. with most min. oils (Cachalot C-50, C-51)

Sol. in VMP naphtha (Adol 52 NF, 54, 520 NF)

Sol. in oils (Crodacol C); misc. (Lipocol C)

Sol. in trichlorethylene (Adol 52 NF, 54, 520 NF)

Sol. in turpentine (Adol 52 NF, 54, 520 NF)

Misc. with most veg. oils (Cachalot C-50, C-51)

Insol. in water (generic—Vista; Alfol 16; Cycetyl Alcohol NF; Cyclogol Cetyl Alcohol NF; Lipocol C; Niox EO-10)

Misc. with most waxes (Cachalot C-50, C-51)

Ionic Nature:

Nonionic (Cyclogol Cetyl Alcohol NF; Emery 1787; Lanol C)

M.W.:

246 (Adol 520 NF)

247 (Adol 52 NF)

249 (Adol 54)

Sp.gr.:

0.813 (Alfol 16)

0.815 (60/25 C) (Adol 52 NF, 520 NF)

0.816 (60/25 C) (Adol 54)

Density:

6.77 lb/gal (Alfol 16)

M.P.:

45–50 C (closed tube) (Adol 52 NF, 520 NF; Cyclogol Cetyl Alcohol NF)

45–51 C (Cetal)

Acid No.:

Nil (Cyclogol Cetyl Alcohol NF)

1.0 max. (Adol 52 NF, 54, 520 NF; Cetal)

Iodine No.:

1.0 max. (Adol 520 NF)

2.0 max. (Adol 52 NF, 54; Cetal)

Saponification No.:

2.0 max. (Adol 520 NF; Cetal)

3.0 max. (Adol 52 NF, 54)

Hydroxyl No.:

218–238 (Cetal)

220–230 (Adol 54)

220–235 (Adol 52 NF)

221–235 (Adol 520 NF)

TOXICITY/HANDLING:

Avoid inhaling vapors from the hot liquid; wash body parts contacting the alcohol with water (Alfol 16 Alcohol)

41

Cetyl alcohol (cont'd.)

STORAGE/HANDLING:
Combustible; treat with same precautions as high boiling hydrocarbons; store in dry carbon steel, aluminum, epoxy or heresite-lined steel (Alfol 16 Alcohol)

Cetyl lactate (CTFA)

SYNONYMS:
n-Hexadecyl-2-hydroxypropanoate
n-Hexadecyl lactate
Propanoic acid, 2-hydroxy-, hexadecyl ester

EMPIRICAL FORMULA:
$C_{19}H_{38}O_3$

STRUCTURE:

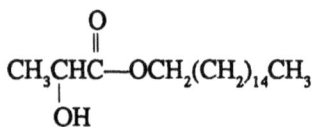

CAS No.:
35274-05-6

TRADENAME EQUIVALENTS:
Cegesoft C19 [Henkel KGaA]
Ceraphyl 28 [VanDyk]
Crodamol CL [Croda]
Cyclochem CL [Alcolac]
Lipal LC [Aquatec Quimica]
Liponate CL [Lipo]
Schercemol CL [Scher]

CATEGORY:
Emollient, thickener, viscosity agent, penetrant, solubilizer

APPLICATIONS:
Bath products: bubble bath (Liponate CL); bath oils (Cyclochem CL; Liponate CL)

Cleansers: cleansing creams (Liponate CL); cleansing lotions (Cyclochem CL; Liponate CL)

Cosmetic industry preparations: (Lipal LC); creams and lotions (Schercemol CL); hair preparations (Schercemol CL); makeup (Ceraphyl 28; Crodamol CL; Cyclochem CL; Schercemol CL); personal care products (Crodamol CL); shampoos (Ceraphyl 28)

Pharmaceutical applications: sunscreens (Schercemol CL)

PROPERTIES:
Form:
   Soft solid (Schercemol CL)
   Soft solid to liquid (Liponate CL)
   Solid (Cegesoft C 19; Crodamol CL; Cyclochem CL; Lipal LC)
Color:
   White (Liponate CL)
   Off-white (Cyclochem CL)
   Colorless to pale straw (Crodamol CL)
   Gardner 2.0 max. (Schercemol CL)
Odor:
   Slight, typical (Schercemol CL)
Composition:
   100% conc. (Lipal LC)
Solubility:
   Sol. in aq. alcoholic sol'ns. (Crodamol CL)
   Sol. in alcohols (Schercemol CL)
   Sol. in esters (Schercemol CL)
   Sol. in glycol ethers (Schercemol CL)
   Sol. in glycols (Crodamol CL; Schercemol CL)
   Sol. in min. oil (Ceraphyl 28; Schercemol CL)
   Sol. in natural oils and fats (Schercemol CL)
   Partly sol. in water (Ceraphyl 28); insol. in water (Schercemol CL)
M.W.:
   314 (theoret.) (Schercemol CL)
Sp.gr.:
   0.890 ± 0.01 (35 C) (Schercemol CL)
Density:
   7.4 (35 C) (Schercemol CL)
M.P.:
   24 C (Cyclochem CL)
Flash Pt.:
   > 160 C (OC) (Schercemol CL)
Acid No.:
   3.0 max. (Liponate CL; Schercemol CL)
   3.5 max. (Cyclochem CL)
Iodine No.:
   Nil (Schercemol CL)
Saponification No.:
   174–190 (Schercemol CL)
   174–195 (Liponate CL)
   180 (Cyclochem CL)

Cetyl myristate (CTFA)

SYNONYMS:
Tetradecanoic acid, hexadecyl ester
EMPIRICAL FORMULA:
$C_{30}H_{60}O_2$
STRUCTURE:

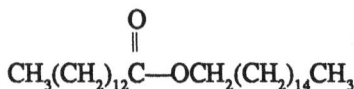

CAS No.:
2599-01-1
TRADENAME EQUIVALENTS:
Kessco 654 [Stepan]
Schercemol CM [Scher]
CATEGORY:
Emollient, thickener, substantivity agent
APPLICATIONS:
Cosmetic industry preparations: (Kessco 654); creams and lotions (Schercemol CM); makeup (Schercemol CM); skin preparations (Schercemol CM)
Pharmaceutical applications: sunscreens (Schercemol CM)
PROPERTIES:
Form:
Waxy solid (Schercemol CM)
Color:
White to pale yellow (Schercemol CM)
Odor:
Mild, characteristic (Schercemol CM)
Solubility:
Sol. in aromatic hydrocarbons (Schercemol CM)
Sol. in esters (Schercemol CM)
Disp. in glycerol (Schercemol CM)
Disp. in glycol ethers (Schercemol CM)
Disp. in glycols (Schercemol CM)
Sol. in min. oil (Schercemol CM)
Sol. in min. spirits (Schercemol CM)
M.W.:
456 (theoret.) (Schercemol CM)
Sp.gr.:
0.835 (55 C) (Schercemol CM)
Density:
7.0 lb/gal (55 C) (Schercemol CM)
M.P.:
51 C (Schercemol CM)

Flash Pt.:
> 170 C (OC) (Schercemol CM)
Acid No.:
2.0 max. (Schercemol CM)
Iodine No.:
Nil (Schercemol CM)
Saponification No.:
110–125 (Schercemol CM)

Cetyl palmitate (CTFA)

SYNONYMS:
Hexadecanoic acid, hexadecyl ester
n-Hexadecyl hexadecanoate
EMPIRICAL FORMULA:
$C_{32}H_{64}O_2$
STRUCTURE:

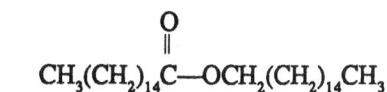

$$\underset{\|}{\overset{\displaystyle O}{}}$$
$CH_3(CH_2)_{14}C-OCH_2(CH_2)_{14}CH_3$
CAS No.:
540-10-3
TRADENAME EQUIVALENTS:
CEP-33 [Hefti Ltd.]
Cutina CP, CP-A [Henkel; Henkel KGaA]
Cyclochem CP [Alcolac]
Kemester CP [Humko/Witco]
Kessco 653 [Stepan]
Nikkol N-SP [Nikko]
Radia 7500 [Synfina-Oleofin]
Schercemol CP [Scher]
Standamul 1616 [Henkel]
Starfol CP [Sherex]
Waxenol 815, 816 [CasChem]
CATEGORY:
Wax, emollient, intermediate, lubricant, binder, thickener, visc. booster, consistency
agent, conditoner, opacifier, sheen additive
APPLICATIONS:
Cleansers: cleansing creams (Cutina CP); cleansing lotions (Cutina CP); liquid soaps
(Kemester CP)

Cetyl palmitate (cont'd.)

Cosmetic industry preparations: (Kessco 653); cosmetic base (CEP-33; Cyclochem CP); creams and lotions (Cutina CP, CP-A; Kemester CP; Schercemol CP; Starfol CP); hair preparations (Schercemol CP); makeup (CEP-33; Cutina CP, CP-A; Cyclochem CP; Kemester CP; Schercemol CP); personal care products (CEP-33; Cyclochem CP; Standamul 1616); pressed powders (Waxenol 815, 816)

Industrial applications: lubricating/cutting oils (Radia 7500); metalworking (Radia 7500; Waxenol 815, 816); polishes and waxes (CEP-33; Nikkol N-SP; Radia 7500); textile/leather processing (Radia 7500)

Pharmaceutical applications: (Kessco 653; Standamul 1616); ointments (Kessco 653)

PROPERTIES:

Form:

Powder (Waxenol 815)

Glistening flakes (Kessco 653)

Flakes (Cutina CP-A; Cyclochem CP; Starfol CP; Waxenol 816)

Waxy flakes (Cutina CP)

Waxy solid (CEP-33; Nikkol N-SP; Schercemol CP; Standamul 1616)

Color:

White (Cutina CP; Kemester CP; Kessco 653; Standamul 1616)

White to pale yellow (Schercemol CP)

Off-white (Cyclochem CP)

Odor:

Mild, characteristic (Schercemol CP)

Faint, characteristic (Standamul 1616)

Composition:

100% conc. (CEP-33; Nikkol N-SP; Standamul 1616; Waxenol 816)

Solubility:

Sol. in alcohols (Cutina CP-A); sol. in boiling alcohol (Kessco 653)

Sol. in aromatic hydrocarbons (Cutina CP-A; Schercemol CP)

Sol. in benzene @ 10% (Radia 7500)

Sol. in castor oil (Standamul 1616); @ 10% (Cutina CP)

Sol. in chloroform (Kessco 653)

Sol. in esters (Cutina CP-A; Schercemol CP)

Sol. in ether (Kessco 653)

Disp. in glycerol (Schercemol CP)

Disp. in glycol ethers (Schercemol CP)

Disp. in glycols (Cutina CP-A; Schercemol CP)

Sol. in hydrocarbons (Kessco 653)

Sol. in isopropyl myristate (Standamul 1616); @ 10% (Cutina CP)

Sol. in ketones (Cutina CP-A)

Sol. in min. oil (Schercemol CP; Standamul 1616; Starfol CP); @ 10% (Cutina CP)

Sol. in min. spirits (Schercemol CP)

Sol. in oils (Kessco 653); sol. in warm oils (Nikkol N-SP); sol. in cosmetic oils (Cutina CP-A)

Sol. in oleyl alcohol (Standamul 1616); @ 10% (Cutina CP)
Sol. in trichlorethylene @ 10% (Radia 7500)
Insol. in water (Cutina CP-A; Kessco 653); @ 10% (Cutina CP)
Sol. in waxes (Kessco 653)
Ionic Nature:
Nonionic (CEP-33; Waxenol 816)
M.W.:
480 (theoret.) (Schercemol CP)
484 avg. (Radia 7500)
Sp.gr.:
0.805 (98.9 C) (Radia 7500)
0.830 (55 C) (Schercemol CP)
Density:
7.0 lb/gal (55 C) (Schercemol CP)
Visc.:
4.30 cps (98.9 C) (Radia 7500)
M.P.:
45–53 C (Kemester CP)
45–55 C (Standamul 1616)
49 C (Radia 7500)
50 C (Cutina CP; Cyclochem CP)
51–55 C (Kessco 653)
54 C (Schercemol CP)
Flash Pt.:
> 170 C (OC) (Schercemol CP)
214 C (COC) (Radia 7500)
HLB:
9 (Cutina CP; Standamul 1616)
Acid No.:
1.0 (Kemester CP)
1.0 max. (Cutina CP)
2.0 max. (Kessco 653; Schercemol CP)
3.0 max. (Cyclochem CP)
Iodine No.:
Nil (Schercemol CP)
1.0 (Kemester CP)
1.0 max. (Cutina CP; Kessco 653)
Saponification No.:
109–117 (Kessco 653)
110–125 (Schercemol CP)
114 (Cyclochem CP)
115–130 (Standamul 1616)
116–126 (Cutina CP)

Cetyl palmitate *(cont'd.)*

STD. PKGS.:
200 lb net fiber drums (Kessco 653)

Cetyl stearate *(CTFA)*

SYNONYMS:
1-Hexadecyloctadecanoate
n-Hexadecyl stearate
Octadecanoic acid, 1-hexadecyl ester
EMPIRICAL FORMULA:
$C_{34}H_{68}O_2$
STRUCTURE:

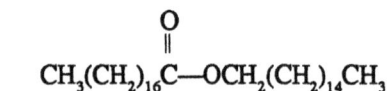

$$\overset{\overset{\textstyle O}{\|}}{CH_3(CH_2)_{16}C}\!\!-\!\!OCH_2(CH_2)_{14}CH_3$$

CAS No.:
1190-63-2
TRADENAME EQUIVALENTS:
Radia 7501 [Synfina-Oleofin]
Schercemol CS [Scher]
CATEGORY:
Lubricant, chemical intermediate, conditioner, opacifier, sheen additive
APPLICATIONS:
Cosmetic industry preparations: conditioners (Schercemol CS); creams and lotions (Schercemol CS); hair preparations (Schercemol CS); makeup (Schercemol CS)
Industrial applications: chemical synthesis (Radia 7501); lamination (Radia 7501); lubricating/cutting oils (Radia 7501); plastics (Radia 7501); rust inhibitors (Radia 7501); textile/leather processing (Radia 7501)
PROPERTIES:
Form:
Flakes (Radia 7501)
Powder (Radia 7501)
Waxy solid (Schercemol CS)
Color:
White (Schercemol CS)
Odor:
Slight, typical (Schercemol CS)
Solubility:
Sol. in aromatic hydrocarbons (Schercemol CS)
Sol. in chlorinated hydrocarbons (Schercemol CS)

Sol. in esters (Schercemol CS)
Insol. in glycerol (Schercemol CS)
Disp. in glycol ethers (Schercemol CS)
Disp. in glycols (Schercemol CS)
Sol. in min. oil (Schercemol CS)
Sol. in min. spirits (Schercemol CS)
Sol. in trichlorethylene @ 10% (Radia 7501)
Disp. in veg. oils (Schercemol CS)
Insol. in water (Schercemol CS)
M.W.:
494 (theoret.) (Schercemol CS)
534 (avg.) (Radia 7501)
Sp.gr.:
0.807 (98.9 C) (Radia 7501)
0.830 ± 0.01 (55 C) (Schercemol CS)
Visc.:
4.60 cps (98.9 C) (Radia 7501)
M.P.:
56 C (Radia 7501)
Setting Pt.:
52–56 C (Schercemol CS)
Flash Pt.:
> 200 C (OC) (Schercemol CS)
216 C (COC) (Radia 7501)
Acid No.:
2.0 max. (Schercemol CS)
Iodine No.:
1.0 max. (Schercemol CS)
Saponification No.:
113 ± 5.0 (Schercemol CS)

Cetyl trimethyl ammonium bromide

SYNONYMS:
CETAB
Cetrimonium bromide (CTFA)
1-Hexadecanaminium, N,N,N-trimethyl-, bromide
N,N,N-Trimethyl-1-hexadecanaminium bromide
EMPIRICAL FORMULA:
$C_{19}H_{42}N \cdot Br$

Cetyl trimethyl ammonium bromide *(cont'd.)*

STRUCTURE:

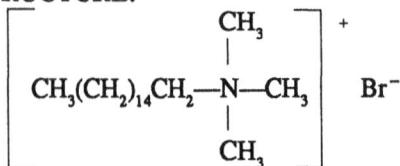

CAS No.:

57-09-0

TRADENAME EQUIVALENTS:

Acetoquat CTAB [Aceto]

Bromat [Hexcel]

Cycloton M242B/99 [Alcolac]

CATEGORY:

Conditioner, germicide, sanitizer, surfactant, emulsifier, antistat

APPLICATIONS:

Cosmetic industry preparations: (Bromat); conditioners (Cycloton M242B/99); hair
preparations (Bromat)

Industrial cleaners: sanitizers/germicides (Acetoquat CTAB)

Pharmaceutical applications: topical preparations (Bromat)

PROPERTIES:

Form:

Solid (Cycloton M242B/99)

Powder (Acetoquat CTAB; Bromat)

Color:

White (Bromat)

Odor:

Characteristic (Bromat)

Composition:

95% active (Acetoquat CTAB)

98% min. assay (Bromat)

100% active (Cycloton M242B/99)

Solubility:

Sol. in alcohols (Bromat)

Sol. in chloroform (Bromat)

Sol. in water (Bromat)

Ionic Nature:

Cationic (Acetoquat CTAB; Bromat; Cycloton M242B/99)

pH:

5.0–8.0 (Bromat)

Surface Tension:

40.0 dynes/cm (Bromat)

Cetyl trimethyl ammonium chloride

SYNONYMS:

Cetrimonium chloride (CTFA)
1-Hexadecanaminium, N,N,N-trimethylchloride
Hexadecyl trimethyl ammonium chloride
Palmityl trimethyl ammonium chloride
N,N,N-Trimethyl-1-hexadecanaminium chloride
Trimethyl hexadecyl ammonium chloride

EMPIRICAL FORMULA:

$C_{19}H_{42}N \cdot Cl$

STRUCTURE:

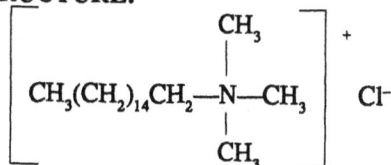

CAS No.:

112-02-7

TRADENAME EQUIVALENTS:

Adogen 444 [Sherex]
Ammonyx CETAC, CETAC-30 [Millmaster-Onyx]
Arquad 16-29, 16-50 [Akzo/Armak; Akzo BV]
Bio-Quat ASH-29 [Bio-Lab; Lonza]
Carsoquat CT-429, CTM-29, CTM-429 [Lonza]
Chemquat 16-50 [Chemax]
Cycloton M242C/29 [Alcolac]
Dehyquart A [Henkel]
Genamin CTAC [Hoechst-Celanese AG]
Incroquat CTC-25, CTC-30 [Croda]
Nissan Cation PB-40, PB-300 [Nippon Oils & Fats]
Querton 16CL29, 16CL50, 24CL35 [Berol Nobel]
Radiaquat 6444 [Synfina-Oleofin]
Synoquart P50 [Aquatec Quimica]
Tequat BC [Tessilchimica SpA]
Variquat E228, E290 [Sherex]

CATEGORY:

Conditioner, softener, emollient, emulsifier, surfactant, corrosion inhibitor, antistat, detergent, foaming agent, dispersant, wetting agent, germicide, coagulant, raw material

APPLICATIONS:

Cleansers: cleansing creams (Ammonyx CETAC); cleansing lotions (Ammonyx CETAC, CETAC-30)

Cosmetic industry preparations: conditioners (Adogen 444; Ammonyx CETAC;

Cetyl trimethyl ammonium chloride (cont'd.)

Arquad 16-29, 16-50; Carsoquat CTM-29, CTM-429; Dehyquart A; Variquat E228); hair preparations (Ammonyx CETAC, CETAC-30; Carsoquat CT-429; Cycloton M242C/29; Dehyquart A; Genamin CTAC; Incroquat CTC-25, CTC-30; Synoquart P50; Tequat BC; Variquat E228); personal care products (Ammonyx CETAC-30; Variquat E290); shampoos (Synoquart P50); skin preparations (Carsoquat CT-429)

Farm products: herbicides (Nissan Cation PB-40, PB-300)

Food applications: (Nissan Cation PB-40, PB-300); fermentation aid (Adogen 444)

Industrial applications: dyes and pigments (Nissan Cation PB-40, PB-300); industrial processing (Querton 16CL29, 24CL35); metalworking (Chemquat 16-50); mineral flotation (Arquad 16-29, 16-50); paint mfg. (Chemquat 16-50); paper mfg. (Nissan Cation PB-40, PB-300); plastics (Bio-Quat ASH-29; Nissan Cation PB-40, PB-300); textile/leather processing (Arquad 16-29, 16-50; Bio-Quat ASH-29; Nissan Cation PB-40, PB-300); water/sewage treatment (Arquad 16-29, 16-50; Nissan Cation PB-40, PB-300)

Industrial cleaners: metal processing surfactants (Arquad 16-29, 16-50)

Pharmaceutical applications: (Carsoquat CTM-29, CTM-429); antibiotics mfg. (Carsoquat CTM-29, CTM-429); deodorant (Dehyquart A)

PROPERTIES:
Form:
Liquid (Adogen 444; Ammonyx CETAC, CETAC-30; Arquad 16-29, 16-50; Carsoquat CTM-29, CTM-429; Cycloton M242C/29; Genamin CTAC; Incroquat CTC-25, CTC-30; Nissan Cation PB-40, PB-300; Querton 16CL29, 24CL35; Radiaquat 6444; Synoquart P50; Tequat BC; Variquat E228)

Clear liquid (Dehyquart A; Variquat E290)

Color:
Off-white (Carsoquat CTM-29, CTM429)

Light/pale yellow (Dehyquart A; Nissan Cation PB-40, PB-300)

APHA 100 max. (Incroquat CTC-25)

Gardner 1 max. (Variquat E290)

Gardner 2 max. (Incroquat CTC-30)

Gardner 6 max. (Adogen 444; Arquad 16-29, 16-50)

Composition:
24–26% active (Dehyquart A; Incroquat CTC-25)

24–26% quaternary; 25% min. solids (Ammonyx CETAC)

27% active min. (Nissan Cation PB-300)

28.5–30.0% in water (Variquat E290)

29% active (Cycloton M242C/29; Radiaquat 6444)

29% active min. (Incroquat CTC-30)

29% active in water (Bio-Quat ASH-29; Carsoquat CTM-29, CTM-429)

29% active in aq. isopropanol (Arquad 16-29)

29% min. quaternary; 30% min. solids (Ammonyx CETAC-30)

35% conc. (Querton 24CL35)

40% active min. (Nissan Cation PB-40)

49–52% quat (Adogen 444)

50% active (Chemquat 16-50; Querton 16CL29; Radiaquat 6444)

50% active in aq. isopropanol (Arquad 16-50)

Solubility:

Sol. in water (Carsoquat CT-429)

Ionic Nature:

Cationic (Adogen 444; Ammonyx CETAC, CETAC-30; Bio-Quat ASH-29; Carsoquat CTM-29, CTM-429; Cycloton M242C/29; Dehyquart A; Incroquat CTC-25, CTC-30; Nissan Cation PB-40, PB-300; Querton 16CL29, 16CL50, 24CL35; Radiaquat 6444; Synoquart P50; Variquat E229, E290)

M.W.:

319 (Adogen 444; Arquad 16-29, 16-50; Carsoquat CTM29, CTM429)

Sp.gr.:

0.88 (Arquad 16-50)

0.93 (25/20 C) (Ammonyx CETAC, CETAC-30)

0.96 (Arquad 16-29)

0.968 (Carsoquat CTM-29, CTM-429)

Density:

8.1 lb/gal (Carsoquat CTM-29, CTM-429)

F.P.:

61 F (Arquad 16-29)

Flash Pt.:

58 F (PM) (Adogen 444)

< 80 F (Arquad 16-50)

> 200 F (Ammonyx CETAC, CETAC-30)

> 212 F (Arquad 16-29)

HLB:

15.8 (Arquad 16-29, 16-50)

Stability:

Good (Arquad 16-29, 16-50)

pH:

3.5–4.0 (2% active) (Carsoquat CTM-429; Incroquat CTC-25)

3.5–4.0 (5%) (Incroquat CTC-30)

7.0–9.0 (10% active) (Carsoquat CTM-29)

Surface Tension:

34 dynes/cm (0.1%) (Arquad 16-50)

Biodegradable: (Arquad 16-29, 16-50)

TOXICITY/HANDLING:

Skin irritant, severe eye irritant; wear protective goggles, gloves (Arquad 16-29, 16-50)

Irritating to skin and eyes in conc. form (Dehyquart A)

Avoid prolonged contact with skin (Incroquat CTC-25, CTC-30)

Cetyl trimethyl ammonium chloride *(cont' d.)*

Corrosive (Arquad 16-29)
STORAGE/HANDLING:
Flammable (Arquad 16-50)
Store in sealed containers below 30 C and protected from frost to obtain 3 yr shelf life
(Dehyquart A)
Store in a cool, dry place (Incroquat CTC-25, CTC-30)
STD. PKGS.:
15-kg Atron can, 180 kg drum (Nissan Cation PB-40)
16-kg Atron can, 170 kg drum (Nissan Cation PB-300)
55-gal (425 lb net) Liquipak container (Incroquat CTC-25)
55-gal (450 lb net) lined drum (Incroquat CTC-30)
242 lb net steel drums (Dehyquart A)

Cholesterol (CTFA)

SYNONYMS:
Cholest-5-en-3-ol (3β)-
EMPIRICAL FORMULA:
$C_{27}H_{46}O$
STRUCTURE:

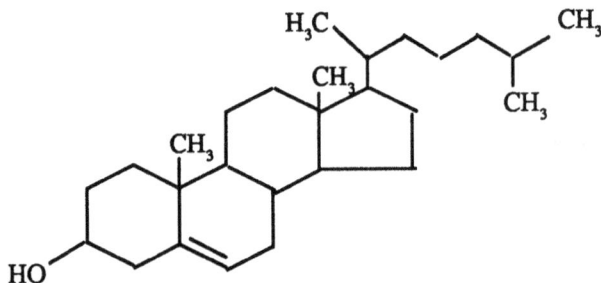

CAS No.:
57-88-5
TRADENAME EQUIVALENTS:
Dastar [Croda Ltd.]
Nimlesterol 1730, 1732 [Henkel/Emery]
Generically sold by:
[Croda (USP); Croda Ltd.; Duphar BV (BP, USP, and tech. grades); RITA]
CATEGORY:
Conditioner, emollient, emulsifier, penetrant, base, raw material, moisturizer

APPLICATIONS:
  Cosmetic industry preparations: (generic—Croda Ltd., Duphar (BP, USP, Tech.));
    hair preparations (Dastar); skin preparations (Dastar; Nimlesterol 1730, 1732)
  Pharmaceutical applications: (generic—Croda Ltd.); steroid hormones (generic—
    Duphar (USP))
PROPERTIES:
Form:
  Liquid (Nimlesterol 1732)
  Powder (generic—RITA, Croda (USP))
  Solid (generic—Croda Ltd., Duphar (BP, USP, Tech.))
  Crystalline wax (Dastar)
Composition:
  70–80% conc. (generic—Duphar (Tech.))
  100% conc. (generic—Croda (USP), Croda Ltd.)
Solubility:
  Sol. in isopropanol (Nimlesterol 1730)
  Sol. in min. oil (Nimlesterol 1730)
  Insol. in water (Nimlesterol 1730)
Ionic Nature:
  Nonionic (generic—Croda (USP), Croda Ltd., Duphar (BP, USP, Tech.))
HLB:
  1.0 (generic—Croda (USP))

Cocamidopropyl dimethylamine (CTFA)

SYNONYMS:
  Amides, coco, N-[3-(dimethylamino) propyl]-
  N-[3-(Dimethylamino) propyl] coco amides
STRUCTURE:

$$RC\overset{\overset{\displaystyle O}{\|}}{-}NH-(CH_2)_3-N\begin{smallmatrix}\nearrow CH_3 \\ \searrow CH_3\end{smallmatrix}$$

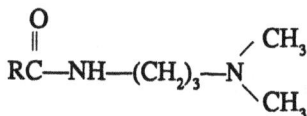

  where RCO⁻ represents the coconut acid radical
CAS No.:
  68140-01-2
TRADENAME EQUIVALENTS:
  Carsamine CB [Lonza]
  Cyclomide CODI [Alcolac]
  Incromine CB [Croda]

Cocamidopropyl dimethylamine *(cont'd.)*

Jordamine DAPL, DMCAPA, SHCFA [PPG-Mazer]
Lexamine C-13 [Inolex]
Mackine 101 [McIntyre]
Schercodine C [Scher]
Tegamine P-7 [Goldschmidt]
Tego-Amid D5040 [Goldschmidt]

CATEGORY:

Emollient, emulsifier, base, foaming agent, intermediate, conditioner, softener, surfactant

APPLICATIONS:

Bath products: (Incromine CB; Schercodine C)

Cosmetic industry preparations: conditioners (Jordamine DAPL, DMCAPA, SHCFA; Lexamine C-13; Mackine 101; Tegamine P-7); cosmetic base (Cyclomide CODI); creams and lotions (Cyclomide CODI; Lexamine C-13; Tego-Amid D5040); hair preparations (Cyclomide CODI; Incromine CB; Lexamine C-13; Mackine 101; Schercodine C; Tego-Amid D5040); shampoos (Jordamine DAPL, DMCAPA, SHCFA; Lexamine C-13); skin preparations (Jordamine DAPL, DMCAPA, SHCFA)

Industrial applications: chemical synthesis (Jordamine DAPL, DMCAPA, SHCFA); lubricating/cutting oils (Jordamine DAPL, DMCAPA, SHCFA)

PROPERTIES:

Form:

Liquid (Cyclomide CODI; Jordamine DMCAPA; Lexamine C-13; Mackine 101)
Liquid/solid (Incromine CB)
Clear liquid (Carsamine CB)
Soft solid (Schercodine C)
Solid (Jordamine DAPL, SHCFA)
Flake (Tegamine P-7)
Waxy (Tego-Amid D5040)

Color:

Amber (Carsamine CB)
Yellow (Cyclomide CODI; Incromine CB)
Tan (Schercodine C)

Composition:

98% amide min. (Schercodine C)
99% active (Jordamine DAPL, DMCAPA, SHCFA)
100% active, 98% amide content (Cyclomide CODI)
100% active (Incromine CB; Lexamine C-13; Mackine 101; Tegamine P-7)

Solubility:

Sol. in acidic media (Lexamine C-13)

Ionic Nature:

Cationic (Cyclomide CODI; Lexamine C-13; Mackine 101; Schercodine C; Tegamine P-7; Tego-Amid D5040)

Cocamidopropyl dimethylamine *(cont'd.)*

Nonionic/cationic (Incromine CB)
M.W.:
304 (Schercodine C)
Density:
7.6 lb/gal (Carsamine CB)
HLB:
10.2 (Lexamine C-13)
Acid No.:
3.0 max. (Incromine CB)
Alkali No.:
172–180 (Incromine CB)
177–187 (Schercodine C)
Equivalent Wt.:
314–326 (Incromine CB)
TOXICITY/HANDLING:
Avoid prolonged contact with skin (Incromine CB)
STORAGE/HANDLING:
Store in a cool, dry place (Incromine CB)
STD. PKGS.:
55-gal (400 lb net) unlined steel drums (Incromine CB)

Cocamidopropyl dimethylamine lactate (CTFA)

SYNONYMS:
Amides, coco, N-[3-(dimethylamino) propyl], lactates
N-[3-(Dimethylamino) propyl cocamide lactate
STRUCTURE:

$$
\begin{array}{c}
\qquad\qquad\qquad\quad CH_3 \\
\quad O \qquad\qquad\quad | \\
\quad \| \qquad\qquad\qquad | \\
RC-NH-(CH_2)_3-N \cdot HOOCCHCH_3 \\
\qquad\qquad\qquad | \qquad\quad | \\
\qquad\qquad\quad CH_3 \qquad OH
\end{array}
$$

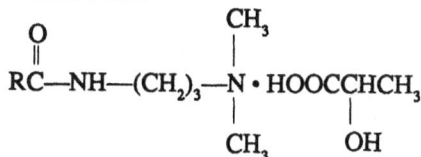

where RCO⁻ represents the coconut acid radical
CAS No.:
68425-42-3
TRADENAME EQUIVALENTS:
Incromate CDL [Croda]
Mackalene 116 [McIntyre]
CATEGORY:
Conditioner, softener, base

57

Cocamidopropyl dimethylamine lactate *(cont' d.)*

APPLICATIONS:

Cosmetic industry preparations: conditioners (Incromate CDL; Mackalene 116); cosmetic base (Incromate CDL); creams and lotions (Incromate CDL); hair preparations (Incromate CDL; Mackalene 116); personal care products (Incromate CDL); shampoos (Incromate CDL)

PROPERTIES:

Form:

Liquid (Mackalene 116)

Viscous liquid (Incromate CDL)

Color:

Yellow (Incromate CDL)

Composition:

95–97% solids (Incromate CDL)

100% conc. (Mackalene 116)

Solubility:

Sol. in water (Incromate CDL)

Ionic Nature:

Cationic (Incromate CDL; Mackalene 116)

pH:

6.0–7.0 (5% sol'n.) (Incromate CDL)

TOXICITY/HANDLING:

Avoid prolonged contact with skin and eyes (Incromate CDL)

STORAGE/HANDLING:

Store in a cool, dry place (Incromate CDL)

STD. PKGS.:

55-gal (450 lb net) polyethylene-lined leverpak (Incromate CDL)

Cocamidopropyl dimethylamine propionate *(CTFA)*

SYNONYMS:

Amides, coco, N-[3-(dimethylamino) propyl], propionates

N-[3-(Dimethylamino) propyl] coco amides, propionates

STRUCTURE:

$$RC\overset{\overset{\textstyle O}{\|}}{-}NH-(CH_2)_3-\underset{\underset{\textstyle CH_3}{|}}{\overset{\overset{\textstyle CH_3}{|}}{N}} \cdot HOOCCH_2CH_3$$

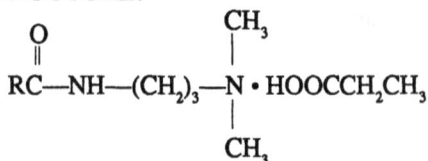

where RCO⁻ represents the coconut acid radical

CAS No.:

68425-43-4

Cocamidopropyl dimethylamine propionate *(cont'd.)*

TRADENAME EQUIVALENTS:
Emcol 1655 [Witco]
Incromate CDP [Croda]
Mackalene 117 [McIntyre]
Mackam CAP [McIntyre]
Richamate 1655, MBF [Richardson]
CATEGORY:
Conditioner, emollient, substantive agent, surfactant, detergent, foaming agent, antistat, dispersant, solubilizer, emulsifier
APPLICATIONS:
Cosmetic industry preparations: (Richamate 1655, MBF); conditioners (Incromate CDP); hair preparations (Richamate 1655, MBF); personal care products (Incromate CDP; Richamate 1655, MBF); shampoos (Emcol 1655; Incromate CDP); skin preparations (Richamate 1655, MBF)
Industrial applications: textile/leather processing (Emcol 1655)
Industrial cleaners: lime soap dispersant (Emcol 1655); textile scouring (Emcol 1655)
PROPERTIES:
Form:
Liquid (Emcol 1655; Incromate CDP; Mackam CAP; Richamate 1655)
Paste (Richamate MBF)
Color:
White (Richamate MBF)
Gardner 4 max. (Incromate CDP)
Gardner 5 (Richamate 1655)
Composition:
30% conc. (Mackam CAP)
33% solids (Richamate MBF)
40% active (Emcol 1655; Incromate CDP)
40% solids (Richamate 1655)
Ionic Nature:
Cationic (Emcol 1655; Incromate CDP; Richamate 1655)
Anionic (Mackam CAP)
Density:
7.9 lb/gal (Richamate MBF)
8.4 lb/gal (Richamate 1655)
pH:
4.0–5.0 (Richamate MBF)
6.0–7.0 (Incromate CDP; Richamate 1655)

Coco amidopropyl betaine

SYNONYMS:
CADG
N-(Carboxymethyl)-N,N-dimethyl-3-[(1-oxococonut) amino]-1-propanaminium hydroxide, inner salt
Cocamidopropyl betaine (CTFA)
Cocamidopropyl dimethyl glycine
Coco amido betaine
Coconut amido betaine
Cocoyl amide propylbetaine
1-Propanaminium, N-(carboxymethyl)-N,N-dimethyl-3-[(1-oxococonut) amino]-, hydroxide, inner salt
Quaternary ammonium compounds, (carboxymethyl) (3-cocoamidopropyl) dimethyl, hydroxides, inner salts

STRUCTURE:

$$RC\overset{O}{\overset{\|}{-}}NH-(CH_2)_3-\overset{\overset{CH_3}{|}}{N^+}-CH_2COO^-$$
$$\underset{CH_3}{|}$$

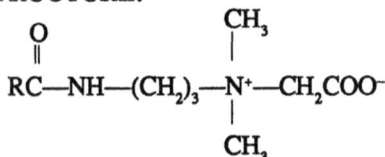

where RCO⁻ represents the coconut acid radical

CAS No.:
61789-40-0

TRADENAME EQUIVALENTS:
Aerosol 30 [Amer. Cyanamid]
Alkateric CAB [Alkaril]
Amonyl 380 BA [Seppic]
Amphosol CA, CG [Stepan; Stepan Europe]
Aremsol A [Ronsheim & Moore]
Carsonam 3, 32, 33-S, 3147 [Lonza]
Chemadene NA-30 [Richardson]
Chimin AX [Tessilchimica]
Cycloteric BET-C30 [Alcolac] (cosmetic grade)
Cycloteric BET-CB [Alcolac] (cosmetic, glycerin-free)
Cycloteric BET-W [Alcolac] (tech.)
Dehyton K [Henkel KGaA]
Deriphat BA, BAW [Henkel]
Emcol DG, NA-30 [Witco]
Emery 5430, 6744, 6748 [Henkel/Emery]
Incronam 30 [Croda; Croda Universal]
Jortaine C, CAB-35, CFA-35 [PPG-Mazer]
Lebon 2000 [Sanyo]
Lexaine C, CG-30, CS [Inolex]
Lonzaine C, CO [Lonza]

Coco amidopropyl betaine (cont'd.)

Mackam 35, 35 HP [McIntyre]
Mafo CAB [PPG-Mazer]
Maprolyte C [Stepan]
Mirataine BD, CB, CBC, CBR [Miranol]
Monateric ADA, ADFA, CAB, COAB, MCB [Mona]
Rewoteric AM-B13 [Rewo GmbH]
Schercotaine CAB, CAB-G [Scher]
Sipoteric CB, COB [Alcolac]
Standapol BAW, BC-35 [Henkel]
Sterling CAB [Canada Packers Ltd.]
Surco Coco Betaine [Stepan]
Tego-Betaine C, L-7, L-10, S, T [Goldschmidt; Goldschmidt AG]
Varion CADG, CADG-HS, CADG-LS, CADG-W [Sherex]
Velvetex BA, BA-35, BC-35, BK-35 [Henkel; Henkel Canada]

CATEGORY:

Conditioner, emulsifier, detergent, solubilizer, raw material, foaming agent, thickener, viscosity builder, viscosity control agent, gelling agent, wetting agent, cloud point depressant, dispersant, surfactant, coupling agent, antistat, emollient, spreading agent, corrosion inhibitor

APPLICATIONS:

Automobile cleaners: car shampoo (Tego-Betaine T)

Bath products: (Aerosol 30; Amphosol CG; Emery 6748; Lexaine C, CG-30, CG-50, CS; Mackam 35; Schercotaine CAB; Standapol BAW, BC-35; Tego-Betaine C, L-7, S; Velvetex BC-35); bubble bath (Mackam 35, 35 HP; Maprolyte C; Mirataine BD; Monateric CAB; Rewoteric AM-B13; Velvetex BA, BA-35, BK-35)

Cleansers: hand cleanser (Amphosol CG; Carsonam 3, 33-S, 3147; Incronam 30; Mafo CAB; Surco Coco Betaine; Belvetex BA, BA-35); skin cleaners (Aerosol 30; Emery 6748; Lexaine C, CG-30, CG-50; Schercotaine CAB; Velvetex BA, BA-35)

Cosmetic industry preparations: (Alkateric CAB; Chemadene NA-30; Cycloteric BET-CB, BET-W; Emcol NA-30; Emery 6748; Incronam 30; Jortaine C, CAB-35, CFA-35; Lonzaine C, CO; Monateric COAB, MCB); conditioners (Lexaine C; Lonzaine C, CO; Monateric CAB; Standapol BAW, BC-35); cosmetic base (Velvetex BK-35); creams and lotions (Carsonam 3, 33-S, 3147; Incronam 30; Tego-Betaine C); hair preparations (Aerosol 30; Emery 6748; Incronam 30; Lexaine C, CG-30, CG-50, CS; Schercotaine CAB; Standapol BAW; Velvetex BC-35, BK-35); personal care products (Aremsol A; Chimin AX; Cycloteric BET-CB, BET-W; Schercotaine CAB-G; Sipoteric CB, COB; Sterling CAB; Tego-Betaine L-10; Varion CADG-HS, CADG-LS, CADG-W; Velvetex BA, BA-35, BC-35); shampoos (Alkateric CAB; Amonyl 380BA; Amphosol CG; Carsonam 3, 33-S, 3147; Cycloteric BET-CB, BET-W; Emery 5430, 6748; Incronam 30; Lebon 2000; Lexaine C, CG-30, CG-50, CS; Lonzaine C, CO; Mackam 35, 35 HP; Mafo CAB; Maprolyte C; Mirataine BD, CB, CBC, CBR; Monateric CAB; Rewoteric AM-B13; Schercotaine CAB; Standapol BC-35; Sterling CAB; Tego-Betaine C, L-7;

Coco amidopropyl betaine (cont'd.)

Varion CADG; Velvetex BA, BA-35, BC-35, BK-35); skin preparations (Incronam 30; Lexaine CS; Standapol BAW, BC-35; Velvetex BC-35, BK-35); toiletries (Emcol NA-30; Incronam 30; Lonzaine C, CO; Sipoteric CB, COB)

Household detergents: (Alkateric CAB; Deriphat BAW; Emcol DG; Jortaine CAB-35; Monateric COAB, MCB; Surco Coco Betaine; Tego-Betaine T); carpet & upholstery shampoos (Carsonam 3, 33-S, 3147; Incronam 30; Mafo CAB); dishwashing (Carsonam 3, 33-S, 3147; Incronam 30; Mackam 35, 35 HP; Mafo CAB; Mirataine CBC, CBR; Surco Coco Betaine; Tego-Betaine C); hard surface cleaner (Tego-Betaine T); heavy-duty cleaner (Amphosol CG; Deriphat BAW; Lonzaine C, CO; Varion CADG-HS, CADG-LS, CADG-W); light-duty cleaners (Lebon 2000; Lexaine CG-30, CG-50; Lonzaine C, CO); liquid detergents (Lexaine CG-30, CG-50)

Industrial applications: (Chemadene NA-30; Cycloteric BET-CB, BET-W; Jortaine CAB-35); cement (Velvetex BA, BA-35); construction (Monateric ADA; Velvetex BA, BA-35); electroplating (Lonzaine C, CO); foam drilling/blanketing (Monateric ADA, ADFA; Velvetex BA, BA-35); metalworking (Lonzaine C, CO; Monateric COAB, MCB); petroleum industry (Monateric COAB, MCB); textile/leather processing (Chemadene NA-30; Lonzaine C, CO)

Industrial cleaners: (Carsonam 32; Deriphat BAW; Monateric COAB, MCB; Surco Coco Betaine; Velvetex BA, BA-35); institutional cleaners (Alkateric CAB); metal processing surfactants (Chemadene NA-30); sanitizers/germicides (Lonzaine C, CO)

Pharmaceutical applications: (Tego-Betaine L-10); antidandruff shampoo (Carsonam 3, 33-S, 3147; Incronam 30); antiseptic soaps (Lonzaine C, CO); disinfectants/germicides (Aerosol 30; Carsonam 3, 33-S, 3147; Lexaine C)

PROPERTIES:
Form:

Liquid (Alkateric CAB; Amonyl 380BA; Aremsol A; Carsonam 3, 33-S, 3147; Chemadene NA-30; Chimin AX; Cycloteric BET C-30; Dehyton K; Deriphat BA, BAW; Emcol DG, NA-30; Emery 5430, 6744, 6748; Jortaine C, CAB-35, CFA-35; Lebon 2000; Lexaine C, CS; Mackam 35, 35 HP; Mafo CAB; Maprolyte C; Mirataine BD, CB, CBC, CBR; Monateric ADFA; Sipoteric CB, COB; Sterling CAB; Surco Coco Betaine; Tego-Betaine C, L-7, L-10, S, T; Varion CADG, CADG-W; Velvetex BA, BC-35)

Clear liquid (Aerosol 30; Amphosol CA, CG; Carsonam 32; Incronam 30; Monateric ADA, CAB, COAB, MCB; Schercotaine CAB, CAB-G; Varion CADG-HS, CADG-LS; Velvetex BA-35, BK-35)

Clear, nonviscous liquid (Lexaine CG-30)

Clear, viscous liquid (Lexaine CG-50)

Nonviscous liquid (Lonzaine C, CO)

Low viscosity liquid (Rewoteric AM-B13)

Color:

Pale (Mirataine CB)

Coco amidopropyl betaine *(cont'd.)*

Straw (Amphosol CA)
Straw to light amber (Lexaine CG-30, CG-50)
Light amber (Standapol BAW, BC-35; Velvetex BC-35)
Amber (Alkateric CAB; Amphosol CG; Carsonam 32; Monateric ADA)
Pale/light yellow (Monateric MCB; Schercotaine CAB, CAB-G; Velvetex BK-35)
Yellow (Cycloteric BET C-30; Monateric CAB, COAB; Varion CADG-HS, CADG-
    LS; Velvetex BA-35)
Gardner 2 max. (Emery 6744; Lonzaine C, CO)
Gardner 3 (Chemadene NA-30)
Gardner 4 (Emery 6748)
Gardner 4 max. (Aerosol 30; Lexaine C)
Gardner 5 (Carsonam 3, 3147)
Gardner 5 max. (Incronam 30)
Gardner 6 max. (Carsonam 33-S)

Odor:

Bland (Alkateric CAB; Lexaine CG-30, CG-50)
Characteristic (Lonzaine C)
Slight, characteristic (Schercotaine CAB)

Composition:

25% conc. (Tego-Betaine T)
28% conc. (Tego-Betaine L-7)
29% active (Lexaine CG-30)
29–31% active (Amphosol CG; Cycloteric BET C-30)
30% active (Amonyl 380BA; Amphosol CA; Chimin AX; Dehyton K; Emery 6748;
    Incronam 30; Lebon 2000; Lexaine C; Lonzaine C, CO; Mirataine BD, CBC, CBR;
    Monateric CAB, MCB; Schercotaine CAB; Tego-Betaine C, L-10)
30% active in water (Carsonam 3, 33-S)
32% active (Monateric COAB; Sterling CAB; Varion CADG)
32% active in water (Alkateric CAB)
33% conc. (Monateric ADFA)
33% active in water (Carsonam 3147)
33–37% solids (Maprolyte C)
34% active (Jortaine C, CAB-35, CFA-35; Standapol BAW)
34–36% solids (Velvetex BK-35)
34–37% solids (Velvetex BA-35)
35% active (Aerosol 30; Lexaine CS; Mackam 35, 35 HP; Mafo CAB; Mirataine CB;
    Rewoteric AM-B13; Sipoteric CB, COB; Standapol BC-35; Velvetex BC-35)
35% solids (Varion CADG-W; Velvetex BC-35)
35–37% solids (Varion CADG-HS, CADG-LS)
37% solids (Chemadene NA-30)
38% solids (Monateric ADA)
40% conc. (Aremsol A)
40–44% solids (Surco Coco Betaine)

Coco amidopropyl betaine *(cont'd.)*

45% active (Deriphat BAW)
45% solids min. (Schercotaine CAB-G)
50% active (Lexaine CG-50)
58–62% solids in ethanol (Carsonam 32)
60% active (Deriphat BA)
60% active in alcohol (Velvetex BA)

Solubility:

Sol. in glycerol trioleate (Emery 6744, 6748)
Sol. in water (Emcol DG, NA-30; Emery 5430, 6748; Sipoteric CB, COB; Velvetex
  BA-35, BK-35); completely sol. (Aerosol 30; Alkateric CAB; Schercotaine CAB);
  sol. in all proportions (Incronam 30); sol. @ 5% (Emery 6744)

Ionic Nature:

Amphoteric (Aerosol 30; Alkateric CAB; Amonyl 380 BA; Amphosol CA; Aremsol
  A; Carsonam 3, 33-S, 3147; Chemadene NA-30; Chimin AX; Cycloteric BET-C30,
  BET-CB, BET-W; Dehyton K; Deriphat BA, BAW; Emcol DG, NA-30; Emery
  6744, 6748; Incronam 30; Jortaine C, CAB-35, CFA-35; Lebon 2000; Lexaine C,
  CG-30, CS; Mackam 35, 35 HP; Mafo CAB; Mirataine BD, CB, CBC, CBR;
  Monateric ADFA, CAB, COAB, MCB; Rewoteric AM-B13; Schercotaine CAB,
  CAB-G; Sipoteric CB, COB; Standapol BAW, BC-35; Sterling CAB; Surco Coco
  Betaine; Tego-Betaine C, L-7, L-10, S, T; Varion CADG, CADG-HS, CADG-LS;
  Velvetex BA, BA-35, BC-35)
Nonionic (Maprolyte C)

M.W.:

386 avg. (Schercotaine CAB)

Sp.gr.:

1.02 (Monateric MCB)
1.039 (Velvetex BK-35)
1.04 (Aerosol 30; Monateric COAB); (25/20 C) (Maprolyte C; Surco Coco Betaine)
1.044 (Lexaine C)
1.045 (Mirataine CB)
1.05 (Velvetex BA-35)
1.05 ± 0.01 (Schercotaine CAB)
1.051 (Lonzaine CO)

Density:

8.2 lb/gal (Velvetex BA)
8.45 lb/gal (Carsonam 3, 33-S, 3147)
8.5 lb/gal (Monateric MCB)
8.7 lb/gal (Chemadene NA-30; Emery 6748; Monateric COAB)
8.71 lb/gal (Mirataine CB)
8.75 lb/gal (Schercotaine CAB)
8.77 lb/gal (Lonzaine CO)
8.8 lb/gal (Emery 6744)

Visc.:

7 cSt (100 F) (Emery 6744)

100 cps max. (Schercotaine CAB; Velvetex BA-35, BK-35)

F.P.:

0 C (Aerosol 30; Lexaine C)

Pour Pt.:

< −2 C (Emery 6744)

Flash Pt.:

> 200 F (Maprolyte C; Surco Coco Betaine)

Cloud Pt.:

Clear to freeze (Mirataine CB)

−2 C max. (Schercotaine CAB)

< 0 C (Velvetex BA-35)

Stability:

Stable in acid and alkaline media (Cycloteric BET C-30; Deriphat BA; Lonzaine C, CO; Velvetex BA-35)

Stable in strong acid and alkaline sol'ns. (Standapol BAW; Velvetex BC-35)

Stable over wide pH range (Amphosol CG; Emery 6748; Schercotaine CAB)

High tolerance to hard water (Emery 6748; Monateric CAB; Velvetex BK-35)

High tolerance to electrolytes (Deriphat BA; Emery 6748; Lonzaine C, CO; Schercotaine CAB)

Good salt tolerance (Monateric ADA)

Good tolerance to high levels of detergent builders (Lonzaine C, CO)

pH:

4.5–5.5 (Cycloteric BET C-30; Lexaine C; Velvetex BK-35); (10% sol'n.) (Lonzaine C)

4.5–6.5 (Amphosol CG)

4.8 (10%) (Monateric MCB)

5.0 (10% sol'n.) (Amphosol CA)

5.0–6.0 (Chemadene NA-30)

5.0–7.0 (Schercotaine CAB-G; Varion CADG-LS)

5.5 ± 1.0 (Schercotaine CAB)

5.5–7.5 (10% sol'n.) (Carsonam 3, 33-S; Incronam 30)

6.0–7.0 (Carsonam 32)

6.0–8.0 (10% sol'n.) (Lonzaine CO)

6.5–8.0 (Lexaine CG-30)

6.5–8.5 (Varion CADG-HS)

7.0 ± 0.5 (Monateric CAB)

7.5 (10% aq.) (Lexaine CG-50)

7.9 (Monateric ADA); (10%) (Monateric COAB)

8.5 (Mirataine CB)

Surface Tension:

33.6 dynes/cm (0.1%) (Lonzaine C)

Coco amidopropyl betaine *(cont'd.)*

33.9 dynes/cm (0.1%) (Lonzaine CO)

35 dynes/cm (Aerosol 30)

Wetting (Draves):

10 s (3-g hook, 1%) (Monateric COAB)

12 s (Velvetex BA-35)

18 s (3-g hook, 1%) (Monateric MCB)

Biodegradable: (Carsonam 3, 33-S, 3147; Chemadene NA-30; Deriphat BA; Incronam 30; Lonzaine C, CO; Schercotaine CAB; Velvetex BA, BA-35, BK-35); partially biodegradable (Aerosol 30)

TOXICITY/HANDLING:

Very mild on skin, eyes (Aerosol 30)

Not believed hazardous; however, it is recommended that eye contact and prolonged skin contact be avoided (Amphosol CG)

Avoid prolonged contact with skin (Incronam 30)

Ocular irritant; moderate skin irritant; relatively harmless by acute oral ingestion (Velvetex BA-35)

Irritating to skin and eyes in conc. form (Velvetex BK-35)

Irritating to the eye; nonirritating to the skin; toxic orally (Varion CADG-LS)

STORAGE/HANDLING:

Store at R.T. in stainless steel storage tanks (Amphosol CA)

Store in closed containers above 7 C (Velvetex BA-35, BK-35)

STD. PKGS.:

55-gal drums (Lexaine C)

55-gal poly-lined drums (Schercotaine CAB)

55-gal (450 lb net) Liquipak containers (Incronam 30)

400 lb net polyethylene-lined fiber drums (Lexaine CG-30)

440 lb net fiber drums, bulk, tank wagons, rail cars (Velvetex BA-35, BK-35)

450 lb net polyethylene-lined fiber drums (Lexaine CG-50)

Coco amine oxide

SYNONYMS:

Amines, coco alkyl dimethyl, oxides

Cocamine oxide (CTFA)

Coco dimethyl amine oxide

Coconut dimethylamine oxide

Dimethyl cocamine oxide

STRUCTURE:

CH$_3$

R—N → O

CH$_3$

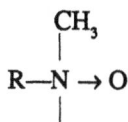

where R represents the coconut radical

CAS No.:

61788-90-7

TRADENAME EQUIVALENTS:

Aromox DMC, DMCD, DMC-W [Akzo/Armak; Akzo BV]

Barlox 12 [Lonza]

Conco XA-Y [Continental Chem.]

Cyclomox C [Alcolac]

Genaminox CS, KC [Hoechst-Celanese AG]

Hartox DMCD [Hart Prod.]

Mackamine CO [McIntyre]

CATEGORY:

Emollient, wetting agent, emulsifier, foaming agent, foam stabilizer, stabilizer, thickener, viscosity modifier, antistat, detergent, dyeing assistant

APPLICATIONS:

Cosmetic industry preparations: (Aromox DMC, DMCD, DMC-W; Conco XA-Y; Hartox DMCD); personal care products (Genaminox CS, KC); shampoos (Aromox DMC, DMC-W)

Farm products: agricultural chemicals (Aromox DMC, DMC-W)

Household detergents: (Aromox DMC, DMCD, DMC-W; Barlox 12; Conco XA-Y)

Industrial applications: dyes and pigments (Aromox DMC, DMC-W; Hartox DMCD); electroplating (Conco XA-Y); lubricating/cutting oils (Aromox DMC, DMC-W); metalworking (Aromox DMC, DMC-W); paint mfg. (Aromox DMC, DMC-W); paper mfg. (Aromox DMC, DMC-W); petroleum industry (Aromox DMC, DMC-W); photography (Conco XA-Y); plastics (Aromox DMC, DMC-W); polymers/polymerization (Conco XA-Y); rubber (Aromox DMC, DMC-W); textile/leather processing (Aromox DMC, DMC-W; Conco XA-Y; Hartox DMCD)

Industrial cleaners: (Conco XA-Y)

PROPERTIES:

Form:

Liquid (Aromox DMCD; Barlox 12; Cyclomox C; Genaminox CS, KC; Hartox DMCD; Mackamine CO)

Clear liquid (Aromox DMC, DMC-W)

Color:

Gardner 1 (Aromox DMC, DMC-W)

Gardner 1 max. (Aromox DMCD)

Coco amine oxide *(cont'd.)*

Composition:
  30% active (Barlox 12; Conco XA-Y; Cyclomox C; Genaminox CS, KC; Mackamine CO)
  30% active in water (Aromox DMC-W)
  39% active min. in isopropanol (Aromox DMCD)
  40% active (Hartox DMCD)
  40% active in aq. isopropanol (Aromox DMC)
Solubility:
  Sol. in ethanol (Conco XA-Y)
  Sol. in hexylene glycol (Conco XA-Y)
  Sol. in isopropanol (Conco XA-Y)
  Sol. in water (Conco XA-Y)
Ionic Nature:
  Cationic (Aromox DMC, DMC-W; Hartox DMCD)
  Cationic/amphoteric (Conco XA-Y)
  Amphoteric (Mackamine CO)
  Nonionic (Barlox 12; Cyclomox C; Genaminox CS, KC)
Sp.gr.:
  0.89 (Aromox DMC, DMCD)
  0.971 (Aromox DMC-W)
Visc.:
  17 cp (Aromox DMC-W)
  18 cp (Aromox DMC)
Pour Pt.:
  < −25 F (Aromox DMC)
  35 F (Aromox DMC-W)
Flash Pt.:
  21 C (APCC) (Aromox DMCD)
  < 80 F (Aromox DMC)
  > 212 F (Aromox DMC-W)
Fire Pt.:
  100 F (Aromox DMC)
Cloud Pt.:
  6 F (Aromox DMC)
  34 F (Aromox DMC-W)
HLB:
  18.6 (Aromox DMCD)
pH:
  6.0–9.0 (Aromox DMCD)
Surface Tension:
  32.5 dynes/cm (0.1%) (Aromox DMC-W)
  33.1 dynes/cm (0.1%) (Aromox DMC)
Biodegradable: (Aromox DMC, DMC-W; Conco XA-Y)

Coco amine oxide *(cont'd.)*

TOXICITY/HANDLING:
Protective clothing, goggles, gloves should be worn (Aromox DMC, DMC-W)
STORAGE/HANDLING:
SS, glass or fiberglass reinforced polyester tanks (Aromox DMC, DMC-W)

Coco-betaine *(CTFA)*

SYNONYMS:
Coco dimethyl glycine
Coconut betaine
Quaternary ammonium compounds, (carboxymethyl) (coco alkyl) dimethyl, hydroxides, inner salt
STRUCTURE:

$$CH_3$$
$$|$$
$$R—N^+—CH_2COO^-$$
$$|$$
$$CH_3$$

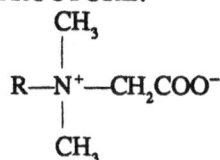

where R represents the coconut radical
CAS No.:
68424-94-2
RD No. 977056-36-2
TRADENAME EQUIVALENTS:
Accobetaine CL [Capital City]
Alkateric BC [Alkaril; Alkaril UK; Alkaril Canada]
Amonyl 265BA [Seppic]
Ampho B11-34 [Capital City]
Amphoram CB A30 [Ceca SA]
Carsonam BCW [Lonza]
Cycloteric BET-C41 [Alcolac]
Dehyton AB-30 [Henkel Canada; Henkel KGaA]
Emcol CC-37-18 [Witco]
Hartaine CB-40 [Hart Prod.]
Jortaine CB40 [PPG-Mazer]
Lonzaine 12C [Lonza]
Mackam CB, CB-35, CB-LS [McIntyre]
Mirataine CDMB [Miranol]
Standapol AB-45 [Henkel]
Velvetex AB-45, BC [Henkel; Henkel Canada]

Coco-betaine *(cont'd.)*

CATEGORY:
Conditioner, detergent, wetting agent, emulsifier, solubilizer, coupling agent, hydrotrope, foaming agent, surfactant, viscosity builder/modifier, gelling agent, foam
stabilizer, antistat, dyeing assistant, bactericide

APPLICATIONS:
Bath products: bubble bath (Carsonam BCW; Cycloteric BET-C41; Lonzaine 12C)

Cleansers: hand cleanser (Accobetaine CL; Cycloteric BET-C41; Velvetex AB-45,
BC)

Cosmetic industry preparations: (Alkateric BC); conditioners (Alkateric BC; Lonzaine
12C; Standapol AB-45); cosmetic base (Accobetaine CL; Ampho B11-34; Dehyton
AB-30; Emcol CC-37-18; Hartaine CB-40); hair preparations (Alkateric BC);
personal care products (Accobetaine CL; Ampho B11-34; Dehyton AB-30; Emcol
CC-37-18; Jortaine CB-40; Standapol AB-45); shampoos (Accobetaine CL; Alkateric BC; Amonyl 265BA; Carsonam BCW; Cycloteric BET-C41; Dehyton AB-
30; Lonzaine 12C; Mirataine CDMB; Standapol AB-45; Velvetex AB-45, BC)

Household detergents: (Alkateric BC; Carsonam BCW); detergent base (Hartaine CB-
40); dishwashing (Accobetaine CL; Mirataine CDMB)

Industrial applications: (Jortaine CB-40; Lonzaine 12C); dyes and pigments (Hartaine
CB-40); industrial processing (Mackam CB, CB-35, CB-LS); plastics (Hartaine
CB-40); polishes and waxes (Hartaine CB-40); textile/leather processing (Alkateric
BC; Hartaine CB-40)

Industrial cleaners: (Carsonam BCW); textile scouring (Accobetaine CL)

PROPERTIES:

Form:
Liquid (Accobetaine CL; Alkateric BC; Amonyl 265BA; Amphoram CB A30;
Carsonam BCW; Dehyton AB-30; Emcol CC-37-18; Hartaine CB-40; Jortaine CB-
40; Lonzaine 12C; Mackam CB, CB-35, CB-LS; Mirataine CDMB; Standapol AB-
45; Velvetex AB-45, BC)
Clear liquid (Ampho B11-34)

Color:
Very light (Hartaine CB-40)
Pale straw (Alkateric BC)
Light amber (Dehyton AB-30)
Light yellow (Ampho B11-34)
Yellow (Accobetaine CL; Standapol AB-45)
Gardner 2 (Carsonam BCW; Velvetex AB-45, BC)

Odor:
Low (Alkateric BC)
Mild (Ampho B11-34)
Pleasant (Accobetaine CL)

Composition:
29–30% conc. (Dehyton AB-30)
30% conc. (Amonyl 265BA; Amphoram CB A30; Mirataine CDMB)

31% active (Alkateric BC)
32% conc. (Mackam CB-LS)
35% active (Accobetaine CL; Ampho B11-34; Lonzaine 12C; Mackam CB-35)
36.5% active min. (Standapol AB-45)
39% active (Jortaine CB-40)
40% conc. (Hartaine CB-40)
43–45% solids (Velvetex AB-45)
44% solids in water (Carsonam BCW)
45% conc. (Mackam CB)
59% active (Velvetex BC)
Solubility:
Sol. in alcohols (Standapol AB-45)
Sol. in alkanolamides (Standapol AB-45)
Sol. in glycols (Standapol AB-45)
Sol. in water (Accobetaine CL; Alkateric BC; Ampho B11-34; Hartaine CB-40)
Ionic Nature:
Amphoteric (Alkateric BC; Amonyl 265BA; Ampho B11-34; Amphoram CB A30; Cycloteric BET-C41; Dehyton AB-30; Emcol CC-37-18; Jortaine CB-40; Lonzaine 12C; Mackam CB, CB-35, CB-LS; Mirataine CDMB; Standapol AB-45; Velvetex AB-45, BC)
Sp.gr.:
0.96 (Ampho B11-34)
1.03 (Velvetex AB-45)
1.044 (Carsonam BCW)
1.06 (Hartaine CB-40)
Density:
7.7 lb/gal (Velvetex BC)
8.0 lb/gal (Ampho B11-34)
8.5 lb/gal (Carsonam BCW)
8.85 lb/gal (Hartaine CB-40)
Visc.:
< 100 cps (Velvetex AB-45)
F.P.:
< 0 C (Accobetaine CL; Alkateric BC)
Cloud Pt.:
< 0 C (Standapol AB-45; Velvetex AB-45)
Stability:
Stable over wide pH range (Mackam CB)
Stable in strong acid and alkaline sol'n. (Standapol AB-45; Velvetex AB-45)
pH:
6.5–7.5 (10% sol'n.) (Standapol AB-45)
6.5–8.5 (Velvetex AB-45)
7.5–8.5 (Alkateric BC)

Coco-betaine *(cont'd.)*

8.0 (Carsonam BCW)
Biodegradable: (Ampho B11-34; Carsonam BCW; Lonzaine 12C; Velvetex AB-45, BC)
TOXICITY/HANDLING:
Nontoxic (Accobetaine CL)
STD. PKGS.:
Drums (Hartaine CB-40)

Coconut oil *(CTFA)*

SYNONYMS:
Copra oil
CAS No.:
8001-31-8
TRADENAME EQUIVALENTS:
Cobee 76 [Stepan/PVO] (refined, bleached, deodorized)
Super Refined Coconut Oil [Croda]
CATEGORY:
Emollient, superfatting agent, clouding agent,
APPLICATIONS:
Cosmetic industry preparations: conditioners (Cobee 76)
Food applications: (Cobee 76)
Pharmaceutical applications: (Cobee 76); sunscreens (Super Refined Coconut Oil)
PROPERTIES:
Form:
Soft solid (Cobee 76)
Color:
APHA 20 (Super Refined Coconut Oil)
Lovibond 20/2 (Cobee 76)
Odor:
None (Cobee 76)
Taste:
None (Cobee 76)
M.P.:
76–80 F (Cobee 76)

Decyl oleate (CTFA)

SYNONYMS:
Decyl-9-octadecenoate
9-Octadecenoate, decyl ester

EMPIRICAL FORMULA:
$C_{28}H_{54}O_2$

STRUCTURE:

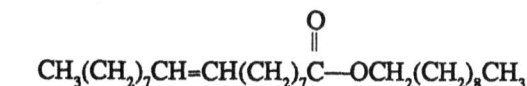

$$CH_3(CH_2)_7CH=CH(CH_2)_7C-OCH_2(CH_2)_8CH_3$$

CAS No.:
3687-46-5

TRADENAME EQUIVALENTS:
Ceraphyl 140 [Van Dyk]
Cetiol V [Henkel; Henkel Canada; Henkel KGaA]
Crodamol DO [Croda Ltd.]
Schercemol DO [Scher]
Standamul CTV [Henkel]

CATEGORY:
Emollient, lubricant, penetrant, base, carrier, dispersant, cosolvent, moisturizer

APPLICATIONS:
Bath products: bubble bath (Cetiol V); bath oils (Ceraphyl 140; Cetiol V; Standamul CTV)

Cosmetic industry preparations: (Crodamol DO; Schercemol DO; Standamul CTV); cosmetic base (Cetiol V; Standamul CTV); creams and lotions (Ceraphyl 140; Cetiol V; Standamul CTV); makeup (Ceraphyl 140; Cetiol V; Crodamol DO; Standamul CTV); personal care products (Cetiol V; Schercemol DO; Standamul CTV); pressed powders (Ceraphyl 140)

Industrial applications: dyes and pigments (Ceraphyl 140)

Pharmaceutical applications: (Cetiol V; Crodamol DO; Standamul CTV)

PROPERTIES:
Form:
Liquid (Ceraphyl 140; Crodamol DO)
Clear liquid (Schercemol DO)
Low viscosity liquid (Cetiol V)
Clear, low viscosity liquid (Standamul CTV)

Decyl oleate *(cont'd.)*

Color:
   White to straw (Ceraphyl 140)
   Pale straw (Crodamol DO)
   Yellow (Cetiol V)
   Gardner 3 max. (Schercemol DO)
Odor:
   Bland characteristic (Schercemol DO)
   Faint characteristic (Standamul CTV)
   Mild characteristic (Crodamol DO)
Taste:
   Bland (Crodamol DO)
Composition:
   87% min. ester content (Schercemol DO)
   100% active (Ceraphyl 140; Standamul CTV)
Solubility:
   Sol. in anhydrous ethanol (Standamul CTV)
   Sol. in aromatic hydrocarbons (Schercemol DO)
   Sol. in castor oil @ 10% (Cetiol V; Standamul CTV)
   Sol. in most esters (Schercemol DO)
   Sol. in 95% ethanol (Ceraphyl 140)
   Sol. in isopropyl myristate (Ceraphyl 140); sol. @ 10% (Cetiol V; Standamul CTV)
   Sol. in min. oil (Ceraphyl 140); sol. @ 10% (Cetiol V; Schercemol DO; Standamul CTV)
   Sol. in min. spirits (Schercemol DO)
   Sol. in oleyl alcohol (Ceraphyl 140); sol. @ 10% (Cetiol V; Standamul CTV)
   Sol. in peanut oil (Ceraphyl 140)
   Disp. in silicone fluid @ 10% (Cetiol V)
   Insol. in water @ 10% (Cetiol V; Schercemol DO); insol. @ 5% (Ceraphyl 140)
M.W.:
   422 (theoret.) (Schercemol DO)
Sp.gr.:
   0.855–0.865 (Ceraphyl 140)
   0.86 ± 0.02 (Schercemol DO)
   0.86–0.87 (20 C) (Standamul CTV)
Density:
   0.86 g/ml (Cetiol V)
F.P.:
   −13 C (Schercemol DO)
Gel Pt.:
   5 C max. (Standamul CTV)
Solidification Pt.:
   < 0 C (Cetiol V)

Flash Pt.:
240–260 C (Cetiol V)
Cloud Pt.:
10 C max. (Cetiol V; Standamul CTV)
HLB:
9 (Cetiol V; Standamul CTV)
Acid No.:
0.5 max. (Standamul CTV)
1 max. (Cetiol V)
5 max. (Ceraphyl 140; Schercemol DO)
Iodine No.:
55–65 (Cetiol V; Standamul CTV)
57 ± 5 (Schercemol DO)
57–60 (Ceraphyl 140)
Saponification No.:
130–140 (Cetiol V)
130–150 (Standamul CTV)
132–142 (Ceraphyl 140)
135 ± 5 (Schercemol DO)
Hydroxyl No.:
5.0 max. (Standamul CTV)
Ref. Index:
1.450 (Cetiol V)
1.4530–1.4555 (Ceraphyl 140)
1.454 (Schercemol DO)
STORAGE/HANDLING:
Subject to oxidation (Standamul CTV)
STD. PKGS.:
385 lb net closed-head steel drums or bulk (Standamul CTV)

Dibutyl adipate (CTFA)

SYNONYMS:
Hexanedioic acid, dibutyl ester
EMPIRICAL FORMULA:
$C_{14}H_{26}O_4$
STRUCTURE:

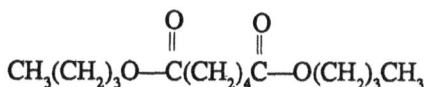

$$CH_3(CH_2)_3O-C(CH_2)_4C-O(CH_2)_3CH_3$$

Dibutyl adipate *(cont' d.)*

CAS No.:
105-99-7
TRADENAME EQUIVALENTS:
Cetiol B [Henkel Canada; Henkel KGaA]
Rilanit DBA [Henkel KGaA]
CATEGORY:
Emollient, lubricant, fatting agent
APPLICATIONS:
Cosmetic industry preparations: creams and lotions (Cetiol B)
Industrial applications: ceramics (Rilanit BS); lubricating/cutting oils (Rilanit BS); metalworking (Rilanit BS); textile/leather processing (Rilanit BS)
PROPERTIES:
Form:
Liquid (Cetiol B; Rilanit BS)
Composition:
100% active (Cetiol B)

Dihydrogenated tallow dimethyl ammonium chloride

SYNONYMS:
Dimethyl di(hydrogenated tallow) ammonium chloride
Dimethyl tallow ammonium chloride, dihydrogenated
Ditallowalkonium chloride
Quaternary ammonium compounds, bis (hydrogenated tallow alkyl) dimethyl, chlorides
Quaternium-18 (CTFA)
STRUCTURE:

$$\left[\begin{array}{c} CH_3 \\ | \\ R-N-R \\ | \\ CH_3 \end{array} \right]^{+} \quad Cl^{-}$$

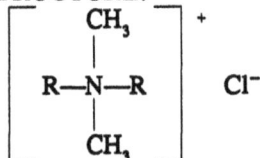

where R represents hydrogenated tallow fatty radicals
CAS No.:
61789-80-8
TRADENAME EQUIVALENTS:
Accosoft 707 [Capital City]
Adogen 442, 442H [Sherex]
Arquad 2HT, 2HT-75, 88 [Akzo/Armak; Akzo BV]
Crapol FU-25 [Pulcra SA]

Dihydrogenated tallow dimethyl ammonium chloride
(cont'd.)

Cycloton D261C/70, D261C/75 [Alcolac]
Jet Quat 2HT-75 [Jetco]
Kemamine Q-9702C [Humko/Witco]
Noramium M2SH [Ceca SA]
Querton 442, 442-11, 442-82 [Berol Nobel]
Radiaquat 6442 [Synfina-Oleofin]
Varisoft 110, 3262, DHT [Sherex]

CATEGORY:

Softener, conditioner, surfactant, antistat, bacteriostat, germicide, emulsifier, foaming agent, wetting agent, corrosion inhibitor, dispersant, viscosifier

APPLICATIONS:

Cosmetic industry preparations: conditioners (Arquad 2HT-75; Crapol FU-25; Cycloton D261C/70, D261C/75; Noramium M2SH; Varisoft DHT); hair rinses (Varisoft DHT)

Household applications: fabric softeners (Accosoft 707; Adogen 442; Varisoft 110)

Industrial applications: asphalt (Jet Quat 2HT-75); dyes and pigments (Kemamine Q-9702C); ore flotation (Arquad 2HT-75); petroleum industry (Adogen 442H; Jet Quat 2HT-75); textile/leather processing (Accosoft 707; Adogen 442; Arquad 2HT, 2HT-75, 88; Crapol FU-25; Cycloton D261C/70, D261C/75; Jet Quat 2HT-75; Kemamine Q-9702C; Noramium M2SH; Varisoft 110); water/sewage treatment (Arquad 2HT-75)

Industrial cleaners: metal processing surfactants (Arquad 2HT-75); sanitizers/germicides (Kemamine Q-9702C)

PROPERTIES:

Form:

Liquid (Jet Quat 2HT-75)

Paste (Accosoft 707; Adogen 442, 442H; Arquad 2HT; Noramium M2SH; Querton 442, 442-11; Radiaquat 6442; Varisoft 110)

Soft paste (Arquad 2HT-75; Cycloton D261C/70)

Firm paste (Cycloton D261C/75)

Solid (Arquad 88; Kemamine Q-9702C; Querton 442-82)

Soft solid (Varisoft DHT)

Color:

White (Arquad 2HT-75)

Gardner 2 max. (Kemamine Q-9702C)

Gardner 3 (Accosoft 707)

Gardner 3 max. (Varisoft DHT)

Gardner 6 max. (Varisoft 110)

Composition:

70% active (Cycloton D261C/70)

74% conc. (Radiaquat 6442)

74–76% solids (Varisoft 110)

74–77% quat in aq. isopropanol (Varisoft DHT)

77

Dihydrogenated tallow dimethyl ammonium chloride
(cont'd.)

75% active (Adogen 442H; Arquad 2HT; Cycloton D261C/75; Jet Quat 2HT-75; Kemamine Q-9702C; Noramium M2SH; Querton 442)

75% active in aq. isopropanol (Accosoft 707; Arquad 2HT-75)

75% solids (Adogen 442)

77% conc. (Querton 442-11)

82% conc. (Querton 442-82)

88% active (Arquad 88)

Solubility:

Disp. in water (Crapol FU-25; Kemamine Q-9702C); disp. in warm water (Varisoft DHT)

Ionic Nature:

Cationic (Accosoft 707; Adogen 442; Arquad 2HT, 2HT-75, 88; Cycloton D261C/70, D261C/75; Jet Quat 2HT-75; Kemamine Q-9702C; Noramium M2SH; Querton 442, 442-11, 442-82; Radiaquat 6442; Varisoft 110, DHT)

M.W.:

573 (of active) (Arquad 2HT-75)

575 (Kemamine Q-9702C)

Sp.gr.:

0.87 (Accosoft 707; Arquad 2HT-75)

Density:

7.12 lb/gal (Varisoft DHT)

7.22 lb/gal (Arquad 2HT-75)

7.3 lb/gal (Accosoft 707)

Visc.:

47.5 cps (120 F) (Arquad 2HT-75)

F.P.:

95 F (Arquad 2HT-75)

Pour Pt.:

90–100 F (Arquad 2HT-75)

Flash Pt.:

65 F (CC) (Adogen 442H)

68 F (PMCC) (Varisoft DHT)

99 F (PM) (Varisoft 110)

112 F (Arquad 2HT-75)

HLB:

9.7 (Arquad 2HT-75)

pH:

6.0–9.0 (5% sol'n. in 1:1 IPA/water) (Varisoft DHT)

9.0 max. (5% sol'n.) (Kemamine Q-9702C)

Surface Tension:

37 dynes/cm (0.1%) (Arquad 2HT-75)

Dihydrogenated tallow dimethyl ammonium chloride

(cont'd.)

TOXICITY/HANDLING:
Nontoxic (Accosoft 707)
Skin and eye irritant; protective gloves and glasses should be worn (Arquad 2HT-75)
Eye irritant, but not corrosive to the skin (Varisoft DHT)

STORAGE/HANDLING:
Combustible (Arquad 2HT-75)
Flammable; supports bacterial growth (Accosoft 707)
Pumping temperature: 110–120 F (Varisoft DHT)

STD. PKGS.:
55-gal nonreturnable epoxy-phenolic lined drums; bulk SS tank trucks (Arquad 2HT-75)
55-gal (390 lb net) lined open-head steel drums; tank wagons, tank cars (Varisoft DHT)

Disodium laneth-5 sulfosuccinate (CTFA)

STRUCTURE:

$$R(OCH_2CH_2)_5O—\overset{\overset{O}{\|}}{C}CHCH_2\overset{\overset{O}{\|}}{C}—ONa$$
$$|$$
$$SO_3Na$$

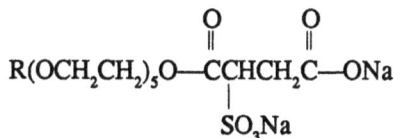

where R represents the lanolin alcohol radical

CAS No.:
68890-92-6 (generic); 977059-29-2

TRADENAME EQUIVALENTS:
Cyclopol SB-5 [Alcolac]
Incrosul LAFS [Croda]

CATEGORY:
Conditioner, moisturizer, surfactant

APPLICATIONS:
Bath products: bubble bath (Cyclopol SB-5)
Cleansers: body cleansers (Cyclopol SB-5)
Cosmetic industry preparations: hair preparations (Cyclopol SB-5); personal care products (Incrosul LAFS); shampoos (Cyclopol SB-5; Incrosul LAFS); skin preparations (Cyclopol SB-5)
Household detergents: dishwashing (Cyclopol SB-5)

PROPERTIES:
Form:
Liquid (Cyclopol SB-5)
Liquid/slurry (Incrosul LAFS)

Disodium laneth-5 sulfosuccinate *(cont'd.)*

Color:
Yellow (Cyclopol SB-5; Incrosul LAFS)
Composition:
48–50% active (Cyclopol SB-5)
50% active (Incrosul LAFS)
Ionic Nature:
Anionic (Cyclopol SB-5; Incrosul LAFS)

Disodium oleamido MEA-sulfosuccinate *(CTFA)*

SYNONYMS:
Butanedioic acid, sulfo-, mono [2-[(1-oxo-9-octadecenyl) amino] ethyl] ester, disodium salt
Disodium monooleamido MEA-sulfosuccinate
Sulfobutanedioic acid, mono [2-[1-(oxo-9-octadecenyl) amino] ethyl] ester-, disodium salt
EMPIRICAL FORMULA:
$C_{24}H_{43}NO_8S \cdot 2Na$
STRUCTURE:

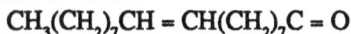

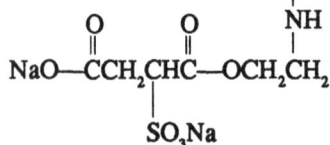

CAS No.:
68479-64-1
TRADENAME EQUIVALENTS:
Incrosul OMS [Croda]
Mackanate OM [McIntyre]
Schercopol OMS-Na [Scher]
CATEGORY:
Conditioner, emollient, surfactant
APPLICATIONS:
Bath products: bath oils (Schercopol OMS-Na)
Cleansers: body soaps (Schercopol OMS-Na); face cleanser (Schercopol OMS-Na); hand cleanser (Incrosul OMS)
Cosmetic industry preparations: hair preparations (Schercopol OMS-Na); shampoos (Incrosul OMS; Mackanate OM; Schercopol OMS-Na); toiletries (Schercopol OMS-Na)

Household detergents: dishwashing (Schercopol OMS-Na)

PROPERTIES:

Form:

Liquid (Mackanate OM)

Clear liquid (Schercopol OMS-Na)

Clear, viscous liquid (Incrosul OMS)

Color:

Yellow (Incrosul OMS; Schercopol OMS-Na)

Odor:

Mild, characteristic (Incrosul OMS; Schercopol OMS-Na)

Composition:

34% solids min. in water (Schercopol OMS-Na)

34–36% active (Incrosul OMS)

35% active (Mackanate OM)

Ionic Nature:

Anionic (Incrosul OMS; Mackanate OM; Schercopol OMS-Na)

Sp.gr.:

1.10 ± 0.05 (Schercopol OMS-Na)

Density:

9.16 lb/gal (Schercopol OMS-Na)

Visc.:

1000 cps max. (Schercopol OMS-Na)

Cloud Pt.:

5 C max. (Schercopol OMS-Na)

Storage Stability:

One year min. shelf life; visc. increases and pH decreases slightly with aging which
does not affect performance (Schercopol OMS-Na)

pH:

6.0 ± 1.0 (Schercopol OMS-Na)

6.0–7.0 (5%) (Incrosul OMS)

Biodegradable: (Incrosul OMS)

TOXICITY/HANDLING:

Avoid prolonged contact with skin and eyes (Incrosul OMS)

STORAGE/HANDLING:

Store in a cool, dry place (Incrosul OMS)

STD. PKGS.:

55-gal (450 lb net) polyethylene-lined Leverpak (Incrosul OMS)

55-gal (500 lb net) Liquipak drums (Schercopol OMS-Na)

Distearyl dimethyl ammonium chloride

SYNONYMS:
Dimethyl dioctadecyl ammonium chloride
Dimethyl distearyl ammonium chloride
N,N-Dimethyl-N-octadecyl-1-octadecanaminium chloride
Distearyldimonium chloride (CTFA)
1-Octadecanaminium, N,N-dimethyl-N-octadecyl-, chloride
Quaternium-5

EMPIRICAL FORMULA:
$C_{38}H_{80}N \cdot Cl$

STRUCTURE:

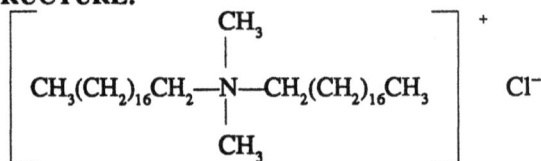

CAS No.:
107-64-2

TRADENAME EQUIVALENTS:
Arosurf TA-100, TA-101 [Sherex]
Cation DS [Sanyo]
Comperlan LD9 [Henkel Canada]
Dehyquart DAM [Henkel KGaA]
Genamin DSAC [Hoechst-Celanese AG]
Querton 442S, 442-Sx [Berol Nobel]

CATEGORY:
Softener, conditioner, antistat, foam builder, viscosity builder, raw material

APPLICATIONS:
Cosmetic industry preparations: conditioners (Arosurf TA-100); hair preparations
(Arosurf TA-100; Dehyquart DAM; Genamin DSAC); personal care products
(Comperlan LD9)
Household detergents: fabric softeners (Arosurf TA-100; Cation DS; Querton 442S,
442-Sx)
Industrial applications: textile/leather processing (Arosurf TA-100, TA-101)

PROPERTIES:

Form:
Paste (Comperlan LD9; Dehyquart DAM)
Solid (Cation DS)
Powder (Arosurf TA-100, TA-101; Genamin DSAC; Querton 442S)

Color:
Gardner 3 max. (Arosurf TA-100)
Gardner 4 max. (Arosurf TA-101)

Composition:
70–80% active (Dehyquart DAM)

82

Distearyl dimethyl ammonium chloride *(cont'd.)*

75% conc. (Comperlan LD9)
90% active (Cation DS)
93% quat min. (Arosurf TA-100)
100% conc. (Querton 442S)
100% solids (Arosurf TA-101)
Solubility:
Disp. in water (Arosurf TA-100, TA-101)
Ionic Nature:
Cationic (Arosurf TA-100, TA-101; Cation DS; Comperlan LD9; Querton 442S, 442-Sx)
M.W.:
583 (Arosurf TA-100)
Density:
24 lb/ft^3 (bulk) (Arosurf TA-101)
Flash Pt.:
> 200 F (PM) (Arosurf TA-100, TA-101)

Ditallow dimethyl ammonium chloride

SYNONYMS:
Dimethyl ditallow ammonium chloride
Ditallowdimonium chloride (CTFA)
Quaternary ammonium compounds, dimethyl ditallow alkyl, chlorides
Quaternium-48
STRUCTURE:

$$\left[\begin{array}{c} CH_3 \\ | \\ R{-}N{-}R \\ | \\ CH_3 \end{array} \right]^{+} \quad Cl^{-}$$

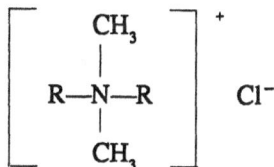

where R represents the tallow radical
CAS No.:
68783-78-8
RD No.: 977068-43-1
TRADENAME EQUIVALENTS:
Adogen 470 [Sherex]
Armosoft L [Akzo/Armak]
Querton 470 [Berol Nobel]

Ditallow dimethyl ammonium chloride *(cont'd.)*

CATEGORY:
Softener, conditioner, lubricant, antistat, bacteriostat
APPLICATIONS:
Household products: fabric softener (Armosoft L; Querton 470); laundry detergent (Armosoft L)
PROPERTIES:
Form:
Liquid (Adogen 470; Querton 470)
Hazy to clear liquid (Armosoft L)
Color:
Gardner 6 max. (Adogen 470)
Composition:
74–77% quat (Adogen 470)
75% active (Querton 470)
75% active in isopropanol (Armosoft L)
Ionic Nature:
Cationic (Armosoft L; Querton 470)
M.W.:
564 (Adogen 470)
Sp.gr.:
0.850 (100 F) (Armosoft L)
Density:
7.09 lb/gal (Armosoft L)
Visc.:
45.5 cps (120 F) (Armosoft L)
Pour Pt.:
70–75 F (Armosoft L)
Flash Pt.:
65 F (PM) (Adogen 470)
Stability:
Stable over wide pH range; not precipitated by hard water (Armosoft L)
pH:
6.0–9.0 (Armosoft L)
TOXICITY/HANDLING:
Skin and eye irritant (Armosoft L)
STORAGE/HANDLING:
Store in bulk between 85–90 F; maintain pH at 6–8; flammable; storage in fiberglass-reinforced furan resin tanks is recommended (Armosoft L)
STD. PKGS.:
55-gal steel drums; 4000-gal tank trucks; 4000- and 8000-gal tank cars (Armosoft L)

EMPIRICAL FORMULA:
$C_{32}H_{62}O_4$
STRUCTURE:

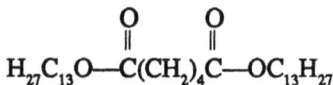

CAS No.:
26401-35-4
TRADENAME EQUIVALENTS:
Kemester 5654 [Humko/Witco]
Plasthall DTDA [C.P. Hall]
Generically sold by:
[Hexagon]
CATEGORY:
Lubricant additive
APPLICATIONS:
Industrial applications: lubricating/cutting oils (generic—Hexagon; Kemester 5654;
Plasthall DTDA); metalworking (Kemester 5654)
Form:
Liquid at ambient temps. (Kemester 5654)
Clear liquid (Plasthall DTDA)
Color:
APHA 35 (Plasthall DTDA)
Gardner 1 (Kemester 5654)
Solubility:
Sol. in acetone (Plasthall DTDA)
Sol. in ASTM oil #1 (Plasthall DTDA)
Sol. in ethanol (Plasthall DTDA)
Sol. in hexane (Plasthall DTDA)
Sol. in kerosene (Plasthall DTDA)
Sol. in min. oil (Plasthall DTDA)
Sol. in petroleum oils (generic—Hexagon)
Sol. in toluene (Plasthall DTDA)
M.W.:
510 (avg.) (Plasthall DTDA)
Sp.gr.:
0.908 (Plasthall DTDA)
F.P.:
< −65 C (Plasthall DTDA)
Flash Pt.:
235 C (Plasthall DTDA)
Acid No.:
0.03 (Plasthall DTDA)

Ditridecyl adipate *(cont'd.)*

    0.1 (Kemester 5654)
Iodine No.:
    Nil (Plasthall DTDA)
    0.1 (Kemester 5654)
Saponification No.:
    220 (Plasthall DTDA)
Ref. Index:
    1.4545 (Plasthall DTDA)

Erucic acid amide

SYNONYMS:
   13-Docosenamide
   Erucamide (CTFA)
EMPIRICAL FORMULA:
   $C_{22}H_{43}NO$
STRUCTURE:

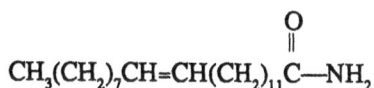

$$CH_3(CH_2)_7CH=CH(CH_2)_{11}C-NH_2$$

CAS No.:
   112-84-5
TRADENAME EQUIVALENTS:
   Armid E [Akzo]
   Armoslip EXP [Akzo/Noury]
   Crodamide E, ER [Croda Ltd.]
   Kemamide E [Humko/Witco]
   Kemester E [Humko/Witco]
   Petrac Eramide [Synthetic Products]
CATEGORY:
   Lubricant, release agent, slip agent, antiblock agent, antitack agent, defoamer, water
      repellent, corrosion inhibitor, dispersant, intermediate, foam stabilizer
APPLICATIONS:
   Household detergents: (Kemamide E)
   Food applications: food contact applications (Petrac Eramide)
   Industrial applications: (Kemamide E); asphalts (Kemamide E); dyes and pigments
      (Kemamide E); lamination (Petrac Eramide); metalworking (Kemamide E); paint
      mfg. (Kemamide E; Petrac Eramide); petroleum industry (Kemamide E); plastics
      (Armid E; Armoslip EXP; Crodamide E, ER; Kemamide E; Kemester E; Petrac
      Eramide); printing inks (Kemamide E); rubber (Kemester E); textile/leather proc-
      essing (Kemamide E)
PROPERTIES:
Form:
   Powder (Petrac Eramide)
   Solid (Armid E; Kemamide E; Kemester E)
   Flake (Armid E; Armoslip EXP; Petrac Eramide)
   Pellets (Armoslip EXP; Petrac Eramide)

Erucic acid amide *(cont'd.)*

Wax (Crodamide E, ER)
Color:
Yellow (Armid E)
APHA 10 (Petrac Eramide)
Gardner 5 max. (Kemamide E)
Odor:
Characteristic mild (Petrac Eramide)
Composition:
90% active (Armid E)
99.5% total amide (Petrac Eramide)
100% conc. (Kemamide E)
Solubility:
Sol. 25 g/100 g in chloroform (Kemamide E)
Sol. 8 g/100 g in isopropanol (Kemamide E)
Sol. 4 g/100 g in MEK (Kemamide E)
Sol. 3 g/100 g in methanol (Kemamide E)
Insol. in water (Armid E; Kemamide E)
Sol. 3 g/100 g in toluene (Kemamide E)
M.W.:
335 (Kemamide E)
M.P.:
76–86 C (Kemamide E)
79 C (Armid E)
80 C (Petrac Eramide)
Flash Pt.:
230 C (COC) (Armid E)
Acid No.:
4.0 max. (Kemamide E)
Iodine No.:
70–80 (Kemamide E)
72–80 (Armid E)
STORAGE/HANDLING:
Avoid contact with strong oxidizing agents (Armid E)
STD. PKGS.:
50 kg fiberboard keg (Armid E)

Ethylene bisoleamide

SYNONYMS:
N,N´-1,2-Ethanediylbis-9-octadecenamide
Ethylene dioleamide (CTFA)
9-Octadecenamide, N,N´-1,2-ethanediylbis-

EMPIRICAL FORMULA:

$C_{38}H_{72}N_2O_2$

STRUCTURE:

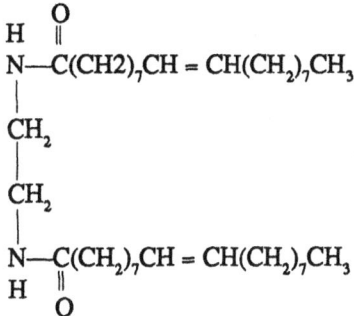

CAS No.:

110-31-6

TRADENAME EQUIVALENTS:

Advawax 240 [Morton Int'l./Carstab]

Kemamide W-20 [Humko/Witco]

CATEGORY:

Lubricant, release agent, antistat, m.p. modifier, dispersant, coupling agent, slip agent, antiblocking agent, defoamer, water repellent, corrosion inhibitor

APPLICATIONS:

Food applications: food packaging (Advawax 240)

Household products: (Kemamide W-20); detergents (Kemamide W-20)

Industrial applications: (Kemamide W-20); adhesives (Advawax 240); asphalts (Advawax 240; Kemamide W-20); dyes and pigments (Advawax 240; Kemamide W-20); latex (Kemamide W-20); lubricating/cutting oils (Kemamide W-20); metalworking (Kemamide W-20); paint mfg. (Advawax 240; Kemamide W-20); paper mfg. (Kemamide W-20); petroleum products (Kemamide W-20); plastics (Advawax 240; Kemamide W-20); polishes and waxes (Advawax 240; Kemamide W-20); printing inks (Kemamide W-20); rubber (Kemamide W-20); textile/leather processing (Kemamide W-20)

PROPERTIES:

Form:

Small beads (Advawax 240)

Flake (Kemamide W-20)

Powder (Kemamide W-20)

Color:

Gardner 6 max. (Kemamide W-20)

D1500 3 (Advawax 240)

Solubility:

Sol. > 10% with heat in benzene (Advawax 240)

Ethylene bisoleamide (cont'd.)

Sol. > 10% with heat in 1,3-butylene glycol (Advawax 240)
Sol. > 10% with heat in carbon tetrachloride (Advawax 240)
Sol. > 10% with heat in Cellosolve (Advawax 240)
Sol. 4 g/100 g @ 35 C in dichloroethane (Kemamide W-20)
Sol. > 10% with heat in dimethyl formamide (Advawax 240)
Sol. > 10% with heat in ethanol (Advawax 240)
Sol. > 10% with heat in heptane (Advawax 240)
Sol. 12 g/100 g @ 35 C in isopropanol (Kemamide W-20)
Sol. > 10% with heat in kerosene (Advawax 240)
Sol. > 10% with heat in MEK (Advawax 240); sol. 3 g/100 g @ 35 C (Kemamide W-20)
Sol. > 10% with heat in MIBK (Advawax 240)
Sol. > 10% with heat in naphtha (Advawax 240)
Sol. > 10% with heat in a-pinene (Advawax 240)
Sol. > 20 g/100 g @ 35 C in toluene (Kemamide W-20)
Sol. > 10% with heat in xylene (Advawax 240)

Density:
5 lb/gal (bulk) (Advawax 240)
M.P.:
113–118 C (Advawax 240)
120 C (Kemamide W-20)
Flash Pt.:
270 C (COC) (Advawax 240)
296 C (COC) (Kemamide W-20)
Fire Pt.:
310 C (COC) (Advawax 240)
315 C (COC) (Kemamide W-20)
Acid No.:
10 (Advawax 240)
10 max. (Kemamide W-20)
Amine No.:
10 (Kemamide W-20)
Stability:
Good chemical resistance to practically all solvents, acids, and alkalis (Advawax 240)

TOXICITY/HANDLING:

Nontoxic; handle powders to prevent eye irritation; wear protective goggles and gloves when handling (Kemamide W-20)

STORAGE/HANDLING:

Flammable; fine powder in air may be combustible or explosive (Advawax 240)
Handle powders to prevent dust explosions; not classified as flammable, but will burn if ignited; keep away from strong oxidizing agents; should not remain at temps. higher than 175–185 C for long periods (Kemamide W-20)

STD. PKGS.:
200-lb fiber drums (Advawax 240)

N,N´-Ethylene bisstearamide

SYNONYMS:
N,N´-1,2-Ethanediylbisoctadecanamide
Ethylene distearamide (CTFA)
Octadecanamide, N,N´-1,2-ethanediylbis-
EMPIRICAL FORMULA:
$C_{38}H_{72}N_2O_2$
STRUCTURE:

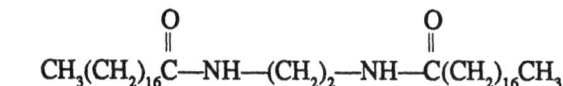

$$CH_3(CH_2)_{16}C—NH—(CH_2)_2—NH—C(CH_2)_{16}CH_3$$

CAS No.:
110-30-5
TRADENAME EQUIVALENTS:
Acrawax C, C DF #1, C DF #2, C SG [Lonza]
Advawax 275, 280, 290 [Morton Int'l./Carstab]
Alkamide STEDA [Alkaril]
EBS Wax [Borg-Warner]
Interstab GT-8257 [Akzo/Interstab]
Kemamide W-35, W-39, W-40, W-40/300, W-40DF, W-45 [Humko/Witco]
Lipowax C [Lipo]
Nopcowax 22-DS [Henkel/Process]
CATEGORY:
Lubricant, processing aid, plasticizer, defoamer, antistat, flow improver, dispersant, detackifier, release agent, antiblocking agent, m.p. modifier, wax, water repellent, corrosion inhibitor, intermediate, binder, thickener
APPLICATIONS:
Food applications: food packaging (Advawax 290; Interstab GT-8257; Lipowax C)
Household detergents: (Kemamide W-39, W-40, W-40/300, W-40DF, W-45)
Industrial applications: (Kemamide W-39, W-40, W-40/300, W-40DF, W-45); adhesives (Acrawax C; Advawax 275, 280, 290; Nopcowax 22-DS); asphalt (Advawax 275, 280, 290; Kemamide W-35, W-39, W-40, W-40/300, W-40DF, W-45); dyes and pigments (Acrawax C; Alkamide STEDA; Kemamide W-39, W-40, W-40/300, W-40DF, W-45; Lipowax C); latex processing (EBS Wax; Kemamide W-39, W-40, W-40/300, W-40DF, W-45; Nopcowax 22-DS); lubricating/cutting oils (Kemamide W-39, W-40, W-40/300, W-40DF, W-45); metalworking (Alkamide

N, N´-Ethylene bisstearamide *(cont'd.)*

STEDA; Kemamide W-39, W-40, W-40/300, W-40DF, W-45; Nopcowax 22-DS); paints and coatings (Acrawax C; Advawax 275, 280, 290; Kemamide W-39, W-40, W-40/300, W-40DF, W-45; Lipowax C; Nopcowax 22-DS); paper mfg. (Advawax 275, 280, 290; Alkamide STEDA; EBS Wax; Kemamide W-35; Lipowax C); petroleum industry (Kemamide W-39, W-40, W-40/300, W-40DF, W-45); plastics (Acrawax C, C DF #1, C DF #2, C SG; Advawax 275, 280, 290; Alkamide STEDA; EBS Wax; Interstab GT-8257; Kemamide W-35, W-39, W-40, W-40/300, W-40DF, W-45; Lipowax C); polishes and waxes (Kemamide W-39, W-40, W-40/300, W-40DF, W-45); potting/dipping compounds for electronics (Kemamide W-35); protective coatings (Kemamide W-39, W-40, W-40/300, W-40DF, W-45); resins (Advawax 275, 280, 290; Alkamide STEDA; Kemamide W-39, W-40, W-40/300, W-40DF, W-45; Lipowax C); printing inks (Kemamide W-39, W-40, W-40/300, W-40DF, W-45); rubber (Kemamide W-35; Lipowax C); textile/leather processing (EBS Wax; Kemamide W-39, W-40, W-40/300, W-40DF, W-45; Lipowax C); wood pulping (Alkamide STEDA; Kemamide W-35)

PROPERTIES:
Form:

Solid (Alkamide STEDA; Kemamide W-35)
Small beads (Advawax 275, 280, 290)
Beads (EBS Wax)
Flakes (Kemamide W-39, W-40, W-40DF, W-45)
Atomized (Acrawax C DF #1, C DF #2; Lipowax C)
Micronized (Kemamide W-40/300)
Powder (Acrawax C DF #1, C DF #2; Kemamide W-40, W-40DF, W-45; Lipowax C)
Prilled (Acrawax C DF #1, C DF #2, C SG; Lipowax C)
Wax (Interstab GT-8257)
Hard waxy solid (Acrawax C)

Fineness:

10% max. on 10 mesh (Acrawax C beads)
2% max. on 40 mesh (Acrawax C prilled)
1% max. on 100 mesh (Acrawax C powder)
0.1% max. on 325 mesh (Acrawax C atomized)

Color:

Off-white (EBS Wax)
Light cream (Lipowax C)
Cream (Acrawax C)
Light tan (Alkamide STEDA)
Gardner 3 (Acrawax C DF #1, C DF #2, C SG)
Gardner 3 max. (Kemamide W-40, W-40DF, W-45)
Gardner 6 max. (Kemamide W-35)
Gardner 18 max. (Kemamide W-39)

Composition:

95% amide, 100% conc. (Alkamide STEDA)

100% active (Kemamide W-35; Lipowax C)

Solubility:

Insol. in aromatics @ 10% (Alkamide STEDA)

Sol. in benzene (Advawax 275, 280, 290)

Sol. in Cellosolve (Advawax 275, 280, 290)

Sol. 1.6 g/100 g in dichloroethane @ 70 C (Kemamide W-39, W-40, W-40/300, W-40DF, W-45)

Sol. in heptane (Advawax 275, 280, 290)

Sol. 1.4 g/100 g in isopropanol @ 70 C (Kemamide W-39, W-40, W-40/300, W-40DF, W-45)

Sol. in kerosene (Advawax 275, 280, 290; Lipowax C)

Sol. 0.9 g/100 g in MEK @ 70 C (Kemamide W-39, W-40, W-40/300, W-40DF, W-45)

Sol. in MIBK (Advawax 275, 280, 290)

Insol. in min. oil @ 10% (Alkamide STEDA)

Insol. in min. spirits @ 10% (Alkamide STEDA)

Sol. in naphtha (Advawax 275, 280, 290; Lipowax C)

Insol. in perchloroethylene @ 10% (Alkamide STEDA)

Sol. in toluene (Lipowax C); sol. 2 g/100 g @ 70 C (Kemamide W-39, W-40, W-40/300, W-40DF, W-45)

Sol. in turpentine (Lipowax C)

Insol. in water (Acrawax C; Kemamide W-35; Nopcowax 22-DS); insol. @ 10% (Alkamide STEDA)

Sol. in xylene (Advawax 275, 280, 290)

Ionic Nature:

Nonionic (Alkamide STEDA; Lipowax C)

M.W.:

580 (EBS Wax)

Sp.gr.:

0.97 (Acrawax C)

Density:

4.75 lb/gal (Advawax 290)

4.9 lb/gal (Advawax 280)

5.3 lb/gal (Advawax 275)

M.P.:

134–137 C (Advawax 275)

139–143 C (Advawax 280)

140 C (Kemamide W-35, W-39, W-40, W-40/300, W-40DF)

140–145 C (Acrawax C; EBS Wax; Lipowax C)

141–145 C (Alkamide STEDA)

143 C (Acrawax C DF #1, C SG)

143–146 C (Advawax 290)

145 C (Acrawax C DF #2; Kemamide W-45)

Flash Pt.:
280 C (COC) (Advawax 275, 280)
285 C (Acrawax C); (COC) (Lipowax C)
290 C (COC) (Advawax 290)
299 C (COC) (Kemamide W-39, W-40, W-40/300, W-40DF)
304 C (COC) (Kemamide W-45)

Fire Pt.:
315 C (Kemamide W-39, W-40, W-40/300, W-40DF)
322 C (Kemamide W-45)

Acid No.:
5 (Acrawax C DF #1, C DF #2, C SG)
5 max. (Kemamide W-40DF)
7.0 max. (EBS Wax)
8 (Advawax 275, 280, 290)
8 max. (Acrawax C)
10 max. (Kemamide W-35, W-39, W-40, W-45; Lipowax C)

Amine No.:
5 (Kemamine W040DF)
10 (Kemamide W-39, W-40, W-45)

Neutralization No.:
3 max. (EBS Wax)

Stability:
Resistant to water, acids, and alkalies (Lipowax C)

pH:
Neutral (Kemamide W-35)

TOXICITY/HANDLING:
Nontoxic; handle powders to prevent eye irritation and dust explosions; wear protective goggles and gloves when handling (Kemamide W-39, W-40, W-40/300, W-40DF, W-45)

STORAGE/HANDLING:
Not classified as flammable, but will burn if ignited; keep away from strong oxidizing agents; should not remain at temps. higher than 175–185 C for long periods (Kemamide W-39, W-40, W-40/300, W-40DF, W-45)

STD. PKGS.:
50 lb net bags and 200 lb net fiber drums (Lipowax C prill)
160 lb net fiber drums (Lipowax C powder and atomized powder)

Glyceryl isostearate (CTFA)

SYNONYMS:
Glyceryl monoisostearate
Isooctadecanoic acid, monoester with 1,2,3-propanetriol

EMPIRICAL FORMULA:
$C_{21}H_{42}O_4$

STRUCTURE:

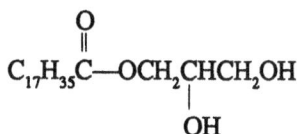

CAS No.:
32057-14-0; 66085-00-5
RD No.: 977064-41-7

TRADENAME EQUIVALENTS:
Emerest 2410 [Henkel/Emery]
Imwitor 780 [Hüls AG]
Schercemol GMIS [Scher]

CATEGORY:
Emollient, lubricant, emulsifier, pearling agent

APPLICATIONS:
Cosmetic industry preparations: creams and lotions (Emerest 2410; Schercemol GMIS); hair preparations (Emerest 2410); makeup (Emerest 2410)

PROPERTIES:

Form:
Liquid (Emerest 2410; Imwitor 780)
Clear liquid to soft solid (Schercemol GMIS)

Color:
Light amber (Schercemol GMIS)
Gardner 2 (Emerest 2410)

Odor:
Slight, typical (Schercemol GMIS)

Composition:
100% active (Emerest 2410; Imwitor 780)

Solubility:
Sol. in alcohols (Schercemol GMIS)

Glyceryl isostearate *(cont'd.)*

Sol. in aliphatic hydrocarbons (Schercemol GMIS)
Sol. in aromatic hydrocarbons (Schercemol GMIS)
Sol. in chlorinated hydrocarbons (Schercemol GMIS)
Sol. in esters (Schercemol GMIS)
Disp. in glycerin @ 5% (Emerest 2410)
Sol. in glycerol trioleate @ 5% (Emerest 2410)
Sol. in glycol ethers (Schercemol GMIS)
Disp. in glycols (Schercemol GMIS)
Sol. in isopropanol @ 5% (Emerest 2410)
Disp. in isopropyl myristate @ 5% (Emerest 2410)
Sol. in ketones (Schercemol GMIS)
Sol. in min. oil (Schercemol GMIS); sol. @ 5% (Emerest 2410)
Disp. in polyols (Schercemol GMIS)
Disp. in triols (Schercemol GMIS)
Sol. in veg. oil (Schercemol GMIS)
Insol. in water (Schercemol GMIS); insol. @ 5% (Emerest 2410)
Ionic Nature:
Nonionic (Emerest 2410; Imwitor 780)
M.W.:
385 (theoret.) (Schercemol GMIS)
Sp.gr.:
0.960 (Schercemol GMIS)
Density:
7.8 lb/gal (Emerest 2410)
8.0 lb/gal (Schercemol GMIS)
F.P.:
6 C (Schercemol GMIS)
Pour Pt.:
5 C (Emerest 2410)
Flash Pt.:
> 170 C (OC) (Schercemol GMIS)
400 F (Emerest 2410)
HLB:
2.9 (Emerest 2410)
3.7 (Imwitor 780)
Acid No.:
5.0 max. (Schercemol GMIS)
Iodine No.:
13.0 max. (Schercemol GMIS)
Saponification No.:
150–170 (Schercemol GMIS)
Ref. Index:
1.4715 (Schercemol GMIS)

SYNONYMS:
Glyceryl monomyristate
Monomyristin
Tetradecanoic acid, monoester with 1,2,3-propanetriol
EMPIRICAL FORMULA:
$C_{17}H_{34}O_4$
STRUCTURE:

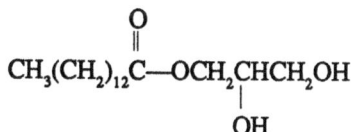

CAS No.:
589-68-4; 27214-38-6
TRADENAME EQUIVALENTS:
GMM-33 [Hefti Ltd.]
Grindtek MM 90 [Grindsted]
CATEGORY:
Lubricant, antistat, antifogging agent, emulsifier, base
APPLICATIONS:
Cosmetic industry preparations: (GMM-33); creams and lotions (GMM-33; Grindtek MM 90)
Industrial applications: plastics (Grindtek MM 90)
Pharmaceutical applications: (GMM-33)
PROPERTIES:
Form:
Flakes (GMM-33)
Block (Grindtek MM 90)
Color:
Whitish (Grindtek MM 90)
Composition:
100% conc. (GMM-33)
Solubility:
Sol. warm in ethanol (Grindtek MM 90)
Sol. warm in propylene glycol (Grindtek MM 90)
Sol. warm in toluene (Grindtek MM 90)
Ionic Nature:
Nonionic (GMM-33)
HLB:
5.0 (Grindtek MM 90)
5.5 (GMM-33)

Glyceryl tribehenate (CTFA)

SYNONYMS:
Docosanoic acid, 1,2,3-propanetriyl ester
Tribehenin

EMPIRICAL FORMULA:
$C_{69}H_{134}O_6$

STRUCTURE:

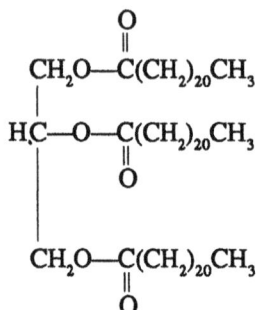

CAS No.:
18641-57-1

TRADENAME EQUIVALENTS:
Syncrowax HR-C [Croda]

CATEGORY:
Emollient, conditioner, emulsifier, wax, gloss improver

APPLICATIONS:
Cosmetic industry preparations: hair preparations (Syncrowax HR-C); makeup (Syncrowax HR-C); skin preparations (Syncrowax HR-C)

Pharmaceutical applications: antiperspirant/deodorant (Syncrowax HR-C)

PROPERTIES:

Form:
Pastille/powder (Syncrowax HR-C)
Wax (Syncrowax HR-C)

Color:
Off-white (Syncrowax HR-C)

Solubility:
Sol. in oils (Syncrowax HR-C)

M.P.:
60–65 C (Syncrowax HR-C)

Glyceryl triisostearate

SYNONYMS:
  Isooctadecanoic acid, 1,2,3-propanetriol ester
  Isostearyl triglyceride
  Triisostearin (CTFA)
EMPIRICAL FORMULA:
  $C_{57}H_{110}O_6$
CAS No.:
  RD No.: 977042-88-8
TRADENAME EQUIVALENTS:
  Cyclochem GTIS [Alcolac]
CATEGORY:
  Emollient, coemulsifier, softener
APPLICATIONS:
  Cosmetic industry preparations: creams and lotions (Cyclochem GTIS); skin preparations (Cyclochem GTIS)
PROPERTIES:
Form:
  Liquid (Cyclochem GTIS)
Color:
  Yellow (Cyclochem GTIS)
M.P.:
  < 0 C (Cyclochem GTIS)
Acid No.:
  8.0 max. (Cyclochem GTIS)
Saponification No.:
  192 (Cyclochem GTIS)

Glyceryl trilaurate

SYNONYMS:
  Dodecanoic acid, 1,2,3-propanetriyl ester
  1,2,3-Propanetriol tridodecanoate
  Trilaurin (CTFA)
EMPIRICAL FORMULA:
  $C_{39}H_{74}O_6$

Glyceryl trilaurate *(cont'd.)*

STRUCTURE:

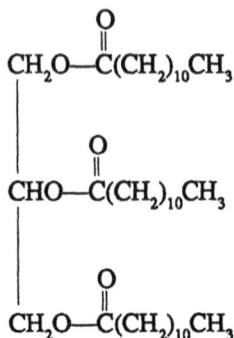

CH$_2$O—C(CH$_2$)$_{10}$CH$_3$

CHO—C(CH$_2$)$_{10}$CH$_3$

CH$_2$O—C(CH$_2$)$_{10}$CH$_3$

CAS No.:
538-24-9

TRADENAME EQUIVALENTS:
Cyclochem GTL [Alcolac]
Softisan 100 [Huls; Huls AG]

CATEGORY:
Emollient

APPLICATIONS:
Cosmetic industry preparations: (Softisan 100); cosmetic base (Softisan 100); creams and lotions (Cyclochem GTL; Softisan 100); makeup (Softisan 100)
Pharmaceutical applications: (Softisan 100); ointment base (Softisan 100)

PROPERTIES:

Form:
Liquid/paste (Cyclochem GTL)
Solid (Softisan 100)

Color:
White (Softisan 100)

Odor:
Neutral (Softisan 100)

Taste:
Neutral (Softisan 100)

Solubility:
Sol. in benzene (Softisan 100)
Sol. in carbon tetrachloride (Softisan 100)
Sol. in chloroform (Softisan 100)
Sol. in dichlorethylene (Softisan 100)
Sol. in dioxane (Softisan 100)
Sol. in ether (Softisan 100)
Sol. in toluene (Softisan 100)
Sol. in xylene (Softisan 100)

Ionic Nature:
Nonionic (Softisan 100)
Sp.gr.:
0.950–0.980 (20 C) (Softisan 100)
Visc.:
≈ 30 cps (40 C) (Softisan 100)
M.P.:
27 C (Cyclochem GTL)
33–35 C (Softisan 100)
Solidification Pt.:
29–33 C (Softisan 100)
Acid No.:
0.5 max. (Softisan 100)
30.0 max. (Cyclochem GTL)
Iodine No.:
3 max. (Softisan 100)
Saponification No.:
230–240 (Softisan 100)
245 (Cyclochem GTL)
Hydroxyl No.:
15 max. (Softisan 100)
Ref. Index:
1.4490–1.4510 (40 C) (Softisan 100)
STD. PKGS.:
20-kg net cartons, in pellet form (Softisan 100)

Glyceryl trioleate

SYNONYMS:
Glycerol trioleate
9-Octadecenoic acid, 1,2,3-propanetriyl ester
Olein
Triester of glycerin and oleic acid
Triglyceride of oleic acid
Triolein (CTFA)
EMPIRICAL FORMULA:
$C_{57}H_{104}O_6$

101

Glyceryl trioleate *(cont'd.)*

STRUCTURE:

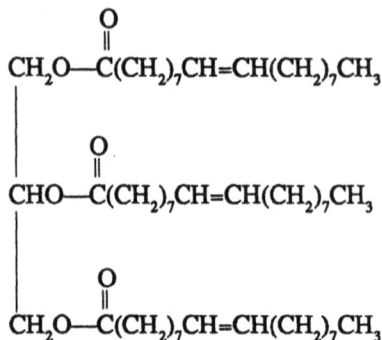

CAS No.:

122-32-7

TRADENAME EQUIVALENTS:

Acconon GTO [Capital City]

Aldo TO [Lonza]

Alkamuls GTO [Alkaril]

Cyclochem GTO [Alcolac]

Emerest 2423 [Henkel/Emery]

Grocor 1000, 1200 [A. Gross]

Hodag GTO [Hodag]

Kemester 1000 [Humko/Witco]

Radia 7161, 7163, 7303, 7363 [Synfina-Oleofin]

Rilanit GTO [Henkel KGaA]

CATEGORY:

Lubricant, emollient, fatting agent, emulsifier, wetting agent, surfactant base, chemical intermediate, opacifier, stabilizer

APPLICATIONS:

Cosmetic industry preparations: (Acconon GTO; Hodag GTO); creams and lotions (Cyclochem GTO)

Industrial applications: ceramics (Rilanit GTO); lubricating/cutting oils (Radia 7161, 7163, 7303, 7363; Rilanit GTO); metalworking (Alkamuls GTO; Emerest 2423; Rilanit GTO); textile/leather processing (Acconon GTO; Alkamuls GTO; Emerest 2423; Grocor 1000, 1200; Kemester 1000; Radia 7161, 7163, 7303, 7363; Rilanit GTO)

Pharmaceutical applications: (Hodag GTO)

PROPERTIES:

Form:

Liquid (Aldo TO; Alkamuls GTO; Cyclochem GTO; Emerest 2423; Grocor 1000, 1200; Hodag GTO; Kemester 1000; Rilanit GTO)

Color:

Amber (Alkamuls GTO; Grocor 1200)

Yellow (Aldo TO)
Gardner 3 (Emerest 2423)
Gardner 6 max. (Grocor 1000)
Odor:
Mild (Grocor 1000, 1200)
Composition:
97% active min. (Grocor 1000, 1200)
100% active (Aldo TO; Alkamuls GTO; Cyclochem GTO; Emerest 2423; Hodag GTO; Kemester 1000)
Solubility:
Sol. in aromatic solvent @ 10% (Alkamuls GTO)
Sol. in benzene (Radia 7161, 7163, 7303, 7363)
Sol. in hexane (Radia 7161, 7163, 7303, 7363)
Sol. in hydrocarbons (Grocor 1000, 1200)
Sol. in isopropanol (Radia 7161, 7163, 7303, 7363)
Sol. in min. oil (Emerest 2423; Radia 7161, 7163, 7303, 7363); (@ 10%) (Alkamuls GTO)
Sol. in min. spirits @ 10% (Alkamuls GTO)
Sol. in oils (Acconon GTO; Grocor 1000, 1200)
Sol. in organic solvents (Acconon GTO)
Sol. in perchloroethylene @ 10% (Alkamuls GTO)
Sol. in toluol (Emerest 2423)
Sol. in trichlorethylene (Radia 7161, 7163, 7303, 7363)
Sol. in veg. oils (Radia 7161, 7163, 7303, 7363)
Insol. in water @ 10% (Alkamuls GTO)
Ionic Nature:
Nonionic (Aldo TO; Alkamuls GTO; Cyclochem GTO; Emerest 2423; Grocor 1000, 1200)
M.W.:
880 avg. (Radia 7161, 7163, 7303, 7363)
Sp.gr.:
0.906 (37.8 C) (Radia 7163, 7363)
0.908 (37.8 C) (Radia 7161)
0.910 (37.8 C) (Radia 7303)
0.9145 (Grocor 1000)
0.945 (Grocor 1200)
Density:
0.95 g/ml (Alkamuls GTO)
7.6 lb/gal (Emerest 2423)
Visc.:
41.20 cps (37.8 C) (Radia 7363)
42.30 cps (37.8 C) (Radia 7163)
45.20 cps (37.8 C) (Radia 7303)

Glyceryl trioleate *(cont'd.)*

47 cs (Emerest 2423)
49.20 cps (37.8 C) (Radia 7161)
190–205 SS (100 F) (Grocor 1000, 1200)
F.P.:
9 C (Emerest 2423)
M.P.:
< 0 C (Aldo TO)
Flash Pt.:
274 C (COC) (Radia 7161)
275 C (COC) (Radia 7303)
283 C (COC) (Radia 7163)
289 C (COC) (Radia 7363)
293 C (Emerest 2423)
Cloud Pt.:
–13 C (Radia 7363)
–12 C (Radia 7163)
–9 C (Radia 7161)
5 C (Radia 7303)
5 F max. (Grocor 1200)
18 F max. (Grocor 1000
HLB:
0.6 (Emerest 2423)
0.8 (Aldo TO; Alkamuls GTO)
1.0 (Hodag GTO)
Acid No.:
5 (Emerest 2423)
Iodine No.:
85 (Emerest 2423)
Saponification No.:
188–195 (Aldo TO)
197 (Emerest 2423)
Ref. Index:
1.4670 (Grocor 1000)
1.4685 (Radia 7161)
1.4688 (Radia 7163)
1.4691 (Radia 7363)
1.4709 (Radia 7303)
STD. PKGS.:
SS drums or aluminum tank cars (Grocor 1000)
SS drums, aluminum tank cars and tank wagons (Grocor 1200)

SYNONYMS:
  Octadecanoic acid, 1,2,3-propanetriyl ester
  1,2,3-Propanetriol trioctadecanoate
  Tristearin (CTFA)
EMPIRICAL FORMULA:
  $C_{57}H_{110}O_6$
STRUCTURE:

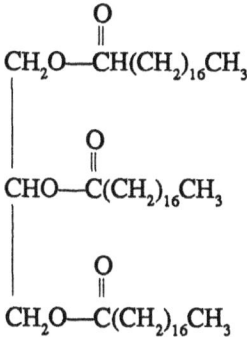

CAS No.:
  555-43-1

TRADENAME EQUIVALENTS:
  Dynasan 118 [Huls]
  GTS-33 [Hefti Ltd.]
  Neobee 62 [Stepan/PVO]
  Rilanit GTS [Henkel KGaA]

CATEGORY:
  Lubricant, filler, solubilizer, stabilizer, fatting agent

APPLICATIONS:
  Cosmetic industry preparations: (Dynasan 118; GTS-33); powders and cakes (Dynasan 118)
  Food applications: (GTS-33; Neobee 62)
  Industrial applications: ceramics (Rilanit GTS); lubricating/cutting oils (Rilanit GTS); metalworking (Rilanit GTS); textile/leather processing (Rilanit GTS)
  Pharmaceutical applications: (GTS-33); tablet mfg. (Dynasan 118)

PROPERTIES:
Form:
  Solid (Rilanit GTS)
  Flakes (GTS-33; Neobee 62)
  Microcrystalline powder (Dynasan 118)
Composition:
  100% conc. (GTS-33)

Glyceryl tristearate (cont' d.)

Solubility:
  Sol. in benzene (Dynasan 118)
  Sol. in ether (Dynasan 118)
Ionic Nature:
  Nonionic (GTS-33)
Setting Pt.:
  60 C (Neobee 62)
HLB:
  1.0 (GTS-33)
Iodine No.:
  3.0 (Neobee 62)
Saponification No.:
  189–195 (Neobee 62)

Hexyl laurate (CTFA)

SYNONYMS:
Dodecanoic acid, hexyl ester
Lauric acid, hexyl ester

EMPIRICAL FORMULA:
$C_{18}H_{36}O_2$

STRUCTURE:

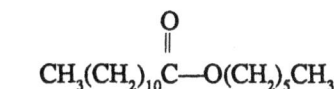

$$CH_3(CH_2)_{10}C-O(CH_2)_5CH_3$$

CAS No.:
34316-64-8

TRADENAME EQUIVALENTS:
Cetiol A [Henkel; Henkel Canada; Henkel KGaA]
Dermalcare HL [Alcolac]
Rilanit HL [Henkel KGaA]
Standamul CTA [Henkel]

CATEGORY:
Emollient, skin lubricant and penetrant, carrier, vehicle, base, spreading agent, dispersant, fatting/refatting agent

APPLICATIONS:
Bath products: bath oils (Standamul CTA)
Cleansers: cleansing creams (Cetiol A); toilet soaps (Cetiol A)
Cosmetic industry preparations: (Standamul CTA); body oils (Standamul CTA); creams and lotions (Cetiol A; Dermalcare HL; Standamul CTA); makeup (Cetiol A; Dermalcare HL; Standamul CTA); shaving preparations (Cetiol A); skin preparations (Cetiol A; Dermalcare HL; Standamul CTA)
Industrial applications: ceramics (Rilanit HL); lubricating/cutting oils (Rilanit HL); metalworking (Rilanit HL); textile/leather processing (Rilanit HL)
Pharmaceutical applications: ointments (Cetiol A); sunscreens (Cetiol A); topical preparations (Standamul CTA)

PROPERTIES:
Form:
Liquid (Rilanit HL)
Oily liquid (Dermalcare HL)
Clear oily liquid (Cetiol A)
Clear, low viscosity oil (Standamul CTA)

Hexyl laurate *(cont'd.)*

Color:
Colorless (Standamul CTA)
Odor:
Odorless (Cetiol A)
Practically odorless (Standamul CTA)
Composition:
100% conc. (Standamul CTA)
Solubility:
Sol. in castor oil (Standamul CTA); (@ 10%) (Cetiol A)
Miscible with essential oils (@ 10%) (Cetiol A)
Miscible with esters (@ 10%) (Cetiol A)
Sol. in ethyl alcohol (anhydrous) (Standamul CTA); sol. in 95% SD40 (@ 10%) (Cetiol A)
Miscible with fatty alcohols (@ 10%) (Cetiol A)
Sol. in isopropyl myristate (Standamul CTA); (@ 10%) (Cetiol A)
Sol. in min. oil (Standamul CTA); (@ 10%) (Cetiol A)
Miscible with commonly used cosmetic oils and fats (@ 10%) (Cetiol A)
Sol. in oleyl alcohol (Standamul CTA); (@ 10%) (Cetiol A)
Sol. in silicone fluid (Standamul CTA); (@ 10%) (Cetiol A)
Insol. in water (@ 10%) (Cetiol A)
Sp.gr.:
0.85–0.87 (20 C) (Standamul CTA)
Density:
0.847 g/ml (Cetiol A)
Gel Pt.:
–5 C max. (Standamul CTA)
Solidification Pt.:
< 0 C (Cetiol A)
Flash Pt.:
> 165 C (Cetiol A)
Cloud Pt.:
0 C max. (Standamul CTA)
< 5 C (Cetiol A)
HLB:
9 (Standamul CTA)
12 (Cetiol A)
Acid No.:
0.2 max. (Cetiol A)
0.5 max. (Standamul CTA)
Iodine No.:
1.0 max. (Cetiol A; Standamul CTA)
Saponification No.:
190–205 (Cetiol A; Standamul CTA)

Hexyl laurate *(cont'd.)*

Hydroxyl No.:
2.0 max. (Standamul CTA)
Ref. Index:
1.435 (Cetiol A)
TOXICITY/HANDLING:
Nonirritating (Cetiol A; Dermalcare HL)
STD. PKGS.:
385 lb net closed-head steel drums or bulk (Standamul CTA)

Hydrogenated lanolin *(CTFA)*

SYNONYMS:
Lanolin, hydrogenated
CAS No.:
8031-44-5
TRADENAME EQUIVALENTS:
Fanchem HL [Fanning]
Fancol HL [Fanning]
Lipolan, S, Distilled Lipolan [Lipo]
Satulan [Croda; Croda Ltd.]
Super-Sat [RITA]
CATEGORY:
Emollient, conditioner, moisturizer, lubricant, plasticizer, chemical intermediate, humectant, release agent
APPLICATIONS:
Cosmetic industry preparations: (Fanchem HL; Fancol HL; Lipolan, S, Distilled Lipolan; Satulan; Super-Sat); conditioners (Lipolan, S, Distilled Lipolan); creams and lotions (Lipolan, S, Distilled Lipolan; Satulan; Super-Sat); hair preparations (Fancol HL; Lipolan, S, Distilled Lipolan); makeup (Lipolan, S, Distilled Lipolan; Satulan; Super-Sat); nail polish removers (Satulan); shaving preparations (Super-Sat); skin preparations (Fancol HL; Satulan); toiletries (Lipolan, S, Distilled Lipolan)
Industrial applications: (Fancol HL)
Pharmaceutical applications: (Fanchem HL; Satulan; Super-Sat); topical preparations (Lipolan, S, Distilled Lipolan; Satulan)
PROPERTIES:
Form:
Paste (Lipolan)
Soft paste (Distilled Lipolan)
Firm paste (Lipolan S)
Soft solid (Satulan)

Hydrogenated lanolin *(cont'd.)*

Solid (Fancol HL; Super-Sat)
Color:
White (Fancol HL; Super-Sat)
White/off-white (Lipolan, S, Distilled Lipolan)
Gardner 2 max. (Satulan)
Odor:
Trace (Fancol HL; Super-Sat)
Low (Satulan)
Mild, characteristic (Lipolan, S, Distilled Lipolan)
Composition:
100% active (Satulan; Super-Sat)
Solubility:
Sol. in acetone @ 75 C (Fancol HL)
Sol. in alcohols (Fanchem HL; Fancol HL)
Sol. in castor oil (Fancol HL)
Sol. in ethyl acetate @ 75 C (Fancol HL)
Insol. in glycerin (Fancol HL)
Sol. in most hydrocarbon oils (Fanchem HL)
Sol. in isopropyl myristate (Fancol HL)
Sol. in min. oil (Fancol HL)
Miscible warm with common oil phase ingredients (Lipolan, S, Distilled Lipolan)
Sol. in olive oil (Fancol HL)
Insol. in propylene glycol (Fancol HL)
Sol. in propylene glycol laurate (Fancol HL)
Insol. in water (Fancol HL; Lipolan, S, Distilled Lipolan)
Ionic Nature:
Nonionic (Satulan; Super-Sat)
Sp.gr.:
0.85–0.86 (Super-Sat)
0.855–0.865 (Fancol HL)
M.P.:
34–43 C (Distilled Lipolan)
37–45 C (Lipolan)
45–53 C (Lipolan S)
47–54 C (Satulan)
48–52 C (Super-Sat)
48–53 C (Fancol HL)
HLB:
1.0 (Satulan)
Acid No.:
1 max. (Lipolan, S, Distilled Lipolan; Satulan)
Iodine No.:
6–16 (Wijs) (Satulan)

Hydrogenated lanolin *(cont'd.)*

12–15 (Fancol HL)
16 (Lipolan)
20 (Lipolan S, Distilled Lipolan)
Saponification No.:
3 (Distilled Lipolan)
5 (Lipolan, S)
6 max. (Fancol HL)
7 max. (Satulan)
Hydroxyl No.:
130–160 (Satulan)
Stability:
Stable over entire pH range (Lipolan, S, Distilled Lipolan)
Ref. Index:
1.460–1.469 (50 C) (Fancol HL)

Hydrolyzed animal protein (CTFA)

SYNONYMS:
Collagen hydrolysates
Hydrolyzed collagen
Protein hydrolysate from collagen
Proteins, collagen, hydrolysate
CAS No.:
9015-54-7
TRADENAME EQUIVALENTS:
Collagen Hydrolyzate Cosmetic 55, N-55, SD [Amerchol]
Cropepsol [Croda Ltd.]
Cropeptone [Croda Ltd.]
Crotein A, C, O, SPA, SPA 55, SPC, SPO [Croda; Croda Ltd.]
Lamequat L [Henkel KGaA]
Lexein X250, X300, X350, X400, X450 [Inolex]
Nutrilan H, I, L [Grunau GmbH]
Polypeptide 37, LSN [Stepan]
Sterling CH-grade, PH-grade, SH-grade [Canada Packers]
CATEGORY:
Conditioner, emollient, humectant, thickener, protective colloid, film-former, foam
stabilizer/booster, leveling agent, texturizer, binder, excipient, encapsulant, surfac-
tant
APPLICATIONS:
Bath products: bubble bath (Nutrilan I, L)

Hydrolyzed animal protein (cont'd.)

Cosmetic industry preparations: (Sterling CH-grade, PH-grade, SH-grade); conditioners (Collagen Hydrolyzate Cosmetic 55, N-55, SD; Cropepsol; Cropeptone; Crotein SPA, SPC, SPO; Lexein X250, X350; Nutrilan I, L); creams and lotions (Crotein A, C, O; Lexein X250, X350); hair preparations (Collagen Hydrolyzate Cosmetic 55, N-55, SD; Cropepsol; Cropeptone; Crotein A, C, O, SPA, SPC, SPO; Lamequat L; Lexein X250; Nutrilan H, I, L); makeup (Crotein A, C, O); personal care products (Crotein SPA, SPC, SPO; Polypeptide 37, LSN; Lexein X450); shampoos (Crotein A, C, O, SPA, SPC, SPO; Lexein X350; Nutrilan I, L); skin preparations (Collagen Hydrolyzate Cosmetic 55, N-55, SD; Cropepsol; Cropeptone; Crotein SPA, SPC, SPO; Lamequat L; Lexein X250, X350; Nutrilan I, L)

Food applications: (Sterling CH-grade, PH-grade, SH-grade)

Pharmaceutical applications: (Crotein SPC); depilatories (Crotein A, C, O, SPA, SPC, SPO)

PROPERTIES:
Form:
Liquid (Crotein SPA 55; Lamequat L; Nutrilan H, I, L; Sterling CH-grade, PH-grade)

Clear liquid (Polypeptide 37, LSN)

Aq. sol'n. (Collagen Hydrolyzate Cosmetic 55, N-55; Cropepsol; Cropeptone)

Spray-dried powder (Collagen Hydrolyzate Cosmetic SD)

Powder (Crotein A, C, O, SPA, SPC, SPO; Lexein X300; Nutrilan I, L; Sterling SH-grade)

Color:
White (Crotein SPA, SPC, SPO)

Very low (Crotein A, C, O)

Amber (Crotein SPA 55; Polypeptide LSN)

Dark amber (Polypeptide 37)

Odor:
Very low (Crotein A, C, O)

Low (Crotein SPA 55; Lexein X450)

Bland, pleasant (Crotein SPA, SPC, SPO)

Composition:
35% active (Lamequat L)

40–42% solids (Polypeptide LSN)

48–50% solids (Polypeptide 37)

55% active (Crotein SPA 55)

93% active (Crotein SPA, SPC, SPO)

100% active (Crotein A, C, O)

Solubility:
Sol. in aq. alcohol sol'ns. (Lexein X250, X350)

Sol. in 50% ethanol @ 1% (Crotein SPA); sol. in 65% ethanol @ 1% (Crotein SPC); sol. in 20% ethanol @ 1% (Crotein SPO)

Sol. in water (Crotein A, C, O, SPA, SPC; Lexein X250, X300, X350, X400, X450); miscible with water (Polypeptide 37, LSN)

Hydrolyzed animal protein *(cont'd.)*

Ionic Nature:
   Cationic (Lamequat L)
   Anionic (Polypeptide 37, LSN)
M.W.:
   1000 (Crotein SPO)
   4000 (Crotein SPA, SPA 55)
   10,000 (Crotein SPC)
Sp.gr.:
   1.20–1.23 (Polypeptide 37)
Visc.:
   15–20 mps (10% sol'n.) (Crotein SPO)
   20–25 mps (10% sol'n.) (Crotein SPA)
   40–50 mps (10% sol'n.) (Crotein SPC)
   50 cps max. (Polypeptide 37)
   100 cps max. (Polypeptide LSN)
Stability:
   Resistant to degradation and hydrolysis under both alkaline and acid conditions (pH
      3–12) (Crotein SPA, SPC, SPO)
pH:
   5.5–6.5 (10% sol'n.) (Crotein SPA, SPC, SPO)
   6.6–7.4 (Polypeptide LSN)
   7.0–7.4 (Polypeptide 37)

Hydroxylated lanolin *(CTFA)*

SYNONYMS:
   Lanolin, hydroxylated
CAS No.:
   68424-66-8
TRADENAME EQUIVALENTS:
   Hydroxylan [Fanning]
   OHlan [Amerchol; Amerchol Europe]
   Ritahydrox [RITA]
CATEGORY:
   Conditioner, emollient, emulsifier, stabilizer, dispersant, tackifier, superfatting agent
APPLICATIONS:
   Cleansers: soaps (OHlan)
   Cosmetic industry preparations: (OHlan); hair preparations (OHlan); makeup
      (OHlan); shaving preparations (OHlan); skin preparations (OHlan)
   Pharmaceutical applications: hypoallergenic products (Ritahydrox)

Hydroxylated lanolin (cont'd.)

PROPERTIES:
Form:
   Paste (Ritahydrox)
   Waxy solid (OHlan)
Color:
   Yellow-amber to light tan (OHlan)
Odor:
   Mild (OHlan)
Composition:
   100% conc. (OHlan; Ritahydrox)
Solubility:
   Slightly sol. in castor oil (OHlan)
   Slightly sol. in isopropyl myristate (OHlan)
   Miscible with common oil phase ingredients (OHlan)
Ionic Nature:
   Nonionic (OHlan; Ritahydrox)
M.P.:
   41–46 C (OHlan)
HLB:
   4.0 (Ritahydrox)
Acid No.:
   10 max. (OHlan)
Iodine No.:
   15–23 (OHlan)
Saponification No.:
   95–110 (OHlan)
Hydroxyl No.:
   38–48 (OHlan)

Hydroxystearic acid (CTFA)

SYNONYMS:
   12-Hydroxyoctadecanoic acid
   12-Hydroxystearic acid
   Octadecanoic acid, 12-hydroxy-
EMPIRICAL FORMULA:
   $C_{18}H_{36}O_3$
STRUCTURE:

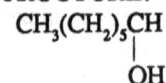

$$CH_3(CH_2)_5\underset{|}{\overset{}{CH}}$$
$$OH$$

CAS No.:
106-14-9; 1330-70-7
TRADENAME EQUIVALENTS:
Loxiol G21 [Henkel KGaA]
Radiacid 200 [Synfina-Oleofin]
CATEGORY:
Lubricant
APPLICATIONS:
Industrial applications: plastics (Loxiol G21; Radiacid 200)
PROPERTIES:
Form:
Solid (Loxiol G21)
Flake (Radiacid 200)
Powder (Radiacid 200)
Sp.gr.:
0.885–0.891 (80 C) (Loxiol G21)
Visc.:
30–35 mPa•s (80 C) (Loxiol G21)
Drop Pt.:
72–78 C (Loxiol G21)
Flash Pt.:
> 210 C (Loxiol G21)
Acid No.:
172–180 (Loxiol G21)
Iodine No.:
< 10 (Loxiol G21)
Ref. Index:
1.440–1.442 (80 C) (Loxiol G21)

Isobutyl stearate (CTFA)

SYNONYMS:
2-Methylpropyl octadecanoate
Octadecanoic acid, 2-methylpropyl ester
Stearic acid, 2-methylpropyl ester

EMPIRICAL FORMULA:
$C_{22}H_{44}O_2$

STRUCTURE:

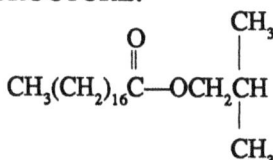

CAS No.:
646-13-9

TRADENAME EQUIVALENTS:
Emerest 2324 [Henkel/Emery]
Kemester 5415 [Humko]
Kessco IBS, Isobutyl Stearate [Stepan]
Radia 7240 [Synfina-Oleofin]
Rilanit IBS [Henkel KGaA]
Uniflex IBYS [Union Camp]

CATEGORY:
Lubricant, emollient, fatting agent, solvent, plasticizer, chemical intermediate, gloss promoter

APPLICATIONS:
Cosmetic industry preparations: creams and lotions (Kessco Isobutyl Stearate); hair preparations (Kessco Isobutyl Stearate); makeup (Kessco Isobutyl Stearate)
Industrial applications: ceramics (Rilanit IBS); chemical synthesis (Radia 7240); lubricating/cutting oils (Radia 7240; Rilanit IBS); metalworking (Emerest 2324; Kemester 5415; Kessco IBS; Rilanit IBS); paint/lacquer mfg. (Radia 7240); paper mfg. (Radia 7240); rust inhibitors (Radia 7240); textile/leather processing (Emerest 2324; Kemester 5415; Kessco IBS, Isobutyl Stearate; Radia 7240; Rilanit IBS)

PROPERTIES:
Form:
Liquid (Emerest 2324; Kemester 5415; Kessco IBS, Isobutyl Stearate; Rilanit IBS)

Color:
APHA 35 max. (Kessco Isobutyl Stearate)
APHA 40 (Uniflex IBYS)
Gardner 1 (Emerest 2324)
Composition:
100% active (Emerest 2324; Kessco IBS)
Solubility:
Sol. in isopropanol (Emerest 2324)
Sol. in min. oil (Emerest 2324; Radia 7240)
Sol. in most solvents (Radia 7240)
Sol. in toluene (Emerest 2324)
Sol. in veg. oil (Radia 7240)
Insol. in water (Emerest 2324)
Sol. in xylene (Emerest 2324)
Ionic Nature:
Nonionic (Emerest 2324)
M.W.:
325 (avg.) (Radia 7240)
Sp.gr.:
0.844 (37.8 C) (Radia 7240)
0.849–0.855 (25/20 C) (Kessco Isobutyl Stearate)
0.855 (Uniflex IBYS)
Density:
7.1 lb/gal (Emerest 2324; Kessco Isobutyl Stearate)
Visc.:
6.20 cps (37.8 C) (Radia 7240)
8.5 cps (Kessco Isobutyl Stearate)
10 cSt (100 F) (Emerest 2324)
11 cSt (Uniflex IBYS)
F.P.:
15 C (Kessco Isobutyl Stearate)
B.P.:
200 C (mid, 4 mm Hg) (Kessco Isobutyl Stearate)
M.P.:
16 C (Uniflex IBYS)
Pour Pt.:
10 C (Emerest 2324)
Flash Pt.:
181 C (COC) (Radia 7240)
190 C (COC) (Uniflex IBYS)
360 F (COC) (Kessco Isobutyl Stearate)
370 F (Emerest 2324)

Isobutyl stearate *(cont'd.)*

Fire Pt.:
216 C (COC) (Uniflex IBYS)
Cloud Pt.:
17 C (Radia 7240)
Acid No.:
0.1 (Uniflex IBYS)
Iodine No.:
< 1.0 (Uniflex IBYS)
1.0 max. (Kessco Isobutyl Stearate)
Ref. Index:
1.441 (Kessco Isobutyl Stearate)
1.4421 (Radia 7240)
Surface Tension:
30.20 dynes/cm (Radia 7240)

Isocetyl alcohol (CTFA)

SYNONYMS:
Isohexadecanol
Isohexadecyl alcohol
EMPIRICAL FORMULA:
$C_{16}H_{34}O$
CAS No.:
36311-34-9
TRADENAME EQUIVALENTS:
Ceraphyl ICA [Van Dyk]
Standamul G-16 [Henkel]
CATEGORY:
Emollient, binder, conditioner, softener, extender, solvent
APPLICATIONS:
Cosmetic industry preparations: conditioners (Ceraphyl ICA); creams and lotions (Ceraphyl ICA); hair preparations (Ceraphyl ICA); makeup (Standamul G-16); perfumery (Standamul G-16); pressed powders (Ceraphyl ICA; Standamul G-16); skin preparations (Ceraphyl ICA; Standamul G-16)
PROPERTIES:
Form:
Clear liquid (Ceraphyl IC; Standamul G-16)
Color:
Water-white (Standamul G-16)

Solubility:
Sol. in most organic solvents (Ceraphyl ICA)
Insol. in water (Ceraphyl ICA)
Visc.:
100 cps (Standamul G-16)
Cloud Pt.:
≈ -50 C (Standamul G-16)
Acid No.:
0.5 max. (Standamul G-16)
Iodine No.:
8.0 max. (Standamul G-16)
Saponification No.:
5.0 max. (Standamul G-16)
Hydroxyl No.:
210–230 (Standamul G-16)
Stability:
Stable to acids and alkalies (Standamul G-16)
Storage Stability:
Stable on long storage (Standamul G-16)

Isocetyl stearate (CTFA)

SYNONYMS:
Ester of isocetyl alcohol and stearic acid
Octadecanoic acid, isohexadecyl ester
EMPIRICAL FORMULA:
$C_{34}H_{68}O_2$
STRUCTURE:

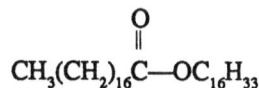

$$CH_3(CH_2)_{16}C-OC_{16}H_{33}$$

CAS No.:
25339-09-7
TRADENAME EQUIVALENTS:
Ceraphyl 494 [Van Dyk]
Cetiol G16S [Henkel Canada]
Crodamol ICS [Croda]
Kemester 5822 [Humko/Witco]
Kessco ICS, Isocetyl Stearate [Stepan]
Nikkol ICS-R [Nikko]

Isocetyl stearate (cont'd.)

Schercemol ICS [Scher]
Standamul 7061 [Henkel]
CATEGORY:
Emollient, conditioner, lubricant, spreading agent, solvent, base, carrier, viscosity
modifier, plasticizer
APPLICATIONS:
Bath products: bath oils (Crodamol ICS)
Cosmetic industry preparations: (Cetiol G16S; Kemester 5822; Nikkol ICS-R);
creams and lotions (Crodamol ICS; Kessco Isocetyl Stearate; Schercemol ICS);
hair preparations (Ceraphyl 494; Crodamol ICS); makeup (Cetiol G16S; Crodamol
ICS); personal care products (Standamul 7061); skin preparations (Ceraphyl 494;
Crodamol ICS; Kessco ICS)
Industrial applications: metalworking (Kessco ICS); textile/leather processing
(Kessco ICS)
Pharmaceutical applications: (Standamul 7061)
PROPERTIES:
Form:
Liquid (Kemester 5822; Kessco ICS, Isocetyl Stearate; Nikkol ICS-R)
Clear liquid (Schercemol ICS)
Clear oily liquid (Cetiol G16S)
Low viscosity liquid (Crodamol ICS)
Clear low viscosity oil (Standamul 7061)
Color:
Water-white (Crodamol ICS)
APHA 200 max. (Kessco Isocetyl Stearate)
Gardner 2 max. (Schercemol ICS)
Odor:
Slight, typical (Schercemol ICS)
Faint, characteristic (Standamul 7061)
Composition:
100% active (Kessco ICS; Nikkol ICS-R; Standamul 7061)
Solubility:
Sol. in alcohols (Cetiol G16S)
Sol. in aromatic hydrocarbons (Schercemol ICS)
Sol. in castor oil (Standamul 7061)
Sol. in esters (Cetiol G16S; Schercemol ICS)
Disp. in glycerol (Schercemol ICS)
Disp. in glycol ethers (Schercemol ICS)
Disp. in glycols (Schercemol ICS)
Sol. in isopropyl myristate (Standamul 7061)
Sol. in min. oil (Cetiol G16S; Schercemol ICS: Standamul 7061)
Sol. in oils (Ceraphyl 494)
Sol. in oleyl alcohol (Standamul 7061)

Sol. in most organic solvents (Ceraphyl 494)
Insol. in water (Ceraphyl 494; Cetiol G16S; Schercemol ICS)
Ionic Nature:
Nonionic (Kessco Isocetyl Stearate)
M.W.:
494 (theoret.) (Schercemol ICS)
Sp.gr.:
0.850 (Schercemol ICS)
0.853–0.859 (25/20 C) (Kessco Isocetyl Stearate)
0.865 (Standamul 7061)
Density:
7.1 lb/gal (Kessco Isocetyl Stearate; Schercemol ICS)
Visc.:
25 cps (Standamul 7061)
32 cps (Kessco Isocetyl Stearate)
F.P.:
–5 C (Schercemol ICS)
0 C (Kessco Isocetyl Stearate)
Set Pt.:
0 C (Crodamol ICS)
Flash Pt.:
> 180 C (Schercemol ICS)
450 F (COC) (Kessco Isocetyl Stearate)
Cloud Pt.:
0 C max. (Standamul 7061)
HLB:
8 (Standamul 7061)
Acid No.:
0.5 max. (Standamul 7061)
1.0 max. (Schercemol ICS)
3.0 max. (Kessco Isocetyl Stearate)
Iodine No.:
Nil (Schercemol ICS)
5.0 max. (Kessco Isocetyl Stearate)
Saponification No.:
105–115 (Standamul 7061)
112 ± 5.0 (Schercemol ICS)
Ref. Index:
1.452 (Kessco Isocetyl Stearate)
STD. PKGS.:
400 lb net closed-head steel drums or bulk (Standamul 7061)

Isopropyl isostearate (CTFA)

EMPIRICAL FORMULA:

$C_{21}H_{42}O_2$

STRUCTURE:

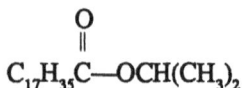

$C_{17}H_{35}C-OCH(CH_3)_2$

CAS No.:

31478-84-9; 68171-33-5

TRADENAME EQUIVALENTS:

Emerest 2310 [Henkel/Emery]

Lan-O-Derm [Alcolac]

Schercemol 318 [Scher]

Wickenol 131 [CasChem]

CATEGORY:

Emollient, lubricant, moisturizer, binder, solubilizer

APPLICATIONS:

Bath products: bath oils (Emerest 2310; Schercemol 318)

Cosmetic industry preparations: (Emerest 2310); creams and lotions (Emerest 2310; Schercemol 318); hair preparations (Emerest 2310); makeup (Emerest 2310; Schercemol 318); pressed powders (Emerest 2310); shampoos (Emerest 2310); skin preparations (Lan-O-Derm)

PROPERTIES:

Form:

Clear liquid (Emerest 2310; Schercemol 318)

Color:

Light lemon yellow (Schercemol 318)

Gardner 1 (Emerest 2310)

Odor:

Bland (Schercemol 318)

Composition:

99% active min. (Schercemol 318)

Solubility:

Sol. in acetone (Wickenol 131)

Sol. in alcohols (Lan-O-Derm)

Sol. in aliphatic hydrocarbons (Lan-O-Derm)

Sol. in aromatic hydrocarbons (Lan-O-Derm)

Sol. in castor oil (Emerest 2310; Wickenol 131)

Sol. in corn oil (Wickenol 131)

Sol. in esters (Lan-O-Derm)

Sol. in ethanol (Emerest 2310; Wickenol 131)

Sol. in ethyl acetate (Wickenol 131)

Insol. in glycerin (Emerest 2310; Wickenol 131)

Sol. in isopropanol (Emerest 2310)
Sol. in isopropyl myristate (Emerest 2310)
Sol. in ketones (Lan-O-Derm)
Sol. in lanolin (Emerest 2310)
Sol. in min. oil (Emerest 2310; Wickenol 131)
Sol. in peanut oil (Emerest 2310)
Insol. in propylene glycol (Emerest 2310; Wickenol 131)
Sol. in silicone (Emerest 2310)
Sol. in triolein (Emerest 2310)
Insol. in water (Emerest 2310; Lan-O-Derm; Wickenol 131)
M.W.:
326 (theoret.) (Schercemol 318)
Sp.gr.:
0.855 ± 0.01 (Schercemol 318)
Density:
7.1 lb/gal (Emerest 2310)
7.12 lb/gal (Schercemol 318)
Visc.:
7 cSt (100 F) (Emerest 2310)
F.P.:
< –20 C (Schercemol 318)
Pour Pt.:
< 2 C (Emerest 2310)
Flash Pt.:
> 170 C (Schercemol 318)
345 F (Emerest 2310)
Acid No.:
1.0 max. (Schercemol 318)
Iodine No.:
3.0 max. (Schercemol 318)
Saponification No.:
170 ± 10 (Schercemol 318)
Ref. Index:
1.422 ± 0.001 (Schercemol 318)

Isopropyl lanolate (CTFA)

SYNONYMS:
Fatty acids, lanolin, isopropyl esters
Lanolin fatty acids, isopropyl esters

Isopropyl lanolate (cont'd.)

CAS No.:
63393-93-1
TRADENAME EQUIVALENTS:
Amerlate P, W [Amerchol; Amerchol Europe]
Crodalan IPL [Croda; Croda Ltd.]
Emerest 1723 [Henkel/Emery]
Lanesta L, S, SA 30 [Westbrook Lanolin]
Rewolan LP [Rewo GmbH]
Ritasol [RITA]
Trisolan 1720 [Henkel/Emery]
CATEGORY:
Conditioner, softener, penetrant, spreading agent, lubricant, moisturizer, emollient, emulsifier, stabilizer, opacifier, dispersant, wetting agent, plasticizer, film-former
APPLICATIONS:
Bath products: bath oils (Crodalan IPL)
Cosmetic industry preparations: (Amerlate P); creams and lotions (Amerlate W); hair preparations (Crodalan IPL; Trisolan 1720); makeup (Amerlate W; Crodalan IPL; Ritasol; Trisolan 1720); personal care products (Amerlate W); skin preparations (Crodalan IPL; Lanesta L, S, SA 30; Rewolan LP); toiletries (Rewolan LP)
Industrial applications: dyes and pigments (Amerlate P); waxes (Amerlate P)
Pharmaceutical applications: (Amerlate P, W)
PROPERTIES:
Form:
Liquid (Lanesta L; Rewolan LP)
Thin opalescent liquid (Crodalan IPL)
Soft paste (Crodalan IPL)
Solid (Ritasol)
Buttery solid (Amerlate P, W)
Soft wax (Lanesta S, SA 30)
Color:
Pale yellow (Crodalan IPL)
Yellow (Amerlate P, W)
Odor:
Faint, characteristic (Amerlate P, W)
Composition:
100% active (Amerlate P, W; Rewolan LP; Ritasol)
Solubility:
Sol. in alcohols (Trisolan 1720)
Miscible with castor oil (Amerlate W)
Sol. in ethanol (Amerlate W)
Miscible with isopropyl myristate (Amerlate W)
Miscible with isopropyl palmitate (Amerlate W)
Miscible with warm min. oil (Amerlate W)

Isopropyl lanolate (cont'd.)

Sol. in oils (Trisolan 1720); sol. in warm oils (Emerest 1723); miscible with oil
  (Amerlate P)
Insol. in water (Amerlate W; Trisolan 1720)
Ionic Nature:
  Nonionic (Amerlate P, W)
HLB:
  9 (Amerlate P, W)
Acid No.:
  18 max. (Amerlate P, W)
Saponification No.:
  130–155 (Amerlate P)
  135–165 (Amerlate W)
Hydroxyl No.:
  35–55 (Amerlate W)

Isopropyl myristate (CTFA)

SYNONYMS:
  Ester of isopropyl alcohol and myristic acid
  IPM
  1-Methylethyl tetradecanoate
  Tetradecanoic acid, 1-methylethyl ester
EMPIRICAL FORMULA:
  $C_{17}H_{34}O_2$
STRUCTURE:

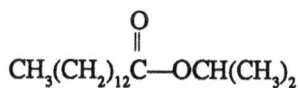

CAS No.:
  110-27-0
TRADENAME EQUIVALENTS:
  Crodamol IPM [Croda]
  Deltyl Extra [Givaudan]
  Emerest 2314 [Henkel/Emery]
  Exceparl IPM [Kao]
  I.P.M. [Akzo Chemie Italia]
  Kessco IPM, Isopropyl Myristate [Stepan]
  Kesso-Isopropyl-myristat [Akzo BV]
  Lexol IPM [Inolex]
  Liponate IPM [Lipo]

Isopropyl myristate (cont'd.)

Promyr [Amerchol]
Radia 7190 [Synfina-Oleofin]
Rilanit IPM [Henkel KGaA]
Schercemol IPM [Scher]
Starfol IPM [Sherex]
Stepan D-50 [Stepan]
Unimate IPM [Union Camp]
Wickenol 101 [CasChem]

CATEGORY:

Emollient, lubricant, conditioner, spreading agent, solvent, cosolvent, solubilizer, plasticizer, carrier, vehicle, penetrant, cloud point depressant, thickener, viscosity control agent, chemical intermediate, auxiliary emulsifier

APPLICATIONS:

Bath products: (Liponate IPM); bath oils (Crodamol IPM; Emerest 2314; Kessco Isopropyl Myristate; Lexol IPM; Liponate IPM; Schercemol IPM; Stepan D-50; Unimate IPM; Wickenol 101)

Cosmetic industry preparations: (Crodamol IPM; Deltyl Extra; Emerest 214; I.P.M.; Exceparl IPM; Kessco IPM; Lexol IPM; Promyr; Radia 7190; Schercemol IPM; Wickenol 101); body oils/powders (Liponate IPM); creams and lotions (Emerest 2314; Kessco Isopropyl Myristate; Liponate IPM; Stepan D-50; Unimate IPM; Wickenol 101); hair preparations (Crodamol IPM; Kesso-Isopropyl-myristat; Kessco Isopropyl Myristate; Schercemol IPM); makeup (Crodamol IPM; Emerest 2314; Kessco Isopropyl Myristate; Liponate IPM; Promyr; Schercemol IPM); perfumery (Schercemol IPM); shaving preparations (Crodamol IPM; Emerest 2314; Kessco Isopropyl Myristate; Lexol IPM; Liponate IPM; Stepan D-50); skin preparations (Crodamol IPM; Kessco IPM; Unimate IPM); toiletries (Promyr; Wickenol 101)

Industrial applications: ceramics (Rilanit IPM); industrial processing (Radia 7190); lamination (Radia 7190); lubricating/cutting oils (Radia 7190; Rilanit IPM); metalworking (Kessco IPM; Rilanit IPM); textile/leather processing (Emerest 2314; Kessco IPM, Isopropyl Myristate; Radia 7190; Rilanit IPM)

Pharmaceutical applications: (Radia 7190); antiperspirant/deodorant (Crodamol IPM; Emerest 2314; Stepan D-50; Unimate IPM); hygiene products (Crodamol IPM); sunscreens (Crodamol IPM; Kessco Isopropyl Myristate; Schercemol IPM)

PROPERTIES:

Form:

Liquid (Crodamol IPM; Emerest 2314; Exceparl IPM; I.P.M.; Kessco IPM, Isopropyl Myristate; Kesso-Isopropyl-myristat; Liponate IPM; Rilanit IPM; Starfol IPM)

Clear liquid (Lexol IPM; Schercemol IPM)

Clear oily liquid (Stepan D-50)

Low viscosity liquid (Promyr)

Color:

Colorless (Lexol IPM; Liponate IPM)

Water-white (Crodamol IPM; Promyr; Schercemol IPM; Stepan D-50)
APHA 20 max. (Kessco Isopropyl Myristate)
APHA 30 max. (Starfol IPM)
Gardner 1 max. (Emerest 2314)
Odor:
Odorless (Crodamol IPM)
Essentially odorless (Lexol IPM; Schercemol IPM)
Practically odorless (Promyr)
Bland, very slightly fatty (Stepan D-50)
Composition:
92% conc. min. (Crodamol IPM)
99% active min. (Schercemol IPM)
> 99% active (I.P.M.)
100% active (Emerest 2314; Kessco IPM; Starfol IPM; Stepan D-50)
Solubility:
Sol. in acetone (Lexol IPM)
Sol. in alcohols (Deltyl Extra; Wickenol 101)
Sol. in aliphatic hydrocarbons (Schercemol IPM)
Sol. in animal oils (Wickenol 101)
Sol. in benzene (Lexol IPM)
Sol. in carbon tetrachloride (Lexol IPM)
Sol. in castor oil (Lexol IPM)
Sol. in chloroform (Lexol IPM)
Sol. in esters (Schercemol IPM)
Sol. in ethanol (Lexol IPM; Liponate IPM); sol. in 95% ethanol (Unimate IPM)
Sol. in heptane (Lexol IPM)
Sol. in isopropanol (Emerest 2314; Lexol IPM)
Sol. in min. oil (Deltyl Extra; Emerest 2314; Liponate IPM; Radia 7190; Schercemol IPM; Unimate IPM)
Sol. in min. spirits (Schercemol IPM)
Sol. in oils (Stepan D-50)
Sol. in most organic solvents (Schercemol IPM)
Sol. in most solvents (Schercmol IPM)
Sol. in toluol (Emerest 2314)
Sol. in veg. oil (Liponate IPM; Radia 7190; Wickenol 101); sol. in peanut, sesame, olive, almond oils (Deltyl Extra)
Insol. in water (Deltyl Extra; Liponate IPM; Schercemol IPM; Unimate IPM)
Sol. in white oils (Wickenol 101)
Ionic Nature:
Nonionic (Emeresst 2314; I.P.M.; Starfol IPM)
M.W.:
270 (theoret.) (Lexol IPM; Schercemol IPM)
271 avg. (Radia 7190)

Isopropyl myristate (cont'd.)

Sp.gr.:
  0.840 (37.8 C) (Radia 7190)
  0.847–0.853 (Crodamol IPM)
  0.847–0.854 (Lexol IPM)
  0.849–0.855 (25/20 C) (Kessco Isopropyl Myristate)
  0.850 (I.P.M.; Schercemol IPM)
Density:
  7.1 lb/gal (Emerest 2314; Kessco Isopropyl Myristate; Lexol IPM)
Visc.:
  3.5 cps (37.8 C) (Radia 7190)
  4.8 cps (Kessco Isopropyl Myristate; Lexol IPM)
  6 cs (38 C) (Emerest 2314)
  6.2–6.4 cs (20 C) (I.P.M.)
F.P.:
  –5 C (Emerest 2314)
  –4 C (Schercemol IPM)
  –3 C (Kessco Isopropyl Myristate)
  3 C (Lexol IPM)
  7 C (I.P.M.)
B.P.:
  160 C (4 mm Hg) (Kessco Isopropyl Myristate)
  170 C (4 mm Hg) (Lexol IPM)
Flash Pt.:
  152 C (Emerest 2314)
  157 C (COC) (Radia 7190)
  > 170 C (OC) (Schercemol IPM)
  305 F (Lexol IPM); (COC) (Kessco Isopropyl Myristate)
Cloud Pt.:
  –3.5 C (Radia 7190)
Acid No.:
  0.22 (Stepan D-50)
  1 (Emerest 2314)
  1 max. (Kessco Isopropyl Myristate; Promyr; Schercemol IPM; Starfol IPM)
  2 max. (Crodamol IPM; Liponate IPM)
Iodine No.:
  0.02 (Stepan D-50)
  1.0 max. (Kessco Isopropyl Myristate; Lexol IPM; Promyr; Starfol IPM)
Saponification No.:
  202–211 (Liponate IPM; Promyr)
  202–212 (Lexol IPM; Schercemol IPM; Starfol IPM)
  210 (Emerest 2314)
Stability:
  Good oxidative stability (Lexol IPM; Promyr)

Ref. Index:
  1.432–1.434 (Crodamol IPM)
  1.4325 (Stepan D-50)
  1.433 (Kessco Isopropyl Myristate; Lexol IPM)
  1.4330–1.4350 (Schercemol IPM)
Surface Tension:
  27.50 dynes/cm (Radia 7190)
STD. PKGS.:
  180-kg steel drums (I.P.M.)
  55-gal steel drums (Lexol IPM)
  400-lb net bung-head steel drums (Schercemol IPM)

Isopropyl oleate (CTFA)

SYNONYMS:
  1-Methylethyl-9-octadecenoate
  9-Octadecenoic acid, 1-methylethyl ester
STRUCTURE:

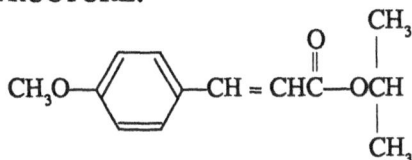

$$CH_3O-\bigcirc-CH=CHC(=O)-OCH(CH_3)CH_3$$

CAS No.:
  112-11-8
TRADENAME EQUIVALENTS:
  Radia 7231 [Synfina-Oleofin]
  Schercemol IPO [Scher]
CATEGORY:
  Lubricant, emollient, chemical intermediate, spreading agent
APPLICATIONS:
  Bath products: bath oils (Schercemol IPO)
  Cosmetic industry preparations: (Schercemol IPO); creams and lotions (Schercemol IPO); makeup (Schercemol IPO); skin preparations (Schercemol IPO)
  Industrial applications: chemical synthesis (Radia 7231); lamination (Radia 7231); lubricating/cutting oils (Radia 7231); rust inhibitors (Radia 7231); textile/leather processing (Radia 7231)
PROPERTIES:
Form:
  Clear liquid (Schercemol IPO)

Isopropyl oleate *(cont'd.)*

Color:
Gardner 2 max. (Schercemol IPO)
Odor:
Typical, slight (Schercemol IPO)
Solubility:
Sol. in alcohols (Schercemol IPO)
Sol. in aliphatic hydrocarbons (Schercemol IPO)
Sol. in aromatic hydrocarbons (Schercemol IPO)
Sol. in chlorinated hydrocarbons (Schercemol IPO)
Sol. in esters (Schercemol IPO)
Sol. in glycol ethers (Schercemol IPO)
Sol. in higher glycols (Schercemol IPO)
Sol. in min. oils (Radia 7231; Schercemol IPO)
Sol. in most solvents (Radia 7231)
Sol. in veg. oils (Radia 7231; Schercemol IPO)
M.W.:
313 (avg.) (Radia 7231)
324 (theoret.) (Schercemol IPO)
Sp.gr.:
0.853 (37.8 C) (Radia 7231)
0.861 ± 0.005 (Schercemol IPO)
Density:
7.17 lb/gal (Schercemol IPO)
Visc.:
5.05 cps (37.8 C) (Radia 7231)
F.P.:
< −15 C (Schercemol IPO)
Flash Pt.:
169 C (COC) (Radia 7231)
> 180 C (OC) (Schercemol IPO)
Cloud Pt.:
−17 C (Radia 7231)
Acid No.:
2.0 max. (Schercemol IPO)
Iodine No.:
85 max. (Schercemol IPO)
Saponification No.:
175–190 (Schercemol IPO)
Ref. Index:
1.4472 ± 0.001 (Schercemol IPO)
1.4479 (Radia 7231)
Surface Tension:
31 dynes/cm (Radia 7231)

Isopropyl palmitate (CTFA)

SYNONYMS:
Hexadecanoic acid, 1-methylethyl ester
IPP
Isopropyl n-hexadecanoate
1-Methylethyl hexadecanoate

EMPIRICAL FORMULA:
$C_{19}H_{38}O_2$

STRUCTURE:

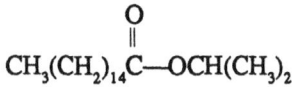

$$CH_3(CH_2)_{14}\overset{\overset{\displaystyle O}{\|}}{C}-OCH(CH_3)_2$$

CAS No.:
142-91-6

TRADENAME EQUIVALENTS:
Crodamol IPP [Croda; Croda Ltd.]
Emerest 2316 [Henkel/Emery]
Exceparl IPP [Kao]
IPP [Henkel KGaA]
Isopropylpalmitat [Henkel Canada]
Kessco IPP, Isopropyl Palmitate [Stepan]
Kessco-Isopropyl-palmitat [Akzo Chemie BV]
Lexol IPP [Inolex]
Liponate IPP [Lipo]
Propal [Amerchol]
Radia 7200 [Synfina-Oleofin]
Rilanit IPP [Henkel KGaA]
Starfol IPP [Sherex]
Stepan D-70 [Stepan]
Unimate IPP [Union Camp]
Wickenol 111 [CasChem]

CATEGORY:
Emollient, lubricant, solvent, cosolvent, wetting agent, conditioner, softener, fatting agent, plasticizer, solubilizer. thickener, viscosity control agent, anticlog agent, dispersant, carrier

APPLICATIONS:
Bath products: (Liponate IPP); bath oils (Emerest 2316; IPP; Isopropylpalmitat; Lexol IPP; Stepan D-70; Unimate IPP; Wickenol 111)

Cosmetic industry preparations: (Crodamol IPP; Emerest 2316; Exceparl IPP; IPP; Isopropylpalmitat; Kessco IPP; Lexol IPP; Propal; Radia 7200; Wickenol 111); conditioners (Lexol IPP); creams and lotions (Emerest 2316; Kessco Isopropyl Palmitate; Lexol IPP; Liponate IPP; Stepan D-70; Unimate IPP; Wickenol 111); hair preparations (Emerest 2316; Kessco-Isopropyl-palmitat; Kessco Isopropyl

Isopropyl palmitate (cont'd.)

Palmitate); makeup (Emerest 2316; Kessco Isopropyl Palmitate; Propal; Wickenol 111); personal care products (IPP; Isopropylpalmitat); pressed powders (Emerest 2316); skin preparations (Kessco IPP); toiletries (Propal; Wickenol 111

Industrial applications: lubricating/cutting oils (Rilanit IPP); metalworking (Kessco IPP; Rilanit IPP); textile/leather processing (Emerest 2316; Kessco IPP; Kessco Isopropyl Palmitate; Rilanit IPP)

Pharmaceutical applications: (Crodamol IPP; Radia 7200); antiperspirant/deodorant (Emerest 2316; Lexol IPP; Stepan D-70; Unimate IPP); sunscreens (Emerest 2316); topical preparations (Lexol IPP)

PROPERTIES:
Form:

Liquid (Crodamol IPP; Emerest 2316; Exceparl IPP; IPP; Isopropylpalmitat; Kessco IPP; Kessco-Isopropyl-palmitat; Kessco Isopropyl Palmitate; Liponate IPP; Propal; Radia 7200; Rilanit IPP; Starfol IPP)

Clear liquid (Lexol IPP; Stepan D-70)

Color:

Colorless (Lexol IPP; Liponate IPP)

Water-white (Crodamol IPP; Propal; Stepan D-70)

APHA 20 max. (Kessco Isopropyl Palmitate)

APHA 30 max. (Starfol IPP)

Gardner 1 (Emerest 2316)

Odor:

Odorless (Crodamol IPP; Lexol IPP)

Essentially odorless (Stepan D-70)

Practically odorless (Propal)

Composition:

100% active (IPP; Kessco IPP; Starfol IPP; Stepan D-70)

Solubility:

Sol. in alcohols (Wickenol 111)

Sol. in animal oils (Wickenol 111)

Sol. in castor oil (Lexol IPP)

Sol. in ethanol (Lexol IPP; Liponate IPP); sol. in 95% ethanol (Unimate IPP)

Insol. in glycerin (Emerest 2316; Lexol IPP; Unimate IPP; Wickenol 111)

Sol. in isopropanol (Emerest 2316; Lexol IPP)

Sol. in isopropyl myristate (Emerest 2316)

Sol. in min. oil (Emerest 2316; Liponate IPP; Radia 7200; Unimate IPP)

Sol. in oil (Stepan D-70)

Insol. in propylene glycol (Lexol IPP; Unimate IPP; Wickenol 111)

Sol. in most solvents (Radia 7200)

Sol. in triolein (Emerest 2316)

Sol. in veg. oil (Liponate IPP; Radia 7200; Wickenol 111)

Insol. in water (Emerest 2316; Lexol IPP; Liponate IPP; Unimate IPP; Wickenol 111)

Sol. in white oils (Wickenol 111)

Ionic Nature:
   Nonionic (Kessco Isopropyl Palmitate; Starfol IPP)
M.W.:
   298 (theoret.) (Lexol IPP)
   300 (avg.) (Radia 7200)
Sp.gr.:
   0.843 (37.8 C) (Radia 7200)
   0.849–0.855 (25/20 C) (Kessco Isopropyl Palmitate)
   0.850–0.855 (Lexol IPP)
Density:
   7.1 lb/gal (Emerest 2316; Kessco Isopropyl Palmitate; Lexol IPP)
Visc.:
   5.05 cps (37.8 C) (Radia 7200)
   6 cSt (100 F) (Emerest 2316)
   6.7 cps (Kessco Isopropyl Palmitate)
   7 cps (Lexol IPP)
F.P.:
   12 C (Lexol IPP)
   13 C (Kessco Isopropyl Palmitate)
B.P.:
   170 C (mid, 4 mm Hg) (Kessco Isopropyl Palmitate)
   172 C (4 mm Hg) (Lexol IPP)
Pour Pt.:
   14 C (Emerest 2316)
Flash Pt.:
   167 C (Radia 7200)
   325 F (COC) (Kessco Isopropyl Palmitate)
   335 F (Lexol IPP)
   340 F (Emerest 2316)
Cloud Pt.:
   11.5 C (Radia 7200)
Acid No.:
   1.0 max. (Kessco Isopropyl Palmitate; Propal)
   2.0 max. (Liponate IPP)
Iodine No.:
   1.0 max. (Kessco Isopropyl Palmitate; Lexol IPP; Propal; Stepan D-70)
Saponification No.:
   182–191 (Lexol IPP; Propal)
   182–192 (Starfol IPP)
   183–190 (Liponate IPP)
   185–1915 (Stepan D-70)
Stability:
   Good oxidative stability (Lexol IPP)

Isopropyl palmitate *(cont'd.)*

Ref. Index:
 1.437 (Kessco Isopropyl Palmitate; Lexol IPP; Stepan D-70)
 1.4373 (Radia 7200)
Surface Tension:
 30.50 dynes/cm (Radia 7200)
STD. PKGS.:
 55-gal steel drums (Lexol IPP)

Isopropyl stearate *(CTFA)*

SYNONYMS:
 1-Methylethyl octadecanoate
 Octadecanoic acid, 1-methylethyl ester
EMPIRICAL FORMULA:
 $C_{21}H_{42}O_2$
STRUCTURE:

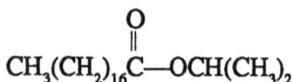

CAS No.:
 112-10-7
TRADENAME EQUIVALENTS:
 IPS [Henkel KGaA]
 Isopropylstearat [Henkel Canada]
 Kessco-Isopropyl-Stearat [Akzo Chemie BV]
 Lipal ST [Aquatec Quimica]
 Radia 7195 [Synfina-Oleofin]
 Rilanit IPS [Henkel KGaA]
 Wickenol 127 [CasChem]
  Generically sold by:
   [Henkel KGaA]
CATEGORY:
 Emollient, lubricant, softener, fatting agent, solvent, wetting agent, chemical interme-
  diate, plasticizer, solubilizer
APPLICATIONS:
 Bath products: bath oils (generic—Henkel; IPS; Isopropylstearat)
 Cosmetic industry preparations: (Lipal ST; Radia 7195); aerosols (generic—Henkel;
  IPS; Isopropylstearat; Kessco-Isopropyl-stearat); decorative cosmetics (generic—
  Henkel; IPS; Isopropylstearat); emulsions (Kessco-Isopropyl-stearat); hair prepa-
  rations (Kessco-Isopropyl-stearat); personal care products (generic—Henkel; IPS;

Isopropylstearat)
Industrial applications: ceramics (Rilanit IPS); chemical synthesis (Radia 7195); lubricating/cutting oils (Radia 7195; Rilanit IPS); metalworking (Rilanit IPS); petroleum industry (Rilanit IPS); rust inhibitors (Radia 7195); textile/leather processing (Radia 7195; Rilanit IPS)
Pharmaceutical applications: (Radia 7195)

PROPERTIES:
Form:
Liquid (generic—Henkel; IPS; Isopropylstearat; Lipal ST; Radia 7195; Rilanit IPS)
Composition:
100% active (generic—Henkel; Lipal ST)
Solubility:
Sol. in acetone (Wickenol 127)
Sol. in alcohols (Wickenol 127)
Sol. in ethyl acetate (Wickenol 127)
Insol. in glycerin (Wickenol 127)
Sol. in min. oils (Radia 7195; Wickenol 127)
Insol. in propylene glycol (Wickenol 127)
Sol. in most solvents (Radia 7195)
Sol. in veg. oils (Radia 7195; Wickenol 127)
Insol. in water (Wickenol 127)
Ionic Nature:
Nonionic (Lipal ST)
M.W.:
315 (avg.) (Radia 7195)
Sp.gr.:
0.842 (37.8 C) (Radia 7195)
Visc.:
5.10 cps (37.8 C) (Radia 7195)
Flash Pt.:
175 C (COC) (Radia 7195)
Cloud Pt.:
18.5 C (Radia 7195)
Ref. Index:
1.4382 (Radia 7195)

Isostearamidopropalkonium chloride (CTFA)

SYNONYMS:
Benzenemethanaminium, N,N-dimethyl-N-[3-[(1-oxoisooctadecyl) amino] propyl]-, chloride

135

Isostearamidopropalkonium chloride *(cont'd.)*

N,N-Dimethyl-N-[3-[(oxoisooctadecyl) amino] propyl] benzenemethanaminium chloride

N-3-Isostearyl amidopropyl) N,N-dimethyl, N-benzyl ammonium chloride

EMPIRICAL FORMULA:

$C_{30}H_{55}N_2O \cdot Cl$

STRUCTURE:

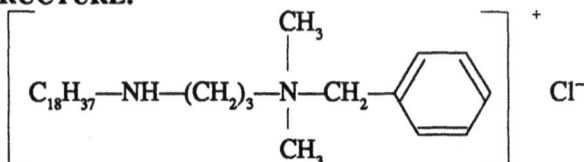

CAS No.:

67633-59-4

TRADENAME EQUIVALENTS:

Incroquat I-85 [Croda]

Schercoquat IB [Scher]

CATEGORY:

Conditioner, foaming agent, bactericide

APPLICATIONS:

Cosmetic industry preparations: conditioners (Schercoquat IB); hair preparations (Incroquat I-85; Schercoquat IB); skin preparations (Incroquat I-85; Schercoquat IB)

PROPERTIES:

Form:

Liquid (Incroquat I-85)

Viscous liquid (Schercoquat IB)

Color:

Amber (Schercoquat IB)

Yellow (Incroquat I-85)

Odor:

Mild almond (Schercoquat IB)

Composition:

85% active (Incroquat I-85)

85% active min. in methanol (Schercoquat IB)

Solubility:

Sol. in water (Incroquat I-85; Schercoquat IB)

Ionic Nature:

Cationic (Schercoquat IB)

M.W.:

494 (avg.) (Schercoquat IB)

Sp.gr.:

0.99 ± 0.01 (40 C) (Schercoquat IB)

Isostearamidopropalkonium chloride *(cont'd.)*

Density:
   8.2 lb/gal (40 C) (Schercoquat IB)
pH:
   4.0–7.0 (5% aq.) (Schercoquat IB)

Isostearamidopropyl betaine *(CTFA)*

SYNONYMS:
   N-Carboxymethyl-N,N-dimethyl-3-[(1-oxoisooctadecyl) amino]-1-propanaminium
      hydroxide, inner salt
   1-Propanaminium, N-(carboxymethyl)-N,N-dimethyl-3-[(1-oxoisooctadecyl)
      amino], hydroxide, inner salt
EMPIRICAL FORMULA:
   $C_{25}H_{50}N_2O_3$
STRUCTURE:

$$C_{17}H_{35}\overset{\displaystyle O}{\overset{\|}{C}}\!-\!NH\!-\!(CH_2)_3\!-\!\underset{\displaystyle CH_3}{\overset{\displaystyle CH_3}{N^+}}\!-\!CH_2COO^-$$

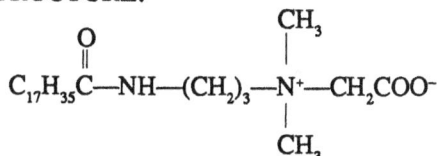

TRADENAME EQUIVALENTS:
   Cycloteric BET-I-30 [Alcolac]
   Incronam I-30 [Croda; Croda Universal]
   Mackam ISA [McIntyre]
   Schercotaine IAB [Scher]
CATEGORY:
   Conditioner, lubricant, detergent, surfactant, thickener, viscosity control agent, foam-
      ing agent
APPLICATIONS:
   Cosmetic industry preparations: (Incronam I-30); conditioners (Incronam I-30);
      creams and lotions (Incronam I-30); hair preparations (Cycloteric BET I-30);
      shampoos (Incronam I-30; Mackam ISA; Schecotaine IAB); skin preparations
      (Cycloteric BET I-30; Schercotaine IAB)
   Industrial applications: textile/leather processing (Schercotaine IAB)

PROPERTIES:
Form:
   Liquid (Cycloteric BET I-30; Mackam ISA)
   Clear viscous liquid (Incronam I-30)
   Viscous liquid (Schercotaine IAB)
   Soft, opaque gel (Schercotaine IAB)

Isostearamidopropyl betaine *(cont'd.)*

Color:
Amber (Schercotaine IAB liquid)
Yellow (Cylcoteric BET I-30; Incronam I-30)
Odor:
Slight, characteristic (Schercotaine IAB)
Composition:
29–31% active (Cycloteric BET I-30)
29.5–31% active (Incronam I-30)
30% active min., 34% dry solids min. (Schercotaine IAB)
35% active (Mackam ISA)
Solubility:
Sol. in aq. alcohol (Schercotaine IAB)
Sol. in glycols (Schercotaine IAB)
Sol. in water (Mackam ISA)
Ionic Nature:
Amphoteric (Cycloteric BET I-30; Incronam I-30; Mackam ISA)
M.W.:
477 (Schercotaine IAB)
Sp.gr.:
1.05 (Schercotaine IAB)
Density:
8.75 lb/gal (Schercotaine IAB)
pH:
5.5 ± 1.0 (Schercotaine IAB)
5.5—6.5 (Cycloteric BET I-30)
6.0–7.5 (5% sol'n.) (Incronam I-30)
TOXICITY/HANDLING:
Avoid prolonged contact with skin and eyes (Incronam I-30)
STORAGE/HANDLING:
Store in a cool, dry place (Incronam I-30)
STD. PKGS.:
55-gal (425 lb net) lined drums (Incronam I-30)

Isostearamidopropyl dimethylamine *(CTFA)*

SYNONYMS:
N-[3-(Dimethylamino) propyl] isooctadecanamide
Isooctadecanamide, N-[3-(dimethylamino) propyl]-
EMPIRICAL FORMULA:
$C_{23}H_{48}N_2O$

STRUCTURE:

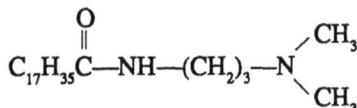

CAS No.:
67799-04-6

TRADENAME EQUIVALENTS:
Cyclomide IODI [Alcolac]
Incromine IB [Croda]
Jordamine DAPI [PPG-Mazer]
Mackine 401 [McIntyre]
Schercodine I [Scher]

CATEGORY:
Lubricant, emollient, conditioner, substantivity agent, softener, intermediate, o/w emulsifier, base

APPLICATIONS:
Cosmetic industry preparations: conditioners (Incromine IB; Mackine 401; Schercodine I); emulsions (Cyclomide IODI; Incromine IB; Schercodine I); hair preparations (Incromine IB; Jordamine DAPI; Schercodine I); personal care products (Incromine IB); skin preparations (Cyclomide IODI; Jordamine DAPI)
Industrial applications: lubricating oils/greases (Jordamine DAPI)

PROPERTIES:

Form:
Liquid (Mackine 401; Schercodine I)
Clear liquid (Cyclomide IODI)
Viscous liquid to soft paste (Incromine IB)
Solid (Jordamine DAPI)

Color:
Light amber (Schercodine I)
Yellow (Incromine IB)

Composition:
98% amide min. (Schercodine I)
99% conc. (Jordamine DAPI)
100% active (Cyclomide IODI; Incromine IB; Mackine 401)

Ionic Nature:
Cationic (Cyclomide IODI; Mackine 401; Schercodine I)
Nonionic/cationic (Incromine IB)

M.W.:
364–384 (Incromine IB)
394 (Schercodine I)

Isostearamidopropyl dimethylamine (cont' d.)

Alkali No.:

146–154 (Incromine IB)

150–160 (Schercodine I)

TOXICITY/HANDLING:

Avoid prolonged contact with skin (Incromine IB)

STORAGE/HANDLING:

Store in a cool, dry place (Incromine IB)

STD. PKGS.:

55-gal (400 lb net) polyethylene-lined Leverpak (Incromine IB)

Isostearamidopropyl dimethylamine lactate (CTFA)

SYNONYMS:

Isostearamidopropyl dimethylamine, lactic acid salt

Propanoic acid, 2-hydroxy-, compd. with N-[3-(dimethylamino) propyl]-16-methylheptadecanamide (1:1)

EMPIRICAL FORMULA:

$C_{23}H_{48}N_2O \cdot C_3H_6O_3$

CAS No.:

55852-15-8

TRADENAME EQUIVALENTS:

Emcol 6613 [Witco]

Incromate IDL [Croda]

Mackalene 416 [McIntyre]

Richamate 6613 [Richardson]

CATEGORY:

Conditioner, softener, emollient, surfactant, solubilizer, emulsifier, base

APPLICATIONS:

Cosmetic industry preparations: (Richamate 6613); conditioners (Incromate IDL; Mackalene 416); emulsions (Incromate IDL); hair preparations (Emcol 6613; Mackalene 416; Richamate 6613); personal care products (Incromate IDL; Richamate 6613); shampoos (Emcol 6613; Incromate IDL); skin preparations (Richamate 6613)

PROPERTIES:

Form:

Liquid (Emcol 6613; Mackalene 416; Richamate 6613)

Viscous liquid (Incromate IDL)

Color:

Yellow (Incromate IDL)

Gardner 12 (Richamate 6613)

140

Isostearamidopropyl dimethylamine lactate *(cont'd.)*

Composition:
  95% solids (Richamate 6613)
  95–97% active (Incromate IDL)
  100% active (Emcol 6613; Mackalene 416)
Ionic Nature:
  Cationic (Emcol 6613; Incromate IDL; Mackalene 416; Richamate 6613)
Density:
  7.8 lb/gal (Richamate 6613)
pH:
  6.0–7.0 (Richamate 6613); (5%) (Incromate IDL)

Isostearamidopropyl ethyldimonium ethosulfate *(CTFA)*

SYNONYMS:
  N-Ethyl-N,N-dimethyl-3-[(1-oxoisooctadecyl) amino]-1-propanaminium ethyl sulfate
  1-Propanaminium, N-ethyl-N,N-dimethyl-3-[(1-oxooctadecyl) amino]-ethyl sulfate
EMPIRICAL FORMULA:
  $C_{25}H_{53}N_2O \cdot C_2H_5O_4S$
STRUCTURE:

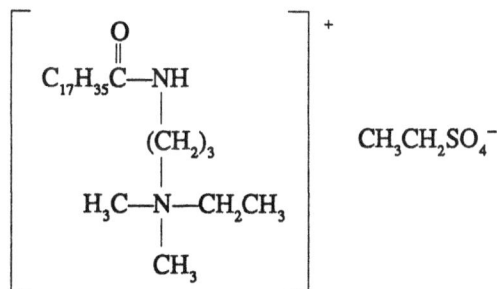

CAS No.:
  67633-63-0
TRADENAME EQUIVALENTS:
  Jordaquat 522 [PPG-Mazer]
  Schercoquat IAS, IAS-LC [Scher]
CATEGORY:
  Conditioner, substantivity agent, bodying agent, antistat
APPLICATIONS:
  Cosmetic industry preparations: conditioners (Jordaquat 522; Schercoquat IAS, IAS-

Isostearamidopropyl ethyldimonium ethosulfate *(cont'd.)*

LC); hair preparations (Jordaquat 522; Schercoquat IAS, IAS-LC); shampoos (Jordaquat 522; Schercoquat IAS); skin preparations (Schercoquat IAS, IAS-LC)

PROPERTIES:

Form:

Clear viscous liquid (Jordaquat 522; Schercoquat IAS, IAS-LC)

Color:

Dark amber (Schercoquat IAS-LC)

Gardner 10 max. (Schercoquat IAS)

Odor:

Slight, mild (Schercoquat IAS, IAS-LC)

Composition:

85% active (Schercoquat IAS-LC)

90% active min. in methanol (Schercoquat IAS)

100% active (Jordaquat 522)

Solubility:

Sol. in water (Jordaquat 522; Schercoquat IAS, IAS-LC)

Ionic Nature:

Cationic (Jordaquat 522; Schercoquat IAS, IAS-LC)

M.W.:

522 (avg.) (Schercoquat IAS)

550 (avg.) (Schercoquat IAS-LC)

Sp.gr.:

0.99 ± 0.01 (Schercoquat IAS, IAS-LC)

Density:

8.2 lb/gal (Schercoquat IAS, IAS-LC)

Flash Pt.:

90 C (OC) (Schercoquat IAS-LC)

pH:

5.0–7.0 (5% aq. sol'n.) (Schercoquat IAS, IAS-LC)

Isostearamidopropyl morpholine (CTFA)

SYNONYMS:

Isooctadecanamide, N-[3-(4-morpholinyl) propyl]-

N-[3-(4-Morpholinyl) propyl- isooctadecanamide

EMPIRICAL FORMULA:

$C_{25}H_{50}N_2O_2$

Isostearamidopropyl morpholine *(cont'd.)*

STRUCTURE:

$$\text{C}_{17}\text{H}_{35}\overset{\displaystyle \text{O}}{\overset{\displaystyle \|}{\text{C}}}\text{—NH—(CH}_2)_3\text{—N}\underset{}{\bigcirc}\text{O}$$

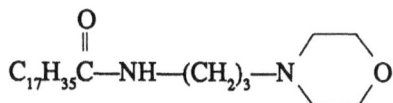

TRADENAME EQUIVALENTS:
Incromine ISM [Croda]
Mackine 421 [McIntyre]

CATEGORY:
Conditioner, softener, intermediate, foaming agent

APPLICATIONS:
Cosmetic industry preparations: conditioners (Incromine ISM; Mackine 421); hair
  preparations (Mackine 421); shampoos (Incromine ISM)

PROPERTIES:

Form:
Liquid (Mackine 421)
Paste to liquid (Incromine ISM)

Color:
Yellow (Incromine ISM)

Composition:
100% active (Incromine ISM; Mackine 421)

Ionic Nature:
Nonionic/cationic (Incromine ISM)
Cationic (Mackine 421)

Alkali No.:
142–148 (Incromine ISM)

TOXICITY/HANDLING:
Avoid prolonged contact with skin and eyes (Incromine ISM)

STORAGE/HANDLING:
Store in a cool, dry place (Incromine ISM)

STD. PKGS.:
55-gal (400 lb net) double lined open-head steel drums (Incromine ISM)

Isostearamidopropyl morpholine lactate (CTFA)

SYNONYMS:
Propanoic acid, 2-hydroxy-, compd. with N-[3-(4-morpholinyl) propyl] isooctadecanamide

EMPIRICAL FORMULA:
$\text{C}_{25}\text{H}_{50}\text{N}_2\text{O}_2 \cdot \text{C}_3\text{H}_6\text{O}_3$

Isostearamidopropyl morpholine lactate *(cont'd.)*

CAS No.:
RD No.: 977010-33-5
TRADENAME EQUIVALENTS:
Emcol ISML [Witco]
Incromate ISML [Croda]
Mackalene 426 [McIntyre]
Richamate ISML [Richardson]
CATEGORY:
Conditioner, base, softener, emollient, viscosity builder, surfactant, solubilizer, emulsifier
APPLICATIONS:
Cosmetic industry preparations: (Richamate ISML); conditioners (Incromate ISML; Mackalene 426); creams and lotions (Incromate ISML); hair preparations (Emcol ISML; Incromate ISML; Mackalene 426; Richamate ISML); personal care products (Richamate ISML); shampoos (Emcol ISML; Incromate ISML); skin preparations (Incromate ISML; Richamate ISML)
PROPERTIES:
Form:
Liquid (Emcol ISML; Mackalene 426; Richamate ISML)
Viscous liquid (Incromate ISML)
Color:
Yellow (Incromate ISML)
Gardner 6 (Richamate ISML)
Composition:
25% solids (Richamate ISML)
25% active (Emcol ISML; Incromate ISML)
100% active (Mackalene 426)
Ionic Nature:
Cationic (Emcol ISML; Mackalene 426)
Density:
8.3 lb/gal (Richamate ISML)
pH:
4–5 (Richamate ISML)

Isostearyl alcohol *(CTFA)*

SYNONYMS:
1-Heptadecanol, 16-methyl-
Isooctadecanol
16-Methyl-1-heptadecanol

EMPIRICAL FORMULA:

$C_{18}H_{38}O$

CAS No.:

27458-93-1; 41744-75-6

TRADENAME EQUIVALENTS:

Adol 66 [Sherex]

Michel XO-146 [M. Michel]

CATEGORY:

Lubricant, coemulsifier, coupler, cosolvent, plasticizer, stabilizer, emollient, intermediate, foam control agent

APPLICATIONS:

Automotive applications: antifreeze (Adol 66)

Household applications: fabric softeners (Adol 66); surfactants (Adol 66)

Industrial applications: industrial processing (Adol 66); lubricating/cutting oils (Adol 66); metalworking (Adol 66); mineral processing (Adol 66); paper mfg. (Adol 66); petroleum industry (Adol 66); printing inks (Adol 66); textile/leather processing (Adol 66)

PROPERTIES:

Color:

Lovibond 5Y/0.5R max. (5 $^1/_4$ in.) (Adol 66)

Solubility:

Sol. in acetone (Adol 66)

Sol. in fatty alcohols (Adol 66)

Sol. in benzene (Adol 66)

Sol. in ethanol/water mixtures (Michel XO-146)

Sol. in ethyl ether (Adol 66)

Sol. in isopropanol (Adol 66)

Sol. in kerosene (Adol 66)

Sol. in light min. oil (Adol 66)

Sol. in VMP naphtha (Adol 66)

Sol. in trichlorethylene (Adol 66)

Sol. in turpentine (Adol 66)

M.W.:

295 (Adol 66)

Sp.gr.:

0.861 (Adol 66)

Cloud Pt.:

8 C max. (AOCS) (Adol 66)

Acid No.:

1.0 max. (Adol 66)

Iodine No.:

1.2 max. (Adol 66)

Isostearyl alcohol (cont' d.)

Saponification No.:
  2.0 max. (Adol 66)
Hydroxyl No.:
  180–200 (Adol 66)
TOXICITY/HANDLING:
  Not considered hazardous (Adol 66)
STORAGE/HANDLING:
  Should be kept as dry as possible; blanket the storage tank with inert gas (Adol 66)

Isostearyl neopentanoate (CTFA)

SYNONYMS:
  2,2-Dimethylpropanoic acid, isooctadecyl ester
  Ester of isostearyl alcohol and neopentanoic acid
  Propanoic acid, 2,2-dimethyl-, isooctadecyl ester
EMPIRICAL FORMULA:
  $C_{23}H_{46}O_2$
STRUCTURE:

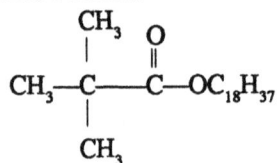

CAS No.:
  588958-60-4
TRADENAME EQUIVALENTS:
  Ceraphyl 375 [Van Dyk]
  Crodamol ISNP [Croda Ltd.]
  Cyclochem INEO [Alcolac]
  Schercemol 185 [Scher]
CATEGORY:
  Emollient, penetrant, cloud and freeze point depressant, emulsion stabilizer, moisturizer
APPLICATIONS:
  Cosmetic industry preparations: (Crodamol ISNP; Schercemol 185); skin preparations
    (Cyclochem INEO; Schercemol 185)
  Pharmaceutical applications: (Crodamol ISNP)
PROPERTIES:
Form:
  Liquid (Cyclochem INEO; Crodamol ISNP)
  Clear liquid (Schercemol 185)
  Low viscosity liquid (Ceraphyl 375)

Color:
Off-white (Cyclochem INEO)
Gardner 2 max. (Schercemol 185)
Odor:
Slight, typical (Schercemol 185)
Solubility:
Sol. in alcohols (Schercemol 185)
Sol. in esters (Schercemol 185)
Sol. in 95% ethanol (Ceraphyl 375)
Sol. in glycol ethers (Schercemol 185)
Sol. in glycols (Schercemol 185)
Sol. in isopropyl myristate (Ceraphyl 375)
Sol. in ketones (Schercemol 185)
Sol. in min. oils (Ceraphyl 375; Schercemol 185)
Sol. in most organic solvents (Schercemol 185)
Sol. in propylene glycol (Ceraphyl 375)
Sol. in veg. oils (Schercemol 185)
Insol. in water (Ceraphyl 375)
M.W.:
368 (theoret.) (Schercemol 185)
Sp.gr.:
0.865 (Schercemol 185)
Density:
7.2 lb/gal (Schercemol 185)
M.P.:
< 0 C (Cyclochem INEO)
Flash Pt.:
> 180 (OC) (Schercemol 185)
Cloud Pt.:
−10 C (Schercemol 185)
Acid No.:
2.0 max. (Cyclochem INEO; Schercemol 185)
Iodine No.:
8.0 (Schercemol 185)
Saponification No.:
144 ± 8.0 (Schercemol 185)
149 (Cyclochem INEO)
Stability:
Highly acid and alkali stable (Cyclochem INEO)
Ref. Index:
1.450 ± 0.002 (Schercemol 185)
TOXICITY/HANDLING:
Very mild with a low level of skin and eye irritation (Schercemol 185)

Jojoba oil (CTFA)

SYNONYMS:
Oils, jojoba
TRADENAME EQUIVALENTS:
Lipovol J [Lipo] (refined)
Nikkol Jojoba Oil N [Nikko]
Nikkol Jojoba Oil S [Nikko] (decolorized, deodorized)
Simchin [RITA]
Wickenol 139 [CasChem] (synthetic)
CATEGORY:
Emollient, conditioner, lubricant, moisturizer
APPLICATIONS:
Bath products: bath oils (Lipovol J)
Cosmetic industry preparations: (Lipovol J); conditioners (Lipovol J); creams and lotions (Lipovol J); hair preparations (Lipovol J; Simchin; Wickenol 139); makeup (Lipovol J); shampoos (Lipovol J); shaving preparations (Lipovol J); skin preparations (Lipovol J; Simchin; Wickenol 139); toiletries (Lipovol J)
PROPERTIES:
Form:
Liquid (Nikkol Jojoba Oil N, S)
Clear liquid (Simchin)
Clear liquid wax (Wickenol 139)
Oil (Lipovol J)
Color:
Yellow (Lipovol J)
Odor:
Characteristic, nut-like (Lipovol J)
Solubility:
Sol. in alcohol (Wickenol 139); insol. in alcohols (Lipovol J)
Sol. in animal oils (Wickenol 139)
Insol. in ethanol (Nikkol Jojoba Oil N, S)
Insol. in glycerin (Wickenol 139)
Sol. in min. oil (Lipovol J; Nikkol Jojoba Oil N, S)
Insol. in propylene glycol (Nikkol Jojoba Oil N, S; Wickenol 139)
Sol. in veg. oil (Lipovol J; Nikkol Jojoba Oil N, S; Wickenol 139)
Insol. in water (Lipovol J; Wickenol 139)
Sol. in white oils (Wickenol 139)

Acid No.:
   5 max. (Lipovol J)
Iodine No.:
   75–95 (Lipovol J)
Saponification No.:
   85–110 (Lipovol J)
Stability:
   Stable at pH levels common to cosmetics and toiletries; not prone to rancidity (Lipovol
   J)
STD. PKGS.:
   400 lb net closed-head steel drums (Lipovol J)

Lanolin (CTFA)

SYNONYMS:
Anhydrous lanolin
Cosmetic lanolin
Lanolin, anhydrous
Lanolin, anhydrous USP
Wool fat
Wool wax

CAS No.:
8006-54-0 (anhydrous); 8020-84-6 (hydrous)

TRADENAME EQUIVALENTS:
Aqualose SLT, SLW [Westbrook Lanolin]
Clearlan 650, 1650, K50 [Henkel/Emery] (anhyd. USP)
Corona Lanolin [Croda Ltd.] (anhyd.)
Coronet Lanolin [Croda Ltd.] (super refined, cosmetic grade, BP)
Cosmetic Lanolin Anhydrous USP [Croda]
Crodapur [Croda Ltd.] (tech.)
Emery 1650, 1656 [Henkel/Emery] (anhyd. USP)
Emery 1660 [Henkel/Emery] (ultra anhyd. USP, cosmetic grade)
Fancor Lanolin [Fanning]
Fancorp Lanolin [Fanning]
Forlan [RITA]
Golden Dawn Grade 1, 2, Superfine [Westbrook Lanolin] (anhyd.)
Golden Dawn Lanolin [Westbrook Lanolin] (pharmaceutical/cosmetic grade)
Golden Fleece DF [Westbrook Lanolin] (extra refined anhyd.)
Golden Fleece P-80, P-95, RA [Westbrook Lanolin] (anhyd.)
Golden Fleece Lanolin [Westbrook Lanolin] (pharmaceutical/cosmetic grade)
Paralan [Croda Ltd.] (tech.)
Pharmaceutical Lanolin Anhydrous USP [Croda]
Sorba [Croda Ltd.]
Sparkelan 656 [Henkel/Emery] (anhyd. USP)
Sparkelan 1656 [Henkel/Emery] (anhyd. USP, pharmaceutical grade)
Super Corona [Croda; Croda Ltd.] (anhyd. USP)
Superfine Lanolin Anhydrous USP [Croda]
White Swan [Croda Ltd.] (anhyd. BP)
Yeoman [Croda Ltd.] (anhyd. BP)

CATEGORY:

Emollient, emulsifier, conditioner, lubricant, moisturizer, penetrant, cosolvent, plasticizer, superfatting agent, wetting agent, dispersant, absorption base, wax crystal inhibitor, binder

APPLICATIONS:

Cleansers: industrial hand cleaners (Emery 1656); soaps (Corona Lanolin; Coronet Lanolin; Super Corona; White Swan; Yeoman)

Cosmetic industry preparations: (Clearlan 1650; Emery 1650, 1656, 1660; Sorba; Super Corona); aerosols (Super Corona); baby preparations (Super Corona); conditioners (Golden Dawn Grade 1, 2, Superfine; Golden Fleece DF, P-80, P-95, RA); creams and lotions (Cosmetic Lanolin Anhydrous USP); emulsions (Paralan); hair preparations (Corona Lanolin; Coronet Lanolin; Golden Dawn Grade 1, 2, Superfine; Golden Fleece DF, P-80, P-95, RA; Super Corona); makeup (Corona Lanolin; Coronet Lanolin; Cosmetic Lanolin Anhydrous USP; Emery 1660; Golden Dawn Grade 1, 2, Superfine; Golden Fleece DF, P-80, P-95, RA; Super Corona); personal care products (White Swan; Yeoman); shampoos (Aqualose SLW; Super Corona); shaving preparations (Super Corona); skin preparations (Clearlan 650, K50; Corona Lanolin; Coronet Lanolin; Sparkelan 656; Super Corona)

Industrial applications: adhesives (Crodapur); metalworking (Fancorp Lanolin); rubber (Fancor Lanolin); textile/leather processing (Fancorp Lanolin)

Pharmaceutical applications: (Pharmaceutical Lanolin Anhydrous USP; Sorba; Sparkelan 1656; Super Corona; White Swan; Yeoman); bases (Corona Lanolin; Coronet Lanolin); ointments/salves (Corona Lanolin; Coronet Lanolin; Emery 1656; Super Corona); sunscreens (Cosmetic Lanolin Anhydrous USP; Emery 1650)

Veterinary products (Emery 1656)

PROPERTIES:

Form:

Paste (Aqualose SLT, SLW; Corona Lanolin; Coronet Lanolin)

Solid (Clearlan 650, K50; Emery 1650, 1656, 1660; Paralan; Sparkelan 656)

Soft solid (Cosmetic Lanolin Anhydrous USP; Golden Fleece DF, RA; Pharmaceutical Lanolin Anhydrous USP; Super Corona; Superfine Lanolin Anhydrous USP)

Soft wax (Golden Dawn Grade 1, 2, Superfine; Golden Fleece Lanolin, P-80, P-95)

Unctuous mass (White Swan; Yeoman)

Color:

Light/pale (Clearlan 1650; Golden Dawn Grade 1, Superfine; Golden Fleece P-80, P-95)

Yellow (White Swan; Yeoman)

Gardner < 5 (Emery 1660)

Gardner 8.5 max. (Super Corona)

Gardner < 9 (Emery 1650)

Gardner 9.5 (Superfine Lanolin Anhydrous USP)

Gardner 10.5 (Cosmetic Lanolin Anhydrous USP)

Lanolin (cont'd.)

Gardner 11.5 (Pharmaceutical Lanolin Anhydrous USP)
Gardner < 12 (Emery 1656)

Odor:
Virtually odorless (Emery 1660)
Practically odorless (slight, characteristic) (Super Corona)

Taste:
Virtually tasteless (Emery 1660)

Composition:
100% active (Clearlan 650, K50; Corona Lanolin; Coronet Lanolin; Golden Dawn Superfine; Golden Fleece DF, RA; Sparkelan 656)

Solubility:
Sparingly sol. in alcohol (Sparkelan 1656)
Insol. in glycerin @ 5% (Emery 1650, 1656, 1660)
Disp. in glycerol trioleate @ 5% (Emery 1650, 1656, 1660)
Sol. in hydrocarbons oils (Fancor Lanolin; Fancorp Lanolin)
Insol. in isopropanol @ 5% (Emery 1650, 1656, 1660)
Sol. in isopropyl myristate @ 5% (Emery 1650, 1656, 1660)
Disp. in min. oil @ 5% (Emery 1650, 1656, 1660)
Sol. in oils (Corona Lanolin)
Sol. in most solvents (Fancor Lanolin; Fancorp Lanolin)
Mixes with twice its wt. in water (Sparkelan 1656); insol. @ 5% (Emery 1650, 1656, 1660)

Ionic Nature:
Nonionic (Clearlan 650, K50; Corona Lanolin; Coronet Lanolin; Golden Dawn Grade 1, 2, Superfine; Golden Fleece DF, P-80, P-95, RA; Paralan; Sparkelan 656)

Density:
7.9 lb/gal (Emery 1650, 1656, 1660)

M.P.:
36–42 C (Emery 1650, 1656)
38–44 C (Cosmetic Lanolin Anhydrous USP; Emery 1660; Pharmaceutical Lanolin Anhydrous USP; Super Corona; Superfine Lanolin Anhydrous USP)

Flash Pt.:
460 F (Emery 1660)
530 F (Emery 1650, 1656)

HLB:
4.0 (Golden Dawn Superfine)
4.5 (Golden Fleece DF)
10.0 (Clearlan 650, K50; Sparkelan 656)

Acid No.:
1.0 max. (Super Corona)

Iodine No.:
18–36 (Hanus) (Super Corona)

152

Lanolin (cont'd.)

Storage Stability:
Refined lanolins will darken upon prolonged heating and are subject to decomposition by excessive heating (Super Corona)

STORAGE/HANDLING:
Avoid excessive melt temps. so as not to exceed 140 F (Super Corona)

Lanolin alcohol (CTFA)

SYNONYMS:
Alcohols, lanolin
Wool wax alcohol

CAS No.:
8027-33-6

TRADENAME EQUIVALENTS:
Argobase 125 [Westbrook Lanolin]
Argowax Distilled, Standard [Westbrook Lanolin]
Ceralan [Amerchol]
Fancol LA [Fanning] (Regular USP and Superfine USP grades)
Hartolan, Super Hartolan [Croda; Croda Ltd.]
Hartolite [Croda; Croda Ltd.]
Lanolin Alcohols LG, LO, THG, THO [Van Schuppen]
Nimco 1780 [Henkel/Emery] (distilled, deodorized)
Ritawax [RITA]

CATEGORY:
Emollient, emulsifier, gelling agent, thickener, stabilizer, plasticizer, superfatting agent, dispersant, chemical intermediate, lubricant, humectant, release agent, conditioner, spreading agent, moisturizer

APPLICATIONS:
Bath products: bath oils (Argobase 125)
Cleansers: soaps (Fancol LA)
Cosmetic industry preparations: (Fancol LA; Hartolan, Super Hartolan; Lanolin Alcohols THG, THO; Ritawax); baby oils (Argobase 125); creams and lotions (Argobase 125; Fancol LA; Hartolite; Nimco 1780); makeup (Fancol LA; Hartolan, Super Hartolan; Nimco 1780); personal care products (Argobase 125); shampoos (Fancol LA); skin preparations (Hartolite; Ritawax)
Industrial applications: (Fancol LA; Lanolin Alcohols THG, THO); paper mfg. (Lanolin Alcohols LG, LO); textile/leather processing (Lanolin Alcohols LG, LO); wood pulping (Lanolin Alcohols LG, LO)
Pharmaceutical applications: (Fancol LA; Hartolan, Super Hartolan); sunscreens (Argobase 125)

Lanolin alcohol (cont'd.)

PROPERTIES:
Form:
Liquid (Argobase 125)
Wax (Argowax Distilled, Standard; Ceralan)
Soft solid (Lanolin Alcohols LO)
Solid (Nimco 1780; Ritawax)
Solid wax (Hartolan, Super Hartolan)
Soft bleached wax (Lanolin Alcohols LG)
Bleached solid (Lanolin Alcohols THG)
Soft waxy solid (Hartolite)
Color:
Light (Argowax Distilled)
Pale amber (Super Hartolan)
Pale yellow (Hartolite)
Tan (Lanolin Alcohols LG, THG)
Brown (Fancol LA Regular USP; Hartolan; Lanolin Alcohols LO)
Gardner-Holdt 8–11 (Fancol LA Superfine USP)
Odor:
Slight (Argowax Distilled)
Composition:
15% conc. (Argobase 125)
100% active (Argowax Distilled, Standard; Ceralan; Fancol LA; Hartolite; Lanolin Alcohols LG, LO, THG; Nimco 1780)
Solubility:
Insol. in acetone (Fancol LA)
Sol. in alcohol (Ritawax); slightly sol. in 90% alcohol (Fancol LA)
Sol. in carbon tetrachloride (Fancol LA)
Sol. in castor oil (Nimco 1780)
Sol. in chlorinated hydrocarbons (Hartolan, Super Hartolan; Lanolin Alcohols LG, LO, THG, THO)
Sol. in chloroform (Fancol LA)
Sol. in ethanol (Nimco 1780)
Insol. in ether (Fancol LA)
Insol. in glycerin (Fancol LA)
Sol. in hydrocarbons (Lanolin Alcohols LG, LO, THG, THO)
Sol. in isopropyl myristate (Nimco 1780; Ritawax); sol. @ 75 C (Fancol LA)
Sol. in min. oil (Hartolan, Super Hartolan; Lanolin Alcohols LG, LO, THG, THO); sol. @ 75 C (Fancol LA)
Sol. in oils (Argobase 125; Ritawax); sol. in warm oils (Nimco 1780)
Sol. in oleyl alcohol (Fancol LA)
Sol. in olive oil @ 75 C (Fancol LA)
Insol. in propylene glycol (Fancol LA)
Sol. in veg. oil (Lanolin Alcohols LG, LO, THG, THO)

Insol. in water (Fancol LA; Hartolan, Super Hartolan; Lanolin Alcohols LG, LO, THG, THO)

Ionic Nature:

Nonionic (Argobase 125; Argowax Distilled, Standard; Ceralan; Hartolite; Lanolin Alcohols LG, LO, THG, THO; Nimco 1780; Ritawax)

M.P.:

56 C min. (Fancol LA)

58 C min. (Hartolan)

60 C min. (Super Hartolan)

HLB:

2.6 (Argowax Distilled)

3.0 (Argobase 125)

8.9 (Nimco 1780)

Acid No.:

1.5 max. (Super Hartolan)

2.0 max. (Fancol LA; Hartolan)

6.0 max. (Lanolin Alcohols LG, THG, THO)

Saponification No.:

5 mg max. (Hartolan, Super Hartolan)

12.0 max. (Fancol LA)

20 max. (Lanolin Alcohols THG, THO)

20–40 (Lanolin Alcohols LG, LO)

Lanolin oil (CTFA)

SYNONYMS:

Dewaxed lanolin

Oils, lanolin

CAS No.:

8038-43-5

TRADENAME EQUIVALENTS:

Argonol 50 Pharmaceutical, 50 Super, 60 [Westbrook Lanolin]

Fluilan [Croda; Croda Ltd.]

Lanogene [Amerchol]

Lantrol 1673, 1674 [Henkel/Emery] (anhyd., cosmetic grade)

Lantrol, 1674 Deodorized [Henkel/Emery]

Lipolan R [Lipo]

Liquilan [Amerchol]

Ritalan [RITA]

Vigilan Regular and Superfine [Fanning]

Lanolin oil (cont'd.)

CATEGORY:
Emollient, lubricant, plasticizer, dispersant, emulsifier, penetrant, conditioner, super-fatting agent, moisturizer, vehicle, spreading agent, cosolvent, solubilizer

APPLICATIONS:
Bath products: bath oils (Lipolan R; Vigilan)

Cleansers: cleansing lotions (Fluilan); germicidal skin cleanser (Fluilan); soaps (Fluilan; Lipolan R)

Cosmetic industry preparations: (Fluilan; Lantrol, 1673, 1674); baby oils (Fluilan); body oils (Lipolan R); conditioners (Vigilan); creams and lotions (Lantrol; Ritalan; Vigilan); hair preparations (Fluilan; Lantrol; Lipolan R; Vigilan); makeup (Fluilan; Lantrol, 1674 Deodorized; Lipolan R; Ritalan; Vigilan); personal care products (Lantrol); shampoos (Fluilan; Lantrol; Lipolan R); shaving preparations (Lipolan R); skin preparations (Lantrol; Lipolan R; Vigilan)

Household detergents: (Lipolan R); dishwashing (Fluilan)

Pharmaceutical applications: (Argonol 50 Pharmaceutical)

PROPERTIES:

Form:
Liquid (Argonol 50 Pharmaceutical, 50 Super, 60; Lanogene; Lantrol 1673, 1674; Lipolan R; Ritalan; Vigilan)

Viscous liquid (Fluilan; Lantrol)

Color:
Amber (Ritalan)

Yellow-amber (Lanogene; Lipolan R)

Pale yellow (Fluilan)

Gardner 6 max. (Vigilan Superfine)

Gardner 8 max. (Vigilan Regular)

Odor:
Low (Lantrol 1674)

Odorless (Argonol 50 Pharmaceutical)

Not completely odorless (Argonol 50 Super)

Slight, characteristic (Lanogene)

Mild, characteristic (Lipolan R)

Pleasant, characteristic (Fluilan)

Composition:
100% active (Fluilan; Lantrol; Ritalan; Vigilan)

Solubility:
Partly insol. in acetone (Fluilan)

Sol. in aerosol propellants (Fluilan; Vigilan)

Insol. in alcohols (Lantrol 1674 Deodorized)

Sol. in castor oil (Lantrol 1674 Deodorized)

Sol. in most cosmetic waxes and oils (Lantrol 1673, 1674)

Sol. in esters (Lanogene)

Insol. in ethanol (Fluilan)

Sol. in fatty alcohol (Fluilan)
Partly insol. in glycerin (Fluilan); insol. (Lipolan R)
Partly insol. in glycols (Fluilan); insol. (Lantrol 1674 Deodorized)
Sol. in hydrocarbons (Fluilan; Lanogene)
Sol. in isopropanol (Fluilan; Lanogene); sol. in anhyd. isopropanol (Lipolan R)
Sol. in isopropyl esters (Lipolan R)
Sol. in min. oil (Argonol 50 Pharmaceutical; Fluilan; Lantrol 1674 Deodorized; Lipolan R; Vigilan)
Sol. in oils (Lanogene; Liquilan)
Insol. in propylene glycol (Lipolan R)
Sol. in veg. oil (Lipolan R; Vigilan)
Partly insol. in water (Fluilan); insol. (Lantrol 1674 Deodorized; Lipolan R)
Ionic Nature:
Nonionic (Fluilan; Lantrol; Vigilan)
Pour Pt.:
8 C max. (Fluilan)
Cloud Pt.:
18 C max. (Fluilan; Lipolan R)
HLB:
9.0 (Lantrol)
Acid No.:
1.0 max. (Vigilan)
2.0 max. (Fluilan; Lanogene; Lipolan R)
Iodine No.:
18–36 (Lanogene; Vigilan); (Hanus) (Lipolan R)
24–40 (Hanus) (Fluilan)
Saponification No.:
80–100 (Fluilan)
85–105 (Lanogene)
85–110 (Lipolan R)
95–110 (Vigilan)
Stability:
Stable at pH levels common in cosmetics and toiletries (Lipolan R)
STD. PKGS.:
400 lb net closed-head steel drums (Lipolan R)

Lanolin wax (CTFA)

SYNONYMS:
De-oiled lanolin
Waxes, lanolin

Lanolin wax *(cont'd.)*

CAS No.:
68201-49-0
TRADENAME EQUIVALENTS:
Albalan [Westbrook Lanolin]
Lanfrax, Lanfrax 1777 Deodorized [Henkel/Emery]
Lanocerin [Amerchol]
CATEGORY:
Emollient, conditioner, emulsifier, stabilizer, waxing agent, thickener
APPLICATIONS:
Cosmetic industry preparations: (Lanfrax 1777 Deodorized); emulsions (Albalan; Lanfrax; Lanocerin); makeup (Lanocerin)
Industrial applications: polishes and waxes (Lanfrax)
PROPERTIES:
Form:
Soft wax (Albalan)
Wax (Lanfrax)
Waxy solid (Lanocerin)
Color:
Yellow-tan (Lanocerin)
Odor:
Faint, pleasant (Lanocerin)
Composition:
100% active (Albalan; Lanfrax)
Solubility:
Insol. in water (Lanfrax 1777 Deodorized; Lanocerin)
Ionic Nature:
Nonionic (Albalan; Lanfrax)
M.P.:
41–51 C (Lanocerin)
HLB:
5.0 (Albalan)
Acid No.:
2.5 max. (Lanocerin)
Iodine No.:
15–35 (Lanocerin)
Saponification No.:
85–115 (Lanocerin)

Lard monoglyceride

SYNONYMS:
Glycerides, lard mono-
Lard glyceride (CTFA)

CAS No.:
61789-10-4

TRADENAME EQUIVALENTS:
Dimodan P, PM, PM 300 [Grindsted] (dist. sat.)
Dimodan S [Grindsted] (dist. unsat.)
Grindtek MOP 90 [Grindsted]
Monomuls 90-10 [Grünau] (dist.)
Myverol 18-40 [Eastman] (dist.)

CATEGORY:
Emulsifier, lubricant, antistat, antifogging agent, stabilizer, dispersant, opacifier

APPLICATIONS:
Cosmetic industry preparations: (Monomuls 90-10; Myverol 18-40); creams and lotions (Grindtek MOP 90)
Food applications: (Monomuls 90-10); food emulsifying (Dimodan P, PM, PM 300, S; Myverol 18-40)
Industrial applications: plastics (Grindtek MOP 90)
Pharmaceutical applications: (Monomuls 90-10)

PROPERTIES:
Form:
Solid (Monomuls 90-10)
Beads (Dimodan PM
Block (Dimodan S; Grindtek MOP 90)
Plastic (Myverol 18-40)
Fine powder (Dimodan PM 300)
Color:
Whitish (Grindtek MOP 90)
Composition:
90% monoester min. (Dimodan P, PM, PM 300, S)
100% active (Monomuls 90-10; Myverol 18-40)
Solubility:
Sol. warm in ethanol (Grindtek MOP 90)
Sol. warm in propylene glycol (Grindtek MOP 90)
Sol. warm in toluene (Grindtek MOP 90)
Sol. warm in white spirit (Grindtek MOP 90)
Ionic Nature:
Nonionic (Monomuls 90-10; Myverol 18-40)
Sp.gr.:
0.91 (80 C) (Myverol 18-40)

Lard monoglyceride *(cont'd.)*

M.P.:
55 C (Dimodan S)
70 C (Dimodan P, PM, PM 300)
HLB:
4.3 (Grindtek MOP 90)

Lauryl amine

SYNONYMS:
1-Dodecanamine
Dodecylamine
Lauramine (CTFA)
EMPIRICAL FORMULA:
$C_{12}H_{27}N$
STRUCTURE:
$CH_3(CH_2)_{10}CH_2NH_2$
CAS No.:
124-22-1
TRADENAME EQUIVALENTS:
Amine 12D [Berol Nobel]
Armeen 12, 12D [Akzo/Armak; Akzo BV]
Kemamine P-690, P-690D [Humko] (dist.)
Lilamin 163, 163D [Berol Nobel] (dist.)
Nissan Amine BB [Nippon Oil & Fats]
Radiamine 6163, 6164 [Synfina-Oleofin] (dist.)
CATEGORY:
Lubricant, release agent, emulsifier, flotation reagent, spinning aid, corrosion inhibitor, dispersant, wetting agent, chemical intermediate, anticaking agent, bactericide, germicide, softener, antistat, antifog agent
APPLICATIONS:
Cosmetic industry preparations: (Armeen 12; Radiamine 6163, 6164)
Farm products: (Nissan Amine BB); fungicides (Lilamin 163, 163D)
Industrial applications: asphalt/concrete (Nissan Amine BB); ceramics (Nissan Amine BB); chemical synthesis (Nissan Amine BB; Radiamine 6163, 6164); dyes and pigments (Nissan Amine BB; Radiamine 6163, 6164); lubricating/cutting oils (Nissan Amine BB); metalworking (Armeen 12D; Kemamine P-690, P-690D); ore flotation (Nissan Amine BB; Radiamine 6163, 6164); petroleum industry (Kemamine P-690, P-690D; Lilamin 163, 163D); plastics (Kemamine P-690, P-690D; Radiamine 6163, 6164); rubber (Kemamine P-690, P-690D; Lilamin 163, 163D; Nissan Amine BB; Radiamine 6163, 6164); textile/leather processing (Nissan Amine BB; Radiamine 6163, 6164); water treatment (Nissan Amine BB)

PROPERTIES:

Form:
Liquid (Amine 12D; Armeen 12D; Kemamine P-690, P-690D)
Liquid in summer, solid in winter (Nissan Amine BB)
Liquid/paste (Armeen 12)
Liquid/solid (Radiamine 6163, 6164)
Solid (Amine 12D)
Flake (Lilamin 163, 163D)

Color:
APHA 120 max. (Nissan Amine BB)
Gardner 2 max. (Armeen 12D)
Gardner 3 max. (Lilamin 163, 163D)

Composition:
95% active (Lilamin 163, 163D)
98% active (Amine 12D; Armeen 12)
98% primary amine min. (Armeen 12D; Nissan Amine BB)
100% active (Kemamine P-690, P-690D; Radiamine 6163, 6164)

Solubility:
Sol. in acetone (Armeen 12D)
Sol. in carbon tetrachloride (Armeen 12D)
Sol. in chloroform (Armeen 12D)
Sol. in ethanol (Armeen 12D)
Sol. in isopropanol (Armeen 12D)
Sol. in kerosene (Armeen 12D)
Sol. in methanol (Armeen 12D)
Sol. in toluene (Armeen 12D)

Ionic Nature:
Cationic (Amine 12; Armeen 12, 12D; Kemamine P-690, P-690D; Lilamin 163, 163D; Radiamine 6163, 6164)

M.W.:
190 (Lilamin 163, 163D)

Sp.gr.:
0.777 (60 C) (Lilamin 163, 163D)
0.801 (25/4 C) (Armeen 12D)

Visc.:
42.2 SSU (Armeen 12D)
1.80 cps (60 C) (Lilamin 163, 163D)

M.P.:
25 C (Lilamin 163, 163D)
79–86 F (Armeen 12D)

Pour Pt.:
80 F (Armeen 12D)

Lauryl amine (cont'd.)

Solidification Pt.:
24–28 C (Nissan Amine BB)
Flash Pt.:
110 C (OC) (Lilamin 163, 163D)
235 F (COC) (Armeen 12D)
Fire Pt.:
250 F (Armeen 12D)
Iodine No.:
1.0 max. (Nissan Amine BB)
2.0 max. (Armeen 12D; Lilamin 163, 163D)
Amine No.:
297 min. (Armeen 12D)

Lauryl dimethyl amine oxide

SYNONYMS:
N,N-Dimethyl-1-dodecanamine-N-oxide
Dimethyl lauryl amine oxide
1-Dodecanamine, N,N-dimethyl, N-oxide
Lauramine oxide (CTFA)
Lauryl amine oxide
EMPIRICAL FORMULA:
$C_{14}H_{31}NO$
STRUCTURE:

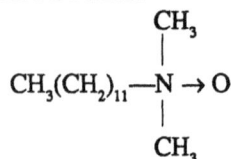

CAS No.:
1643-20-5
TRADENAME EQUIVALENTS:
Alkamox LO [Alkaril; Alkaril UK; Alkaril Canada]
Ammonyx DMCD-40, LO [Millmaster-Onyx]
Aromox DMMC-W [Akzo/Armak; Akzo BV]
Conco XA-L [Continental]
Cyclomox L [Alcolac]
Emcol L [Witco]
Empigen OB [Albright & Wilson/Marchon]
Incromine L-40, Oxide L [Croda]

Lauryl dimethyl amine oxide (cont'd.)

Jordamox LDA [PPG-Mazer]
Loramine L408 [Rewo]
Mackamine LO [McIntyre]
Ninox L [Stepan]
Oxamin LO [ICI Australia]
Rewominox L408 [Rewo GmbH]
Sandoz Amine Oxide XA-L [Sandoz]
Schercamox DML [Scher]
Textamine Oxide LMW [Henkel Canada]
Varox 365, 375 [Vanderbilt]

CATEGORY:

Detergent, wetting agent, foaming agent, emollient, conditioner, softener, lubricant, foam stabilizer, antistat, emulsifier, viscosity builder/modifier, thickener

APPLICATIONS:

Bath products: (Ammonyx LO; Schercamox DML);; bubble bath (Empigen OB; Loramine L408; Ninox L; Rewominox L408)

Cleansers: (Empigen OB; Varox 365)

Cosmetic industry preparations: (Ammonyx DMCD-40; Aromox DMMC-W; Conco XA-L; Emcol L; Incromine L-40, Oxide L; Jordamox LDA); conditioners (Incromine Oxide L; Loramine L408; Ninox L; Rewominox L408; Schercamox DML); personal care products (Oxamin LO; Sandoz Amine Oxide XA-L); shampoos (Alkamox LO; Ammonyx LO; Aromox DMMC-W; Cyclomox L; Empigen OB; Incromine Oxide L; Loramine L408; Oxamin LO; Ninox L; Rewominox L408; Schercamox DML; Varox 365); shaving preparations (Ammonyx LO; Incromine Oxide L; Ninox L); toiletries (Emcol L)

Degreasers: (Ammonyx LO; Incromine Oxide L)

Farm products: (Aromox DMMC-W)

Household detergents: (Alkamox LO; Ammonyx DMCD-40; Aromox DMMC-W; Conco XA-L; Incromine Oxide L; Jordamox LDA; Oxamin LO); dishwashing (Cyclomox L; Incromine Oxide L; Sandoz Amine Oxide XA-L); laundry detergent (Ammonyx LO); light-duty cleaners (Ninox L); liquid detergents (Empigen OB)

Industrial applications: (Incromine L-40); carpet backing (Empigen OB); dyes and pigments (Aromox DMMC-W; Conco XA-L); electroplating (Alkamox LO; Conco XA-L); fire fighting foams (Empigen OB); lubricating/cutting oils (Aromox DMMC-W); metal plating (Aromox DMMC-W); paint mfg. (Aromox DMMC-W); paper mfg. (Alkamox LO; Aromox DMMC-W); petroleum industry (Aromox DMMC-W); photography (Conco XA-L); plastics (Aromox DMMC-W); polymers/polymerization (Conco XA-L); rubber (Aromox DMMC-W); foamed rubber (Alkamox LO); textile/leather processing (Conco XA-L; Sandoz Amine Oxide XA-L)

Industrial cleaners: (Conco XA-L; Emcol L; Jordamox LDA); janitorial cleaners (Ammonyx DMCD-40); textile cleaning (Aromox DMMC-W)

Lauryl dimethyl amine oxide *(cont'd.)*

PROPERTIES:
Form:
> Liquid (Alkamox LO; Ammonyx DMCD-40, LO; Cyclomox L; Emcol L; Incromine L-40; Loramine L408; Mackamine LO; Oxamin LO; Rewominox L408; Sandoz Amine Oxide XA-L; Textamine Oxide LMW; Varox 365, 375); @ 20 C (Empigen OB)
> Clear liquid (Aromox DMMC-W; Incromine Oxide L; Ninox L; Schercamox DML)

Color:
> Colorless (Alkamox LO)
> Nearly colorless (Incromine Oxide L)
> Water-white (Schercamox DML)
> Pale straw @ 20 C (Emigen OB)
> Amber (Sandoz Amine Oxide XA-L)
> Gardner 1 (Aromox DMMC-W; Varox 365)

Odor:
> Faint (Alkamox LO)
> Mild (Schercamox DML)

Composition:
> 27–30% active (Cyclomox L)
> 29–31% amine oxide (Incromine Oxide L)
> 29–31% active (Jordamox LDA)
> 30% active (Conco XA-L; Loramine L408; Mackamine LO; Ninox L; Oxamin LO; Rewominox L408; Sandox Amine Oxide XA-L; Schercamox DML; Textamine Oxide LMW)
> 30% active in water (Alkamox LO; Aromox DMMC-W)
> 30% amine oxide (Ammonyx LO)
> 30% amine oxide in water (Varox 365)
> 30 ± 1.5% active (Empigen OB)
> 40% active (Incromine L-40)
> 40% solids (Varox 375)
> 40–42% amine oxide (Ammonyx DMCD-40)

Solubility:
> Sol. in alcohols (Schercamox DML); sol. in lower alcohols (Sandoz Amine Oxide XA-L)
> Sol. in ethanol (Conco XA-L)
> Sol. in glycol ethers (Schercamox DML)
> Sol. in glycols (Sandoz Amine Oxide XA-L; Schercamox DML)
> Sol. in hexylene glycol (Conco XA-L)
> Sol. in isopropanol (Conco XA-L)
> Sol. in polyols (Schercamox DML)
> Sol. in triols (Schercamox DML)
> Sol. in water (Alkamox LO; Aromox DMMC-W; Conco XA-L; Emcol L; Sandoz Amine Oxide XA-L; Schercamox DML)

Ionic Nature:
Nonionic (Ammonyx DMCD-40, LO; Incromine Oxide L; Loramine L408; Jordamox LDA; Mackamine LO; Rewominox L408)
Nonionic-cationic (Alkamox LO; Ninox L; Sandoz Amine Oxide XA-L; Schercamox DML; Varox 365)
Cationic (Aromox DMMC-W; Conco XA-L; Cyclomox L; Emcol L; Incromine L-40, Oxide L)

M.W.:
235 (avg.) (Schercamox DML)

Sp.gr.:
0.91 (25/20 C) (Ammonyx DMCD-40)
0.96 (Aromox DMMC-W); (25/20 C) (Ammonyx LO)
0.99 ± 0.01 (Schercamox DML0

Density:
0.98 g/cc (20 C) (Empigen OB)

Visc.:
25 cs (20 C) (Empigen OB)
90 cp (Aromox DMMC-W)

F.P.:
< 0 C (Alkamox LO)

Pour Pt.:
36 F (Aromox DMMC-W)

Flash Pt.:
86 F (Ammonyx DMCD-40)
> 200 F (Ammonyx LO)
> 212 F (Aromox DMMC-W)

Cloud Pt.:
22 F (Aromox DMMC-W)

Stability:
Good (Schercamox DML)

pH:
7.0 ± 1.0 (1% sol'n.) (Schercamox DML)
7.0–8.0 (5% aq. sol'n.) (Incromine Oxide L)
7.5 ± 0.5 (5% aq. sol'n.) (Empigen OB)

Surface Tension:
31.2 dynes/cm (0.1%) (Aromox DMMC-W)

Biodegradable: (Conco XA-L; Ninox L; Sandoz Amine Oxide XA-L)

TOXICITY/HANDLING:
Protective clothing, goggles, gloves should be worn (Aromox DMMC-W)
Avoid prolonged contact with skin and eyes (Incromine Oxide L)
Toxic orally, nonirritating to skin, corrosive to eyes (Varox 365)

STORAGE/HANDLING:
SS, glass or fiberglass reinforced polyester tanks (Aromox DMMC-W)

Lauryl dimethyl amine oxide *(cont' d.)*

Store in a cool, dry place (Incromine Oxide L)
STD. PKGS.:
55-gal poly-lined drums (Schercamox DML)
55-gal (425 lb net) lined drums (Incromine Oxide L)

Lauryl lactate (CTFA)

SYNONYMS:
Dodecyl 2-hydroxypropanoate
Ester of lauryl alcohol and lactic acid
2-Hydroxypropanoic acid, dodecyl ester
Propanoic acid, 2-hydroxy-, dodecyl ester
EMPIRICAL FORMULA:
$C_{15}H_{30}O_3$
STRUCTURE:

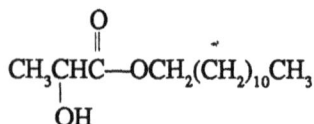

CAS No.:
6283-92-7
TRADENAME EQUIVALENTS:
Ceraphyl 31 [Van Dyk]
Crodamol LL [Croda; Croda Ltd.]
Cyclochem LVL [Alcolac]
Schercemol LL [Scher]
CATEGORY:
Emollient, penetrant, antitack agent, spreading agent, wetting agent, pigment dispersant
APPLICATIONS:
Bath products: bath oils (Schercemol LL)
Cosmetic industry preparations: (Crodamol LL); creams and lotions (Cyclochem LVL; Schercemol LL); hair preparations (Crodamol LL); makeup (Crodamol LL; Cyclochem LVL; Schercemol LL); perfumery (Schercemol LL); powders (Cyclochem LVL); shaving preparations (Crodamol LL; Schercemol LL); skin preparations (Crodamol LL)
Pharmaceutical applications: antiperspirant/deodorant (Ceraphyl 31; Crodamol LL; Cyclochem LVL; Schercemol LL); sunscreens (Schercemol LL)

PROPERTIES:
Form:
   Liquid (Crodamol LL; Cyclochem LVL)
   Clear liquid (Schercemol LL)
Color:
   Off-white (Cyclochem LVL)
   Colorless to pale straw (Crodamol LL)
   Gardner 2 max. (Schercemol LL)
Odor:
   Faint bland (Crodamol LL)
   Typical (Schercemol LL)
Solubility:
   Sol. in alcohols (Scherecemol LL)
   Sol. in aliphatic hydrocarbons (Schercemol LL)
   Sol. in aromatic hydrocarbons (Schercemol LL)
   Sol. in chlorinated hydrocarbons (Schercemol LL)
   Sol. in esters (Schercemol LL)
   Sol. in 95% ethanol (Ceraphyl 31)
   Sol. in natural fats (Schercemol LL)
   Sol. in glycol ethers (Schercemol LL)
   Sol. in glycols (Crodamol LL; Schercemol LL)
   Sol. in isopropyl myristate (Ceraphyl 31)
   Sol. in ketones (Schercemol LL)
   Sol. in min. oil (Ceraphyl 31; Cyclochem LVL; Schercemol LL)
   Sol. in most organic solvents (Schercemol LL)
   Sol. in highly polar solvents (Crodamol LL)
   Disp. in some polyols (Schercemol LL)
   Sol. in propylene glycol (Ceraphyl 31)
   Disp. in water (Schercemol LL); insol. (Ceraphyl 31; Cyclochem LVL)
M.W.:
   275 (theoret.) (Schercemol LL)
Sp.gr.:
   0.900 (Schercemol LL)
Density:
   7.6 lb/gal (Schercemol LL)
F.P.:
   −3 C (Schercemol LL)
M.P.:
   4 C (Cyclochem LVL)
Flash Pt.:
   150 C (OC) (Schercemol LL)
Acid No.:
   3.0 max. (Cyclochem LVL; Schercemol LL)

167

Lauryl lactate *(cont'd.)*

Iodine No.:
   Nil (Schercemol LL)
Saponification No.:
   199 (Cyclochem LVL)
   200–220 (Schercemol LL)
Ref. Index:
   1.4430 (Schercemol LL)
STORAGE/HANDLING:
   Does not attack styrene (Crodamol LL)

Magnesium stearate (CTFA)

SYNONYMS:
Magnesium octadecanoate
Octadecanoic acid, magnesium salt

EMPIRICAL FORMULA:
$C_{18}H_{36}O_2 \cdot \frac{1}{2}Mg$

STRUCTURE:
$[CH_3(CH_2)_{16}COO]_2Mg$

CAS No.:
557-04-0

TRADENAME EQUIVALENTS:
Petrac MG-20, MG-20 NF [Synthetic Products]
Radiastar 1100 [Synfina-Oleofin]
Witco Magnesium Stearate D [Witco]
    Generically sold by:
        [Harwick; Smith]

CATEGORY:
Lubricant, anticaking aid, tableting aid, processing aid, activator

APPLICATIONS:
Cosmetic industry preparations: baby powders (Petrac MG-20 NF); creams and lotions (Petrac MG-20 NF); emulsions (Petrac MG-20 NF)
Food applications: (Petrac MG-20 NF; Radiastar 1100)
Industrial applications: plastics (Witco Magnesium Stearate D); rubber (generic; Witco Magnesium Stearate D)
Pharmaceutical applications: (Petrac MG-20 NF); ointments (Petrac MG-20 NF); tablet mfg. (Petrac MG-20, MG-20 NF)

PROPERTIES:
Form:
Powder (generic; Petrac MG-20; Witco Magnesium Stearate D)
Fluffy powder (Petrac MG-20 NF; Radiastar 1100)
Fineness:
95% through 200 mesh; 85% through 325 mesh (Petrac MG-20 NF)
Color:
White (generic; Petrac MG-20 NF; Radiastar 1100; Witco Magnesium Stearate D)
Odor:
Odorless (generic)

169

Magnesium stearate (cont'd.)

Solubility:
Insol. in alcohol (generic; Witco Magnesium Stearate D)
Sol. in hot aliphatic solvents (Witco Magnesium Stearate D)
Sol. in hot aromatic solvents (Witco Magnesium Stearate D)
Sol. in hot oils (Witco Magnesium Stearate D)
Insol. in water (generic; Witco Magnesium Stearate D)
Sp.gr.:
1.028 (generic)
1.03 (Witco Magnesium Stearate D)
Density:
22 lb/ft^3 (bulk) (Petrac MG-20 NF)
M.P.:
145 C (Witco Magnesium Stearate D)
Softening Pt.:
140 C (Petrac MG-20 NF)

Methyl gluceth-10 (CTFA)

STRUCTURE:
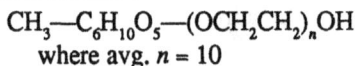
CH_3—$C_6H_{10}O_5$—$(OCH_2CH_2)_n OH$
where avg. $n = 10$
CAS No.:
RD No.: 977065-07-8
TRADENAME EQUIVALENTS:
Glucam E-10 [Amerchol]
CATEGORY:
Solvent, emollient, conditioner, humectant
APPLICATIONS:
Cosmetic industry preparations: (Glucam E-10)
Pharmaceutical applications: (Glucam E-10)
PROPERTIES:
Solubility:
Sol. in ethanol (Glucam E-10)
Sol. in hydroalcoholic systems (Glucam E-10)
Sol. in water (Glucam E-10)

STRUCTURE:
CH_3—$C_6H_{10}O_5$—$(OCH_2CH_2)_n$OH
where avg. $n = 20$

CAS No.:
RD No. 977065-08-9

TRADENAME EQUIVALENTS:
Glucam E-20 [Amerchol]

CATEGORY:
Solvent, solubilizer, emollient, conditioner, humectant, freezing point depressant, foam modifier

APPLICATIONS:
Cleansers: personal cleansers (Glucam E-20)
Cosmetic industry preparations: (Glucam E-20); astringents (Glucam E-20); creams and lotions (Glucam E-20); moisturizers (Glucam E-20); shampoos (Glucam E-20); shaving preparations (Glucam E-20)
Household detergents: (Glucam E-20)
Pharmaceutical applications: (Glucam E-20); topical preparations (Glucam E-20)

PROPERTIES:
Form:
Thin syrup (Glucam E-20)
Color:
Pale yellow (Glucam E-20)
Odor:
Practically odorless (Glucam E-20)
Composition:
99% active min. (Glucam E-20)
Solubility:
Freely sol. in alcohols (Glucam E-20)
Sol. in ethanol (Glucam E-20)
Freely sol. in hydroalcoholic systems (Glucam E-20)
Sol. in all proportions in water (Glucam E-20)
Acid No.:
1.0 max. (Glucam E-20)
Iodine No.:
1.0 max. (Glucam E-20)
Saponification No.:
1.0 max. (Glucam E-20)
Hydroxyl No.:
205–225 (Glucam E-20)

TOXICITY/HANDLING:
Low order of oral toxicity; nonirritating to skin and eyes (Glucam E-20)

Methyl laurate (CTFA)

SYNONYMS:
Dodecanoic acid, methyl ester
Methyl dodecanoate
EMPIRICAL FORMULA:
$C_{13}H_{26}O_2$
STRUCTURE:

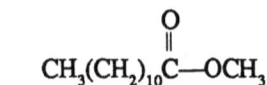

CAS No.:
111-82-0
TRADENAME EQUIVALENTS:
Radia 7118 [Synfina-Oleofin]
Stepan C40, C41, C42, C43 [Stepan]
CATEGORY:
Lubricant, chemical intermediate
APPLICATIONS:
Industrial applications: chemical synthesis (Radia 7118; Stepan C40, C41, C42, C43); lamination (Radia 7118); lubricating/cutting oils (Radia 7118; Stepan C40, C41, C42, C43); plasticizers (Stepan C40, C41, C42, C43); resins (Stepan C40, C41, C42, C43); rust inhibitors (Radia 7118); textile/leather processing (Radia 7118; Stepan C40, C41, C42, C43)
PROPERTIES:
Color:
Water-white (Stepan C40, C41, C42, C43)
Composition:
99.9% active (Stepan C40, C41, C42, C43)
Solubility:
Sol. in min. oils (Radia 7118)
Sol. in most solvents (Radia 7118)
Sol. in veg. oils (Radia 7118)
M.W.:
214 (avg.) (Radia 7118)
Sp.gr.:
0.855 (37.8 C) (Radia 7118)
Visc.:
2.30 cps (37.8 C) (Radia 7118)
M.P.:
1 C (Stepan C41)
5 C (Stepan C40)
Flash Pt.:
132 C (COC) (Radia 7118)

Cloud Pt.:
   1.5 C (Radia 7118)
Acid No.:
   0.5 max. (Stepan C40, C41)
   0.6 max. (Stepan C42, C43)
Iodine No.:
   0.1 max. (Stepan C40)
   0.2 max. (Stepan C41)
   0.3 max. (Stepan C42, C43)
Saponification No.:
   252–256 (Stepan C43)
   254–258 (Stepan C42)
   258–264 (Stepan C41)
   260–264 (Stepan C40)
Ref. Index:
   1.4322 (Radia 7118)

Methyl palmitate (CTFA)

SYNONYMS:
   Hexadecanoic acid, methyl ester
   Methyl hexadecanoate
EMPIRICAL FORMULA:
   $C_{17}H_{34}O_2$
STRUCTURE:

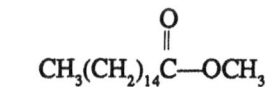

$$CH_3(CH_2)_{14}\overset{\displaystyle O}{\overset{\|}{C}}{-}OCH_3$$

CAS No.:
   112-39-0
TRADENAME EQUIVALENTS:
   Radia 7120 [Synfina-Oleofin]
   Stepan C60 [Stepan]
CATEGORY:
   Lubricant, chemical intermediate
APPLICATIONS:
   Industrial applications: chemical synthesis (Radia 7120; Stepan C60); lamination (Radia 7120); lubricating/cutting oils (Radia 7120; Stepan C60); plasticizers (Stepan C60); resins (Stepan C60); rust inhibitors (Radia 7120); textile/leather processing (Radia 7120; Stepan C60)

Methyl palmitate *(cont'd.)*

PROPERTIES:
Color:
  Gardner 1 max. (Stepan C60)
Composition:
  99.9% active (Stepan C60)
Solubility:
  Sol. in min. oils (Radia 7120)
  Sol. in most solvents (Radia 7120)
  Sol. in veg. oils (Radia 7120)
M.W.:
  267 (avg.) (Radia 7120)
Sp.gr.:
  0.855 (37.8 C) (Radia 7120)
Visc.:
  4.30 cps (37.8 C) (Radia 7120)
M.P.:
  27.5 C (Radia 7120)
  28 C (Stepan C60)
Flash Pt.:
  149 C (COC) (Radia 7120)
Acid No.:
  2.0 max. (Stepan C60)
Iodine No.:
  16 max. (Stepan C60)
Saponification No.:
  203–210 (Stepan C60)

Methyl stearate *(CTFA)*

SYNONYMS:
  Methyl octadecanoate
  Octadecanoic acid, methyl ester
EMPIRICAL FORMULA:
  $C_{19}H_{38}O_2$
STRUCTURE:

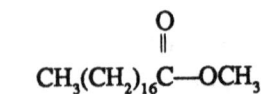

$$CH_3(CH_2)_{16}\overset{\overset{\textstyle O}{\|}}{C}{-}OCH_3$$

CAS No.:
  112-61-8

Methyl stearate (cont'd.)

TRADENAME EQUIVALENTS:
Emery Methyl Stearate [Henkel/Emery]
Grocor 8058 [A. Gross]
Kemester 7018, 9018 [Humko/Witco]
Radia 7110 [Synfina-Oleofin]
Stepan C66 [Stepan]

CATEGORY:
Lubricant, chemical intermediate, detergent, plasticizer, opacifier, viscosity control agent

APPLICATIONS:
Automotive applications: lubricants (Emery Methyl Stearate)
Cosmetic industry preparations: (Kemester 7018)
Household detergents: (Stepan C66)
Industrial applications: chemical synthesis (Radia 7110); lamination (Radia 7110); lubricating/cutting oils (Radia 7110); metalworking (Emery Methyl Stearate; Grocor 8058; Kemester 9018); resins (Stepan C66); rust inhibitors (Radia 7110); textile/leather processing (Emery Methyl Stearate; Grocor 8058; Kemester 7018; Radia 7110; Stepan C66)

PROPERTIES:
Form:
Liquid (Grocor 8058)
Gel (Grocor 8058)
Color:
Water-white (Grocor 8058)
Gardner 1 max. (Kemester 9018; Stepan C66)
Composition:
98% active min. (Grocor 8058)
Solubility:
Sol. in isopropanol (Grocor 8058)
Sol. in min. oils (Radia 7110)
Sol. in most solvents (Radia 7110)
Sol. in veg. oils (Radia 7110)
Ionic Nature:
Nonionic (Grocor 8058)
M.W.:
285 (avg.) (Radia 7110)
Sp.gr.:
0.860 (Grocor 8058)
F.P.:
32 C max. (Grocor 8058)
M.P.:
30 C (Stepan C66)
31 C (Radia 7110)

Methyl stearate (cont'd.)

Flash Pt.:
166 C (OC) (Radia 7110)
Acid No.:
1.5 max. (Stepan C66)
4.0 max. (Kemester 9018)
Saponification No.:
185–192 (Kemester 9018)
195–203 (Stepan C66)
pH:
Neutral (Grocor 8058)
STD. PKGS.:
Bulk or drums (Grocor 8058)

Mica (CTFA)

SYNONYMS:
Muscovite mica
CAS No.:
12001-26-2
TRADENAME EQUIVALENTS:
Huber SM [J.M. Huber] (surface-treated)
Huber WG-1, WG-2 [J.M. Huber] (water-ground)
Mearlmica MMCF, MMSV [Mearl]
Micromesh No. 3 [Mearl] (water-ground)
Micro-Mica C-1000, C-3000 [KMG Minerals]
Generically sold by:
[(water-ground) Harwick; (regular and wet-ground) KMG Minerals]
CATEGORY:
Lubricant, filler, extender, release agent, dusting agent
APPLICATIONS:
Cosmetic industry applications: makeup (Mearlmica MMCF, MMSV); pressed powders (Mearlmica MMCF, MMSV)
Industrial applications: coatings (Huber WG-1, WG-2); pearlescent pigments (Huber WG-1, WG-2); plastics (Huber SM, WG-1, WG-2); rubber (generic; Huber WG-1, WG-2; Micro-Mica C-1000, C-3000)
PROPERTIES:
Form:
Flake (Huber SM, WG-2)
Powder (Huber SM, WG-2; Micro-Mica C-1000)
Platy shaped (Micromesh No. 3)
Powder, platy particle (Micro-Mica C-3000)

Fineness:
    7–8 μ avg. (Mearlmica MMSV)
    26 μ avg. (Mearlmica MMCF)
    82–87% through 325 mesh (Huber SM, WG-2)
    87–92% through 325 mesh (Huber WG-1)
    100% through 100 mesh, 88–94% through 325 mesh (generic—Harwick)
    400 mesh (well below) (Micromesh No. 3)
    Mesh 1000 (theoret.) (Micro-Mica C-1000)
    Mesh 3000 (theoret.) (Micro-Mica C-3000)

Color:
    White (Huber SM; Micro-Mica C-1000, C-3000)
    Nearly white (Micromesh No. 3)
    Off-white (Huber WG-1, WG-2)

Odor:
    Odorless (Micro-Mica C-1000)
    Negligible (Huber SM, WG-2)

Composition:
    42–47% SiO_2; 34–38% Al_2O_3; 9–12% K_2O; 0.5–2.5% Fe_2O_3; 4–5% water (Micromesh No. 3)
    46.2% SiO, 36.9% Al_2O_3, 1.5% Fe_2O_3, 7.9% K_2O, 1.0% Na_2O (generic—Harwick)

Solubility:
    Negligible sol. in water (Huber SM, WG-1, WG-2)

Sp.gr.:
    2.75–2.79 (generic—Harwick)
    2.80 (generic—KMG; Huber SM, WG-1, WG-2; Micro-Mica C-1000, C-3000)
    2.8–2.9 (Micromesh No. 3)

Density:
    7 lb/ft^3 (bulk) (Micromesh No. 3)
    10.80 lb/ft^3 (bulk) (generic—Harwick)
    11.0–13.0 lb/ft^3 (bulk) (Huber SM, WG-1, WG-2)

M.P.:
    ≈ 1000 C (Huber SM, WG-1, WG-2)

Flash Pt.:
    None (Huber SM, WG-1, WG-2)

Ref. Index:
    1.58 (Micromesh No. 3)

pH:
    7.0–8.0 (28% solids) (Huber SM, WG-1, WG-2)

Surface Area:
    6–8 m^2/g (Micromesh No. 3)

TOXICITY/HANDLING:
    Nontoxic (Micro-Mica C-1000)

Microcrystalline wax (CTFA)

SYNONYMS:
Petroleum wax, microcrystalline
Waxes, microcrystalline
CAS No.:
63231-60-7
TRADENAME EQUIVALENTS:
Be Square 175, 185, 195 [Petrolite]
Duron Micro Wax 170/180 (0423) [Dura]
Flexowax C, C Light [Lonza]
Fortex [Petrolite]
Mekon White [Petrolite]
Multiwax 180-M, ML-445, W-445, X-145A (NF) [Witco]
Multiwax W-835 [Witco]
Petrolite C-700, C-1035 [Petrolite]
Polymekon [Petrolite]
Starwax 100 [Petrolite]
Ultraflex [Petrolite]
Victory [Petrolite]
CATEGORY:
Lubricant, solvent, barrier, water repellent, stabilizer, processing aid, plasticizer, modifier, binder, antisoilant, antisunchecking agent, antiozonant, sizing agent, consistency agent, laminating agent, antislip agent, antimar agent
APPLICATIONS:
Cosmetic industry preparations: (Be Square 175; Flexowax C Light; Ultraflex; Victory); creams and lotions (Be Square 185, 195; Fortex; Mekon White; Petrolite C-700, C-1035; Starwax 100); makeup (Be Square 185, 195; Fortex; Mekon White; Petrolite C-700, C-1035; Starwax 100)
Food applications: cheese coating (Multiwax W-445, X-145A); chewing gum (Multiwax 180-M; Ultraflex; Victory)
Household applications: fabric softeners (Be Square 185, 195; Fortex; Mekon White; Petrolite C-700, C-1035; Starwax 100)
Industrial applications: adhesives (Be Square 175, 185, 195; Duron Micro Wax 170/180 (0423); Fortex; Mekon White; Multiwax 180-M, ML-445, W-445, W-835, X-145A; Petrolite C-700, C-1035; Starwax 100; Ultraflex; Victory); ceramics (Be Square 185, 195; Fortex; Mekon White; Petrolite C-700, C-1035; Starwax 100); coatings (Be Square 175, 185, 195; Fortex; Mekon White; Multiwax 180-M, ML-445, W-445, W-835, X-145A; Petrolite C-700, C-1035; Polymekon; Starwax 100; Ultraflex; Victory); electrical/electronic applications (Be Square 175, 185, 195; Fortex; Mekon White; Petrolite C-700, C-1035; Starwax 100; Ultraflex; Victory); explosives (Duron Micro Wax 170/180 (0423)); paint mfg. (Be Square 185, 195; Fortex; Mekon White; Petrolite C-700, C-1035; Starwax 100); paper mfg. (Be Square 185, 195; Fortex; Mekon White; Multiwax ML-445; Petrolite C-700, C-1035; Starwax 100); plastics (Be Square 185, 195; Fortex; Mekon White; Multiwax

W-445, X-145A; Petrolite C-700, C-1035; Starwax 100); polishes (Duron Micro Wax 170/180 (0423)); printing inks (Be Square 185, 195; Fortex; Mekon White; Petrolite C-700, C-1035; Polymekon; Starwax 100); rubber (Be Square 175, 185, 195; Duron Micro Wax 170/180 (0423); Fortex; Mekon White; Petrolite C-700, C-1035; Starwax 100; Ultraflex; Victory); rustproofing compounds (Multiwax ML-445); textile/leather processing (Be Square 175, 185, 195; Flexowax C; Fortex; Mekon White; Petrolite C-700, C-1035; Starwax 100; Ultraflex; Victory); waxy products (candles, crayons) (Be Square 175; Multiwax ML-445; Ultraflex; Victory)

PROPERTIES:

Form:

Solid (Flexowax C, C Light)

Slabs (Duron Micro Wax 170/180 (0423))

Wax (Be Square 175; Fortex; Mekon White; Petrolite C-700, C-1035; Starwax 100; Ultraflex; Victory)

Color:

Cream (Flexowax C Light)

Tan (Flexowax C)

D156 16 (Mekon White)

D1500 0.5 (Be Square 195; Petrolite C-1035)

D1500 1.0 (Be Square 185; Starwax 100)

D1500 1.5 (Fortex; Petrolite C-700); 1.5; also avail. black (Be Square 175); 1.5; also avail. in white (Ultraflex); 1.5; also avail. in white and brown (Victory)

D1500 2.5 (Polymekon)

Saybolt +15 (Duron Micro Wax 170/180 (0423))

Solubility:

Miscible with most animal fats, oils, and waxes (Multiwax 180-M, ML-445, W-445, X-145A)

Miscible with many essential oils (Multiwax 180-M, ML-445, W-445, X-145A)

Very low sol. in organic solvents (Be Square 185, 195; Fortex; Mekon White; Petrolite C-700, C-1035; Starwax 100)

Miscible with petroleum products (Multiwax 180-M, ML-445, W-445, X-145A)

Miscible with most vegetable fats, oils, and waxes (Multiwax 180-M, ML-445, W-445, X-145A)

Sp.gr.:

0.92 (23.9 C) (Be Square 185; Starwax 100)

0.93 (23.9 C) (Be Square 195; Fortex; Mekon White; Petrolite C-700, C-1035)

0.931 (Duron Micro Wax 170/180 (0423))

Density:

0.92 g/cc (Ultraflex)

0.93 g/cc (Be Square 175; Victory)

Visc.:

11 cps (98.8 C) (Be Square 175; Ultraflex)

11.3 cps (98.9 C) (Be Square 185)

Microcrystalline wax *(cont'd.)*

  11.7 cps (98.9 C) (Mekon White; Petrolite C-1035; Starwax 100)
  12 cps (98.8 C) (Victory)
  12.5 cps (98.9 C) (Be Square 195; Petrolite C-700)
  23.5 cps (98.9 C) (Fortex)
  75 SUS (210 F) (Duron Micro Wax 170/180 (0423))
B.P.:
  > 400 F (Duron Micro Wax 170/180 (0423))
M.P.:
  61 C (Flexowax C Light)
  64.4 C (Ultraflex)
  70 C (Flexowax C)
  79.4 C (Victory)
  83.3 C (Be Square 175)
  86.1 C (Starwax 100)
  87.8 C (Be Square 185)
  91.1 C (Be Square 195; Petrolite C-700)
  92.8 C (Mekon White; Petrolite C-1035)
  96.1 C (Fortex; Polymekon)
  150–160 F (Multiwax X-145A)
  165–175 F (Multiwax W-835)
  170–180 F (Duron Micro Wax 170/180 (0423); Multiwax W-445)
  180–190 F (Multiwax 180-M, ML-445)
Flash Pt.:
  230 C (Flexowax C Light)
  255 C (Flexowax C)
  293.3 C (Be Square 175; Ultraflex; Victory)
  ≈ 540 F (Duron Micro Wax 170/180 (0423))
Stability:
  Stable (Duron Micro Wax 170/180 (0423))
STORAGE/HANDLING:
  Store away from hot pipes and boilers (Duron Micro Wax 170/180 (0423))
STD. PKGS.:
  55-lb cartons each containing 5 slabs of 11 lb (Duron Micro Wax 170/180 (0423))

Mineral oil *(CTFA)*

SYNONYMS:
  Heavy mineral oil
  Light mineral oil
  Liquid paraffin
  Paraffin oil
  White mineral oil

CAS No.:
8012-95-1
TRADENAME EQUIVALENTS:
Benol [Witco/Sonneborn] (white NF)
Blandol [Witco/Sonneborn] (white NF)
Carnation [Witco/Sonneborn] (white NF)
Drakeol 5, 6, 7, 8, 9, 10, 13, 15 [Penreco] (light USP)
Drakeol 19, 21, 32, 34, 35 [Penreco] (USP)
Draketex 50 [Penreco] (light USP)
Ervol [Witco/Sonneborn] (white NF)
Gloria [Witco/Sonneborn] (white USP)
Kaydol [Witco/Sonneborn] (white USP)
Klearol [Witco/Sonneborn] (white NF)
Marcol [Exxon] (white)
Orzol [Witco/Sonneborn] (white USP)
Parol 70, 80, 100 [Penreco] (tech.)
Peneteck [Penreco] (tech.)
Protol [Witco/Sonneborn] (white USP)
Sontex 19, 21, 35 [Marathon Morco] (white USP)
Sontex 55, 70 [Marathon Morco]
Sontex 75 [Marathon Morco] (white NF),
Sontex 75T, 85T, 95T [Marathon Morco] (white tech.),
Sontex 100, 150 [Marathon Morco]
CATEGORY:
Lubricant, emollient, release agent, emulsifier, base material, carrier, solvent, dispersant, plasticizer, water repellent, spreading agent, binder
APPLICATIONS:
Bath products: bath oils (Drakeol 6, 13)
Cleansers: cleansing oils (Marcol); soaps (Sontex 19, 21, 35, 55, 70, 75, 75 T, 85, 85 T, 95 T, 100, 150); waterless hand cleanser (Drakeol 7)
Cosmetic industry preparations: (Benol; Carnation; Drakeol 9; Ervol; Gloria; Kaydol; Klearol; Orzol; Protol; Sontex 19, 21, 35, 55, 70, 75, 75 T, 85, 85 T, 95 T, 100, 150); baby oil (Carnation; Drakeol 6, 7); creams and lotions (Drakeol 7; Protol); hair preparations (Drakeol 6); makeup (Drakeol 7)
Farm products: animal feed (Sontex 19, 21, 35, 55, 70, 75, 75 T, 85, 85 T, 95 T, 100, 150)
Food applications: (Benol; Gloria; Orzol); food contact (Ervol); food packaging (Sontex 19, 21, 35, 55, 70, 75, 75 T, 85, 85 T, 95 T, 100, 150); food processing (Sontex 19, 21, 35, 55, 70, 75, 75 T, 85, 85 T, 95 T, 100, 150); preservatives (Marcol); protective coatings (Drakeol 7)
Industrial applications: cellulosics (Drakeol 19, 21, 35); dyes and pigments (Drakeol 7, 9); paper mfg. (Drakeol 9, 19, 35; Sontex 19, 21, 35, 55, 70, 75, 75 T, 85, 85 T, 95 T, 100, 150); plastics (Drakeol 7, 9, 10, 34, 35; Sontex 19, 21, 35, 55, 70, 75, 75

Mineral oil (cont'd.)

T, 85, 85 T, 95 T, 100, 150); rubber (Sontex 19, 21, 35, 55, 70, 75, 75 T, 85, 85 T, 95 T, 100, 150); textile/leather processing (Drakeol 9, 19, 50; Marcol; Sontex 19, 21, 35, 55, 70, 75, 75 T, 85, 85 T, 95 T, 100, 150)

Pharmaceutical applications: (Benol; Gloria; Kaydol; Orzol; Protol; Sontex 19, 21, 35, 55, 70, 75, 75 T, 85, 85 T, 95 T, 100, 150); laxatives (Drakeol 9); ointments (Drakeol 7); sunscreens (Drakeol 7)

Veterinary products (Sontex 19, 21, 35, 55, 70, 75, 75 T, 85, 85 T, 95 T, 100, 150)

PROPERTIES:

Form:

Oil (Sontex 19, 21, 35, 55, 70, 75, 75 T, 85, 85 T, 95 T, 100, 150)

Color:

Water-white (Benol; Gloria; Orzol; Sontex 19, 21, 35, 55, 70, 75, 75 T, 85, 85 T, 95 T, 100, 150)

Odor:

Odorless (Sontex 19, 21, 35, 55, 70, 75, 75 T, 85, 85 T, 95 T, 100, 150)

Taste:

Tasteless (Sontex 95 T, 150

M.W.:

283 (Sontex 55)
298 (Sontex 70)
319 (Sontex 75 T)
335 (Sontex 85, 85 T)
340 (Sontex 75)
364 (Sontex 95 T, 100)
365 (Sontex 150)
391 (Sontex 19)
396 (Sontex 21)
430 (Sontex 35)

Sp.gr.:

0.802–0.811 (Peneteck)
0.817–0.830 (Draketex 50)
0.821–0.833 (Drakeol 5)
0.827–0.836 (Drakeol 6)
0.827–0.844 (Drakeol 7)
0.828–0.849 (Parol 70)
0.830–0.844 (Drakeol 8)
0.830–0.857 (Parol 80)
0.838–0.854 (Drakeol 9)
0.8398 (Sontex 55)
0.8408 (Sontex 75)
0.844–0.864 (Drakeol 10)
0.844–0.865 (Parol 100)
0.845–0.855 (15.5/15.5 C) (Marcol)

0.8473 (Sontex 75 T)
0.848–0.867 (Drakeol 13)
0.8500 (Sontex 100)
0.850–0.870 (Drakeol 15)
0.852–0.871 (Drakeol 19)
0.8524 (Sontex 85 T)
0.853–0.873 (Drakeol 21)
0.8545 (Sontex 85)
0.8550 (Sontex 95 T)
0.856–0.876 (Drakeol 32)
0.858–0.872 (Drakeol 34)
0.8639 (Sontex 70)
0.864–0.876 (Drakeol 35)
0.8720 (Sontex 150)
0.874 (Sontex 19)
0.875 (Sontex 21)
0.881 (Sontex 35)

Density:

6.73–6.81 lb/gal (Peneteck)
6.86–6.96 lb/gal (Draketex 50)
6.89–7.00 lb/gal (Drakeol 5)
6.94–7.02 lb/gal (Drakeol 6)
6.94–7.08 lb/gal (Drakeol 7)
6.94–7.12 lb/gal (Parol 70)
6.97–7.08 lb/gal (Drakeol 8)
6.97–7.19 lb/gal (Parol 80)
6.99 lb/gal (Sontex 55)
7.00 lb/gal (Sontex 75)
7.03–7.16 lb/gal (Drakeol 9)
7.05 lb/gal (Sontex 75 T)
7.08 lb/gal (Sontex 100)
7.08–7.25 lb/gal (Drakeol 10; Parol 100)
7.10 lb/gal (Sontex 85 T)
7.11 lb/gal (Sontex 85)
7.11–7.27 lb/gal (Drakeol 13)
7.12 lb/gal (Sontex 95 T)
7.13–7.30 lb/gal (Drakeol 15)
7.14–7.31 lb/gal (Drakeol 19)
7.15–7.32 lb/gal (Drakeol 21)
7.18–7.35 lb/gal (Drakeol 32)
7.19 lb/gal (Sontex 70)
7.19–7.31 lb/gal (Drakeol 34)
7.25–7.35 lb/gal (Drakeol 35)

Mineral oil *(cont'd.)*

7.26 lb/gal (Sontex 150)
7.28 lb/gal (Sontex 19)
7.29 lb/gal (Sontex 21)
7.34 lb/gal (Sontex 35)

Visc.:

3.4–4.7 cSt (40 C) (Peneteck)
6.5–7.8 cSt (40 C) (Draketex 50)
7.6–8.7 cSt (40 C) (Drakeol 5)
8.8 cs (100 F) (Sontex 55)
9.2–10.6 cSt (40 C) (Drakeol 6)
10.8–12.7 cSt (40 C) (Drakeol 7)
10.8–13.2 cSt (40 C) (Parol 70)
12.7–14.5 cSt (40 C) (Drakeol 8)
13.2–15.7 cSt (40 C) (Parol 80)
13.8 cs (100 F) (Sontex 70)
14.4 cs (100 F) (Sontex 75, 75 T)
14.7–16.8 cSt (40 C) (Drakeol 9)
16.9 cs (100 F) (Sontex 85, 85 T)
17.9–20.0 cSt (40 C) (Drakeol 10; Parol 100)
19.3 cs (100 F) (Sontex 95 T)
20.5 cs (100 F) (Sontex 100)
24.2–26.3 cSt (40 C) (Drakeol 13)
28.2–30.2 cSt (40 C) (Drakeol 15)
32.0 cs (100 F) (Sontex 150)
35.0–37.0 cSt (40 C) (Drakeol 19)
39.7 cs (100 F) (Sontex 19)
39.7–41.7 cSt (40 C) (Drakeol 21)
44.1 cs (100 F) (Sontex 21)
60.0–63.3 cSt (40 C) (Drakeol 32)
65.3–70.0 cSt (40 C) (Drakeol 35)
71.0–78.3 cSt (40 C) (Drakeol 34)
75.5 cs (100 F) (Sontex 35)
50–60 SUS (Saybolt, 100 F) (Klearol)
65–75 SUS (Saybolt, 100 F) (Carnation)
80–90 SUS (Saybolt, 100 F) (Blandol)
95–105 SUS (Saybolt, 100 F) (Benol)
125–135 SUS (Saybolt, 100 F) (Ervol)
180–190 SUS (Saybolt, 100 F) (Protol)
200–210 SUS (Saybolt, 100 F) (Gloria)
320–330 SUS (Saybolt, 100 F) (Orzol)
345–355 SUS (Saybolt, 100 F) (Kaydol)

Pour Pt.:

–40 F (Sontex 70)

184

0 F (Kaydol; Klearol; Orzol; Sontex 19, 21, 35, 150)
10 F (Drakeol 32, 35; Sontex 55, 75, 75 T, 85, 85 T, 95 T, 100)
15 F (Parol 70, 80, 100; Drakeol 5, 6, 7, 8, 9, 10, 13, 15, 19, 21, 34; Gloria)
20 F (Benol; Blandol; Carnation; Draketex 50; Ervol; Peneteck; Protol)
30 F (Peneteck)

Flash Pt.:
265 F (Peneteck)
305 F (Draketex 50)
320 F (Drakeol 5, 6)
325 F (COC) (Sontex 55)
330 F (Parol 70); (COC) (Sontex 70)
340 F (Parol 80)
350 F (Parol 100); (COC) (Sontex 75 T)
355 F (Drakeol 7); (COC) (Sontex 85 T)
360 F (Drakeol 8); (COC) (Sontex 95 T)
365 F (Drakeol 9, 10); (COC) (Sontex 85, 150)
370 F (COC) (Sontex 75, 100)
374 F (COC) (Sontex 19)
385 F (COC) (Sontex 21)
390 F (Drakeol 13, 15)
410 F (Drakeol 19)
415 F (Drakeol 21)
425 F (COC) (Sontex 35)
430 F (Drakeol 32)
435 F (Drakeol 35)
475 F (Drakeol 34)

Ref. Index:
1.4517 (Peneteck)
1.4570 (Draketex 50)
1.4600 (Drakeol 5)
1.4613 (Drakeol 6)
1.4632 (Drakeol 7)
1.4635 (Parol 70)
1.4655 (Drakeol 8)
1.4660 (Parol 80)
1.4665 (Drakeol 9)
1.4688 (Parol 100)
1.4692 (Drakeol 10)
1.4725 (Drakeol 19)
1.4726 (Drakeol 13)
1.4733 (Drakeol 21)
1.4740 (Drakeol 15)
1.4760 (Drakeol 34)

Mineral oil (cont'd.)

1.4770 (Drakeol 32)
1.4785 (Drakeol 35)
pH:
Neutral (Sontex 19, 21, 35, 55, 70, 75, 85, 100, 150)

Myristamidopropyl dimethylamine (CTFA)

SYNONYMS:
Dimethylaminopropyl myristamide
N-[3-(Dimethylamino) propyl] tetradecanamide
Tetradecanamide, N-[3-(dimethylamino) propyl]-
EMPIRICAL FORMULA:
$C_{19}H_{40}N_2O$
STRUCTURE:

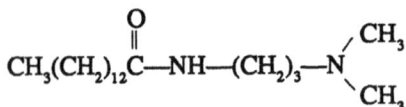

CAS No.:
45267-19-4
TRADENAME EQUIVALENTS:
Schercodine M [Scher]
CATEGORY:
Conditioner, emulsifier, viscosity builder
APPLICATIONS:
Cosmetic industry applications: creams and lotions (Schercodine M)
PROPERTIES:
Form:
Wax (Schercodine M)
Color:
Light tan (Schercodine M)
Composition:
98% amide min. (Schercodine M)
Solubility:
Sol. in organic solvents (Schercodine M)
Ionic Nature:
Cationic (Schercodine M)
M.W.:
312 (Schercodine M)

186

Myristamidopropyl dimethylamine *(cont'd.)*

M.P.:
45–50 C (Schercodine M)
Alkali No.:
180–190 (Schercodine M)

Myristyl lactate (CTFA)

SYNONYMS:
Ester of myristyl alcohol and lactic acid
2-Hydroxypropanoic acid, tetradecyl ester
Propanoic acid, 2-hydroxy-, tetradecyl ester
Tetradecyl 2-hydroxypropanoate
EMPIRICAL FORMULA:
$C_{17}H_{34}O_3$
STRUCTURE:

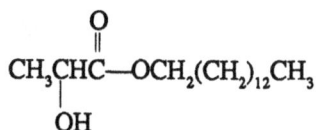

CAS No.:
1323-03-1
TRADENAME EQUIVALENTS:
Cegesoft C17 [Henkel KGaA]
Ceraphyl 50S [Van Dyk]
Crodamol ML [Croda; Croda Ltd.]
Cyclochem ML [Alcolac]
Liponate ML [Lipo]
Schercemol ML [Scher]
Wickenol 506 [CasChem]
CATEGORY:
Emollient, penetrant, antitack agent, lubricant, plasticizer, pigment dispersant, thickener, viscosity control agent, solubilizer
APPLICATIONS:
Bath products: (Liponate ML); bath oils (Liponate ML)
Cosmetic industry preparations: aerosol hair spray (Schercemol ML); creams and lotions (Cyclochem ML; Liponate ML; Schercemol ML); hair preparations (Crodamol ML; Schercemol ML); makeup (Crodamol ML; Cyclochem ML; Liponate ML); shampoos (Schercemol ML); skin preparations (Ceraphyl 50S; Schercemol ML)

Myristyl lactate (cont'd.)

Pharmaceutical applications: antiperspirant/deodorant (Crodamol ML; Cyclochem ML; Schercemol ML)

PROPERTIES:

Form:

Liquid (Crodamol ML; Cyclochem ML)

Liquid to soft solid (Schercemol ML)

Liquid to solid (Liponate ML)

Solid (Cegesoft C17)

Color:

White to pale yellow (Crodamol ML)

White to yellow (Liponate ML)

Off-white (Cyclochem ML)

Gardner 2 max. (Schercemol ML)

Odor:

Typical, slight (Schercemol ML)

Solubility:

Sol. in alcohols (Schercemol ML); soll in aq. alcoholic sol'ns. (Crodamol ML)

Sol. in aromatic hydrocarbons (Schercemol ML)

Sol. in esters (Schercemol ML)

Sol. in ethanol (Liponate ML; Wickenol 506); sol. in 95% ethanol (Ceraphyl 50S)

Sol. in glycol ethers (Schercemol ML)

Sol. in glycols (Crodamol ML; Schercemol ML)

Sol. in isopropyl myristate (Ceraphyl 50S; Wickenol 506)

Sol. in ketones (Schercemol ML)

Sol. in min. oil (Ceraphyl 50S; Liponate ML; Schercemol ML; Wickenol 506)

Sol. in natural oils (Schercemol ML)

Sol. in most organic solvents (Schercemol ML)

Sol. in propylene glycol (Ceraphyl 50S; Liponate ML; Wickenol 506)

Sol. in veg. oil (Liponate ML)

Insol. in water (Ceraphyl 50S; Liponate ML; Schercemol ML; Wickenol 506)

M.W.:

286 (theoret.) (Schercemol ML)

Sp.gr.:

0.90 (30 C) (Schercemol ML)

Density:

7.5 lb/gal (Schercemol ML)

M.P.:

12 C (Cyclochem ML)

28–34 C (Schercemol ML)

Flash Pt.:

> 160 C (OC) (Schercemol ML)

Acid No.:

3.0 max. (Cyclochem ML; Schercemol ML)

Iodine No.:
 Nil (Schercemol ML)
Saponification No.:
 170–190 (Schercemol ML)
 172 (Cyclochem ML)
STORAGE/HANDLING:
 Does not attack polystyrene (Crodamol ML)

Myristyl myristate (CTFA)

SYNONYMS:
 Ester of myristyl alcohol and myristic acid
 Tetradecanoic acid, tetradecyl ester
 Tetradecyl tetradecanoate
EMPIRICAL FORMULA:
 $C_{28}H_{56}O_2$
STRUCTURE:

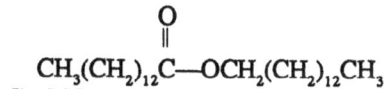

$$CH_3(CH_2)_{12}\overset{\displaystyle O}{\overset{\|}{C}}\!-\!OCH_2(CH_2)_{12}CH_3$$

CAS No.:
 3234-85-3
TRADENAME EQUIVALENTS:
 Ceraphyl 424 [Van Dyk]
 Cetiol MM [Henkel Canada; Henkel KGaA]
 Crodamol MM [Croda; Croda Ltd.]
 Cyclochem MM/M [Alcolac]
 Kemester MM [Humko]
 Kessco Myristyl Myristate [Stepan]
 Liponate MM [Lipo]
 MYM-33 [Hefti Ltd.]
 Schercemol MM [Scher]
 Waxenol 810 [CasChem]
CATEGORY:
 Emollient, thickener, viscosity control agent, spreading agent, superfatting agent,
  pearlescent, bodying agent, lubricant
APPLICATIONS:
 Bath products: (Liponate MM)
 Cleansers: (Liponate MM); cold creams/cleansing creams (Crodamol MM)
 Cosmetic industry preparations: (Crodamol MM; Kemester MM; MYM-33); creams

Myristyl myristate (cont'd.)

and lotions (Ceraphyl 424; Cyclochem MM/M; Liponate MM; Schercemol MM); hair preparations (Crodamol MM; Schercemol MM); makeup (Crodamol MM; Liponate MM; Schercemol MM); shaving preparations (Liponate MM); skin preparations (Schercemol MM)

Pharmaceutical applications: (Crodamo MM); sunscreens (Schercemol MM); suppositories (Crodamol MM)

PROPERTIES:
Form:

Liquid to soft solid (Liponate MM)

Flake (Cyclochem MM/M; MYM-33)

Wax-like (Cetiol MM)

Waxy solid (Crodamol MM; Kessco Myristyl Myristate; Schercemol MM)

Color:

White (Kemester MM; Liponate MM)

Creamy white (Crodamol MM)

Off-white (Cyclochem MM/M)

White to pale yellow (Schercemol MM)

Odor:

Mild, characteristic (Schercemol MM)

Composition:

100% active (Kessco Myristyl Myristate; MYM-33)

Solubility:

Sol. in alcohols (Schercemol MM)

Sol. in aromatic hydrocarbons (Schercemol MM)

Sol. in esters (Schercemol MM)

Disp. in glycerol (Schercemol MM)

Disp. in glycol ethers (Schercemol MM)

Disp. in glycols (Schercemol MM)

Sol. in isopropyl myristate (Ceraphyl 424; Waxenol 810)

Sol. in ketones (Schercemol MM)

Sol. in min. oil (Ceraphyl 424; Cyclochem MM/M; Schercemol MM; Waxenol 810)

Sol. in min. spirits (Schercemol MM)

Sol. in oils (Liponate MM)

Sol. in oleyl alcohol (Waxenol 810)

Partly sol. in water (Schercemol MM); insol. in water (Ceraphyl 424; Cyclochem MM/M; Liponate MM; Waxenol 810)

M.W.:

424 (theoret.) (Schercemol MM)

Sp.gr.:

0.839 (45 C) (Schercemol MM)

Density:

7.0 lb/gal (Schercemol MM)

Myristyl myristate (cont'd.)

M.P.:
   36–40 C (Schercemol MM)
   38 C (Cyclochem MM/M)
   38–40 C (Kemester MM)
Flash Pt.:
   > 170 C (OC) (Schercemol MM)
HLB:
   1.0 (MYM-33)
Acid No.:
   1.0 (Kemester MM)
   2.0 max. (Schercemol MM)
   3.0 max. (Cyclochem MM/M)
Iodine No.:
   Nil (Schercemol MM)
Saponification No.:
   120–135 (Schercemol MM)
   130 (Cyclochem MM/M)

Myristyl propionate (CTFA)

SYNONYMS:
   1-Tetradecanol, propanoate
EMPIRICAL FORMULA:
   $C_{17}H_{34}O_2$
STRUCTURE:

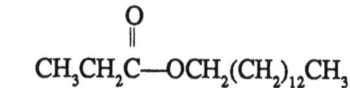

$$CH_3CH_2C—OCH_2(CH_2)_{12}CH_3$$

CAS No.:
   6221-95-0
   RD No.: 977062-14-8
TRADENAME EQUIVALENTS:
   Lonzest 143S [Lonza]
   Schercemol MP [Scher]
CATEGORY:
   Emollient, penetrant, spreading agent, solvent
APPLICATIONS:
   Bath products: bath oils (Lonzest 143S)
   Cleansers: face cleanser (Lonzest 143S); waterless hand cleanser (Lonzest 143S)
   Cosmetic industry preparations: (Lonzest 143S); body oils (Schercemol MP); creams

Myristyl propionate (cont'd.)

and lotions (Schercemol MP); makeup remover (Lonzest 143S); perfumery (Lonzest 143S); shaving preparations (Lonzest 143S)

Pharmaceutical applications: antiperspirant/deodorant (Lonzest 143S; Schercemol MP); sunscreens (Lonzest 143S)

PROPERTIES:

Form:

Clear liquid (Lonzest 143S; Schercemol MP)

Color:

Straw (Schercemol MP)

APHA 5–10 (Lonzest 143S)

Composition:

100% active (Lonzest 143S)

Solubility:

Sol. in acetone (Lonzest 143S)

Sol. in carbon tetrachloride (Lonzest 143S)

Sol. in castor oil (Lonzest 143S)

Sol. in cottonseed oil (Lonzest 143S)

Sol. in ethanol (Lonzest 143S)

Sol. in ethyl acetate (Lonzest 143S)

Sol. in heptane (Lonzest 143S)

Sol. in isopropanol (Lonzest 143S)

Sol. in lanolin oil (Lonzest 143S)

Sol. in peanut oil (Lonzest 143S)

Sol. in silicone oils (Lonzest 143S)

Sol. in toluene (Lonzest 143S)

Ionic Nature:

Nonionic (Lonzest 143S)

Sp.gr.:

0.852 (Lonzest 143S)

Acid No.:

2.0 max. (Schercemol MP)

Iodine No.:

Nil (Schercemol MP)

Saponification No.:

190–210 (Schercemol MP)

Myristyl stearate (CTFA)

SYNONYMS:

Ester of myristyl alcohol and stearic acid

Octadecanoic acid, tetradecyl ester

EMPIRICAL FORMULA:
$C_{32}H_{64}O_2$

STRUCTURE:

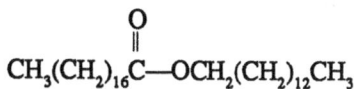

$$CH_3(CH_2)_{16}\overset{\overset{\displaystyle O}{\|}}{C}-OCH_2(CH_2)_{12}CH_3$$

CAS No.:
17661-50-6

TRADENAME EQUIVALENTS:
Cyclochem MST [Alcolac]
Falco S-14S [Fallek]
Hetester MS [Heterene]
Schercemol MS [Scher]

CATEGORY:
Emollient, softener, bodying agent, opacifier, pearlescent

APPLICATIONS:
Cosmetic industry preparations: (Falco S-14S); creams and lotions (Hetester MS; Schercemol MS); emulsions (Cyclochem MST)

PROPERTIES:

Form:
Flake (Cyclochem MST; Hetester MS)

Color:
White to off-white (Hetester MS)
Off-white (Cyclochem MST)

Solubility:
Sol. in min. oil (Cyclochem MST)
Insol. in water (Cyclochem MST)

M.P.:
43–47 C (Hetester MS)
45 C (Cyclochem MST; Schercemol MS)

Acid No.:
2.0 max. (Schercemol MS)
5.0 max. (Cyclochem MST; Hetester MS)

Iodine No.:
Nil (Schercemol MS)
1.0 max. (Hetester MS)

Saponification No.:
109–120 (Hetester MS)
110–120 (Schercemol MS)
115 (Cyclochem MST)

Myristyl trimethyl ammonium bromide

SYNONYMS:

Myrtrimonium bromide (CTFA)

1-Tetradecanaminium, N,N,N-trimethyl-, bromide

N,N,N-Trimethyl-1-, tetradecanaminium bromide

EMPIRICAL FORMULA:

$C_{17}H_{38}N \cdot Br$

STRUCTURE:

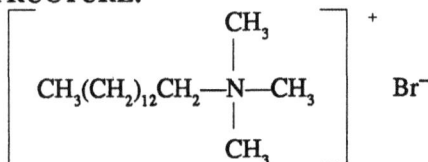

CAS No.:

1119-97-7

TRADENAME EQUIVALENTS:

Cycloton M214B/99, M214C/45, M214C/99 [Alcolac]

Mytab [Hexcel]

Querton 14Br, 14Br-40 [Berol Nobel]

CATEGORY:

Conditioner, softener, antistat, surfactant, emulsifier, antimicrobial

APPLICATIONS:

Cosmetic industry preparations: (Mytab); conditioners (Cycloton M214C/45, M214C/99); hair preparations (Mytab)

Pharmaceutical applications: topical preparations (Mytab)

PROPERTIES:

Form:

Liquid (Querton 14Br-40)

Flake (Cycloton M214B/99)

Powder (Mytab; Querton 14Br)

Color:

White (Mytab)

Odor:

Characteristic (Mytab)

Composition:

40% active (Querton 14Br-40)

45% active (Cycloton M214C/45)

96% active (Querton 14Br)

99% active (Cycloton M214B/99)

100% active (Mytab)

Solubility:

Very sol. in alcohol (Mytab)

Sol. in chloroform (Mytab)

Very sol. in water (Mytab)

Myristyl trimethyl ammonium bromide (cont'd.)

Ionic Nature:

Cationic (Cycloton M214B/99, M214C/45, M214C/99; Querton 14Br, 14Br-40)

M.W.:

336.40 (Mytab)

pH:

5.0–8.0 (1% aq.) (Mytab)

Octyl hydroxystearate (CTFA)

SYNONYMS:
2-Ethylhexyl oxystearate
Hydroxyoctadecanoic acid, 2-ethylhexyl ester
Octadecanoic acid, hydroxy-, 2-ethylhexyl ester
EMPIRICAL FORMULA:
$C_{26}H_{52}O_3$
STRUCTURE:

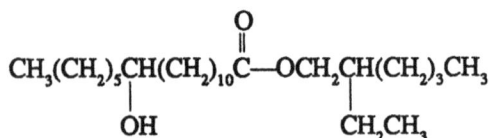

CAS No.:
29383-26-4; 29710-25-6
RD No. 977057-45-6
TRADENAME EQUIVALENTS:
Crodamol OHS [Croda Ltd.]
Naturechem OHS [CasChem]
Wickenol 171 [CasChem]
CATEGORY:
Emollient, lubricant, refatting agent, conditioner

APPLICATIONS:
Cleansers: (Naturechem OHS); soaps (Naturechem OHS)
Cosmetic industry preparations: (Crodamol OHS; Naturechem OHS; Wickenol 171)
Pharmaceutical applications: (Crodamol OHS); topical preparations (Wickenol 171)

PROPERTIES:
Form:
Liquid (Naturechem OHS)
Clear liquid (Crodamol OHS)
Color:
Light yellow (Naturechem OHS)
Composition:
100% active (Naturechem OHS)
Solubility:
Sol. in acetone (Wickenol 171)
Sol. in castor oil (Wickenol 171)

Octyl hydroxystearate *(cont'd.)*

Sol. in chloroform (Wickenol 171)
Sol. in corn oil (Wickenol 171)
Sol. in ethanol (Wickenol 171)
Sol. in ethyl acetate (Wickenol 171)
Sol. in min. oil (Wickenol 171)

Octyl palmitate (CTFA)

SYNONYMS:
2-Ethylhexyl hexadecanoate
2-Ethylhexyl palmitate
Hexadecanoic acid, 2-ethylhexyl ester
EMPIRICAL FORMULA:
$C_{24}H_{48}O_2$
STRUCTURE:

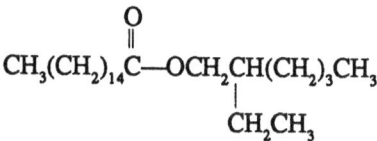

CAS No.:
29806-73-3
TRADENAME EQUIVALENTS:
Cegesoft C25 [Henkel KGaA]
Ceraphyl 368 [Van Dyk]
Crodamol OP [Croda; Croda Ltd.]
Kessco Octyl Palmitate [Stepan]
Lexol EHP [Inolex]
Radia 7129 [Synfina-Oleofin]
Schercemol OP [Scher]
Unimate EHP [Union Camp]
Wickenol 155 [CasChem]

CATEGORY:
Emollient, solubilizer, binder, gloss agent, antitack agent, lubricant, chemical intermediate, suspending agent, anticlogging agent
APPLICATIONS:
Bath products: bath oils (Lexol EHP)
Cosmetic industry preparations: (Lexol EHP; Schercemol OP; Wickenol 155); creams and lotions (Ceraphyl 368; Crodamol OP; Lexol EHP; Schercemol OP; Unimate EHP); emulsions (Cegesoft C25); hair preparations (Crodamol OP; Schercemol

OP); makeup (Ceraphyl 368; Crodamol OP); personal care products (Schercemol OP); pressed powders (Ceraphyl 368); shaving preparations (Schercemol OP); skin preparations (Crodamol OP; Kessco Octyl Palmitate; Unimate EHP)

Industrial applications: chemical synthesis (Radia 7129); lamination (Radia 7129); lubricating/cutting oils (Radia 7129); metalworking (Kessco Octyl Palmitate); rust inhibitors (Radia 7129); textile/leather processing (Kessco Octyl Palmitate; Radia 7129)

Pharmaceutical applications: (Crodamol OP; Lexol EHP; Wickenol 155); antiperspirant/deodorant (Ceraphyl 368; Crodamol OP; Lexol EHP; Schercemol OP); ointments (Crodamol OP); topical preparations (Lexol EHP; Wickenol 155)

PROPERTIES:

Form:
Liquid (Ceraphyl 368; Crodamol OP; Kessco Octyl Palmitate)
Clear liquid (Schercemol OP)
Solid (Cegesoft C25)

Color:
Colorless (Lexol EHP)
Water-white (Crodamol OP)
Gardner 2 max. (Schercemol OP)

Odor:
Odorless (Lexol EHP)
Bland (Schercemol OP)
Mild (Crodamol OP)

Composition:
98% ester content min. (Schercemol OP)
100% active (Kessco Octyl Palmitate)

Solubility:
Sol. in acetone (Lexol EHP)
Sol. in alcohols (Lexol EHP; Wickenol 155)
Sol. in animal oils (Wickenol 155)
Sol. in aromatic hydrocarbons (Schercemol OP)
Sol. in esters (Schercemol OP)
Sol. in 95% ethanol (Ceraphyl 368); sol. in 91% ethanol (Unimate EHP)
Disp. in glycerin (Schercemol OP); insol. (Wickenol 155)
Disp. in glycol ethers (Schercemol OP)
Disp. in glycols (Schercemol OP)
Sol. in isopropyl myristate (Ceraphyl 368)
Sol. in min. oil (Ceraphyl 368; Lexol EHP; Radia 7129; Schercemol OP; Unimate EHP; Wickenol 155)
Sol. in min. spirits (Schercemol OP)
Sol. in most organic solvents (Ceraphyl 368)
Insol. in propylene glycol (Ceraphyl 368; Wickenol 155)
Sol. in most solvents (Radia 7129); sol. in high hydrophobic solvents (Schercemol OP)

Sol. in veg. oil (Lexol EHP; Radia 7129; Wickenol 155)
Insol. in water (Ceraphyl 368; Wickenol 155)
M.W.:
357 (avg.) (Radia 7129)
368 (theoret.) (Schercemol OP)
Sp.gr.:
0.847 (37.8 C) (Radia 7129)
0.855 ± 0.01 (Schercemol OP)
Density:
7.12 lb/gal (Schercemol OP)
Visc.:
7.60 cps (37.8 C) (Radia 7129)
F.P.:
0 C max. (Schercemol OP)
Flash Pt.:
> 170 C (OC) (Schercemol OP)
206 C (COC) (Radia 7129)
Cloud Pt.:
−1 C (Radia 7129)
Acid No.:
3.0 max. (Schercemol OP)
Iodine No.:
Nil (Schercemol OP)
Saponification No.:
155 ± 8.0 (Schercemol OP)
Ref. Index:
1.4460 (Schercemol OP)
1.4467 (Radia 7129)
Surface Tension:
31 dynes/cm (Radia 7129)
STORAGE/HANDLING:
Low order of reaction with polystyrene packaging materials (Crodamol OP)

Octyl pelargonate (CTFA)

SYNONYMS:
2-Ethylhexyl pelargonate
Nonanoic acid, 2-ethylhexyl ester
EMPIRICAL FORMULA:
$C_{17}H_{34}O_2$

Octyl pelargonate *(cont'd.)*

STRUCTURE:

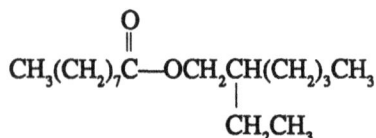

CAS No.:

59587-44-9

RD No.: 977012-73-9

TRADENAME EQUIVALENTS:

Schercemol OPG [Scher]

Wickenol 160 [CasChem]

CATEGORY:

Emollient, penetrant, anticlogging agent

APPLICATIONS:

Cosmetic industry preparations: (Schercemol OPG); makeup (Schercemol OPG); personal care products (Schercemol OPG); shaving preparations (Schercemol OPG); skin preparations (Schercemol OPG)

Pharmaceutical applications: antiperspirant/deodorant (Schercemol OPG)

PROPERTIES:

Form:

Clear liquid (Schercemol OPG)

Color:

Gardner 2 max. (Schercemol OPG)

Odor:

Mild (Schercemol OPG)

Composition:

99% ester content min. (Schercemol OPG)

Solubility:

Sol. in alcohols (Wickenol 160)

Sol. in animal oils (Wickenol 160)

Sol. in aromatic hydrocarbons (Schercemol OPG)

Sol. in esters (Schercemol OPG)

Disp. in glycerin (Schercemol OPG); insol. (Wickenol 160)

Disp. in glycol ethers (Schercemol OPG)

Disp. in glycols (Schercemol OPG)

Sol. in min. oil (Schercemol OPG; Wickenol 160)

Sol. in min. spirits (Schercemol OPG)

Insol. in propylene glycol (Wickenol 160)

Sol. in veg. oils (Wickenol 160)

Insol. in water (Wickenol 160)

M.W.:

270 (theoret.) (Schercemol OPG)

Sp.gr.:
0.857 ± 0.01 (Schercemol OPG)
Density:
7.13 lb/gal (Schercemol OPG)
F.P.:
< −10 C (Schercemol OPG)
Flash Pt.:
> 170 C (OC) (Schercemol OPG)
Acid No.:
1.0 max. (Schercemol OPG)
Iodine No.:
Nil (Schercemol OPG)
Saponification No.:
207 ± 7 (Schercemol OPG)
Ref. Index:
1.4363 ± 0.001 (Schercemol OPG)

Octyl stearate (CTFA)

SYNONYMS:
2-Ethylhexyl octadecanoate
2-Ethylhexyl stearate
Octadecanoic acid, 2-ethylhexyl ester
EMPIRICAL FORMULA:
$C_{26}H_{52}O_2$
STRUCTURE:

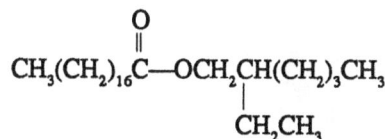

$$CH_3(CH_2)_{16}\overset{\overset{\displaystyle O}{\|}}{C}-OCH_2\underset{\underset{\displaystyle CH_2CH_3}{|}}{C}H(CH_2)_3CH_3$$

CAS No.:
22047-49-0
TRADENAME EQUIVALENTS:
Cetiol 868 [Henkel Canada]
Radia 7131 [Synfina-Oleofin]
Rilanit EHS [Henkel KGaA]
Wickenol 156 [CasChem]
CATEGORY:
Emollient, superfatting agent, solvent, lubricant, plasticizer

Octyl stearate *(cont'd.)*

APPLICATIONS:
Cosmetic industry preparations: (Radia 7131; Wickenol 156)
Industrial applications: ceramics (Rilanit EHS); lubricating/cutting oils (Rilanit EHS); metalworking (Rilanit EHS); plastics (Radia 7131); textile/leather processing (Rilanit EHS)
Pharmaceutical applications: topical preparations (Wickenol 156)

PROPERTIES:

Form:
Liquid (Cetiol 868; Radia 7131; Rilanit EHS)

Solubility:
Sol. in alcohols (Wickenol 156)
Sol. in animal oils (Wickenol 156)
Insol. in glycerin (Wickenol 156)
Sol. in min. oils (Wickenol 156)
Insol. in propylene glycol (Wickenol 156)
Sol. in veg. oils (Wickenol 156)
Insol. in water (Wickenol 156)

Oleamidopropyl dimethylamine *(CTFA)*

SYNONYMS:
N-[3-Dimethylamino) propyl]-9-octadecenamide
Dimethylaminopropyl oleamide
9-Octadecenamide, N-[3-(dimethylamino) propyl]-

EMPIRICAL FORMULA:
$C_{23}H_{46}N_2O$

STRUCTURE:

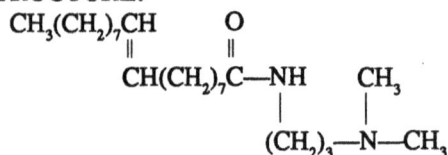

CAS No.:
109-28-4

TRADENAME EQUIVALENTS:
Incromine OPB, OPM [Croda]
Lexamine O13 [Inolex]
Mackine 501 [McIntyre]
Schercodine O [Scher]
Tego-Amid O 18 [Goldschmidt]

Oleamidopropyl dimethylamine (cont'd.)

CATEGORY:
Emollient, conditioner, lubricant, moisturizer, emulsifier
APPLICATIONS:
Cosmetic industry preparations: conditioners (Lexamine O13; Mackine 501); creams
and lotions (Lexamine O13; Tego-Amid O 18); hair preparations (Incromine OPB;
Lexamine O13; Mackine 501; Schercodine O; Tego-Amid O 18); personal care
products (Incromine OPM); shampoos (Lexamine O13); skin preparations (In-
cromine OPB; Schercodine O)
PROPERTIES:
Form:
Liquid (Lexamine O13; Mackine 501; Schercodine O)
Soft paste (Incromine OPB)
Soft solid (Incromine OPM)
Waxy (Tego-Amid O 18)
Color:
Amber (Mackine 501; Schercodine O)
Yellow (Incromine OPB)
Composition:
98% amide min. (Schercodine O)
100% active (Incromine OPB, OPM; Lexamine O13)
Solubility:
Sol. in acidic media (Lexamine O13)
Ionic Nature:
Nonionic/cationic (Incromine OPB, OPM)
Cationic (Lexamine O13; Tego-Amid O 18)
M.W.:
366 (Schercodine O)
HLB:
7.8 (Lexamine O13)
Acid No.:
2.0 max. (Incromine OPB)
Alkali No.:
150–160 (Schercodine O)
Equivalent Wt.:
380–400 (Incromine OPB)
TOXICITY/HANDLING:
Avoid prolonged contact with skin and eyes (Incromine OPB)
STORAGE/HANDLING:
Store in a cool, dry place (Incromine OPB)
STD. PKGS.:
55-gal (450 lb net) polyethylene-lined open-head drums (Incromine OPB)

Oleyl alcohol (CTFA)

SYNONYMS:
9-Octadecen-1-ol

EMPIRICAL FORMULA:
$C_{18}H_{36}O$

STRUCTURE:
$CH_3(CH_2)_7CH=CH(CH_2)_8OH$

CAS No.:
143-28-2

TRADENAME EQUIVALENTS:
Adol 80, 85 NF, 320, 330, 340 [Sherex]
Adol 90 NF [Sherex] (cosmetic/pharmaceutical grade)
Cachalot O-15 [M. Michel]
Fancol OA, OA 50, OA 70, OA 80, OA 90, OA 95 [Fanning]
HD Eutanol [Henkel]
HD-Ocenol [Henkel Canada]
HD-Ocenol 92/96 [Henkel; Henkel Canada]
Lipocol O [Lipo]
Novol [Croda]

CATEGORY:
Emollient, lubricant, stabilizer, plasticizer, coupling agent, coemulsifier, cosolvent, antifoam, conditioner, carrier, release agent, softener, tackifier, penetrant, spreading agent, chemical intermediate, solubilizer, superfatting agent, suspending agent

APPLICATIONS:
Bath products: bath oils (Adol 90 NF; Lipocol O; Novol)
Cleansers: cleansing creams and lotions (Novol)
Cosmetic industry preparations: (Adol 80, 85 NF; Cachalot O-15; Fancol OA 50, OA 70, OA 80, OA 90, OA 95; HD-Ocenol; Novol); conditioners (Novol); creams and lotions (Adol 90 NF; Lipocol O; Novol); hair preparations (Adol 90 NF; Novol); makeup (Adol 90 NF; Lipocol O; Novol); skin preparations (Novol)
Food applications: (Fancol OA 50, OA 70, OA 80, OA 90, OA 95)
Household detergents: (Fancol OA 50, OA 70, OA 80, OA 90, OA 95)
Industrial applications: adhesives (Fancol OA 50, OA 70, OA 80, OA 90, OA 95); aerosols (Adol 90 NF; Fancol OA 50, OA 70, OA 80, OA 90, OA 95); dyes and pigments (Fancol OA, OA 50, OA 70, OA 80, OA 90, OA 95; HD Eutanol); industrial processing (Adol 80, 85 NF, 90 NF); lubricating/cutting oils (Adol 80, 85 NF); metalworking (Fancol OA 50, OA 70, OA 80, OA 90, OA 95; paint mfg. (Fancol OA 50, OA 70, OA 80, OA 90, OA 95); paper mfg. (Adol 80, 85 NF; Fancol OA 50, OA 70, OA 80, OA 90, OA 95); petrochemicals (Fancol OA, OA 50, OA 70, OA 80, OA 90, OA 95; polymers/polymerization (Fancol OA 50, OA 70, OA 80, OA 90, OA 95; printing inks (Adol 80, 85 NF; Fancol OA 50, OA 70, OA 80, OA 90, OA 95; rubber (Fancol OA); textile/leather processing (Adol 80, 85 NF; Fancol OA 50, OA 70, OA 80, OA 90, OA 95); waxes (HD Eutanol)

Pharmaceutical applications: (Fancol OA 50, OA 70, OA 80, OA 90, OA 95); antiperspirant/deodorant (Novol); sunscreens (Novol)

PROPERTIES:

Form:

Liquid (Fancol OA, OA 50, OA 70, OA 80, OA 90, OA 95; Lipocol O)
Clear liquid (Adol 90 NF)
Clear, oily liquid (HD Eutanol; HD-Ocenol 92/96)
Oil (Novol)

Color:

Colorless (Adol 90 NF; HD Eutanol)
Gardner 1 max. (Novol)
Lovibond 3Y/0.3R (5¹/₄ in. max.) (Adol 85 NF)
Lovibond 5Y/0.5R (5¹/₄ in. max.) (Adol 80, 320, 330, 340)

Odor:

Low (Adol 90 NF)
Bland (HD Eutanol)
Mild (Novol)

Composition:

100% active (Fancol OA)

Solubility:

Sol. in acetone (Adol 80, 85 NF, 90 NF, 320, 330, 340; Cachalot O-15); (@ 25–50 C) (Fancol OA 50, OA 70, OA 80, OA 90, OA 95)
Sol. in alcohols (Cachalot O-15)
Sol. in aromatic hydrocarbons (Cachalot O-15)
Sol. in benzene (Adol 80, 85 NF, 90 NF, 320, 330, 340); (@ 25–50 C) (Fancol OA 50, OA 70, OA 80, OA 90, OA 95)
Sol. in carbon disulfide (Cachalot O-15)
Sol. in chloroform (Cachalot O-15)
Sol. in diethyl ether (@ 25–50 C) (Fancol OA 50, OA 70, OA 80, OA 90, OA 95)
Sol. in diglycol ethers (Cachalot O-15)
Sol. in ethanol (Adol 90 NF)
Sol. in ethyl ether (Adol 80, 85 NF, 90 NF, 320, 330, 340)
Miscible with fat, oil, and wax mixtures (Novol); miscible with most fats and waxes (Cachalot O-15)
Sol. in glycol ethers (Cachalot O-15)
Sol. in isopropanol (Adol 80, 85 NF, 90 NF, 320, 330, 340); (@ 25–50 C) (Fancol OA 50, OA 70, OA 80, OA 90, OA 95)
Sol. in kerosene (Adol 80, 85 NF, 90 NF, 320, 330, 340); (@ 25–50 C) (Fancol OA 50, OA 70, OA 80, OA 90, OA 95)
Sol. in light min. oil (Adol 80, 85 NF, 90 NF, 320, 330, 340); (@ 25–50 C) (Fancol OA 50, OA 70, OA 80, OA 90, OA 95); miscible with most min. oils (Cachalot O-15)
Sol. in VM&P naphtha (Adol 80, 85 NF, 90 NF, 320, 330, 340); (@ 25–50 C) (Fancol OA 50, OA 70, OA 80, OA 90, OA 95)

Oleyl alcohol (cont'd.)

Sol. in oils (HD-Ocenol; Lipocol O)

Sol. in trichlorethylene (Adol 80, 85 NF, 320, 330, 340); (@ 25–50 C) (Fancol OA 50, OA 70, OA 80, OA 90, OA 95)

Sol. in turpentine (Adol 80, 85 NF, 90 NF, 320, 330, 340); (@ 25–50 C) (Fancol OA 50, OA 70, OA 80, OA 90, OA 95)

Miscible with most veg. oils (Cachalot O-15)

Insol. in water (Lipocol O)

Ionic Nature:

Nonionic (Fancol OA)

M.W.:

261 (Adol 320, 330, 340)

263 (Adol 80)

267 (Adol 85 NF)

Sp.gr.:

0.840 (Adol 80, 85 NF, 90 NF)

0.845 (Adol 320, 330, 340)

0.845–0.855 (15 C) (Novol)

Visc.:

24–32 cps (Brookfield #1 spindle, 20 rpm) (Novol)

Cloud Pt.:

5 C (Adol 90 NF)

5 C max. (Fancol OA 95)

6 C max. (Fancol OA 90)

10 C max. (Adol 85 NF)

12–18 C max. (Fancol OA 80)

13 C max. (Adol 80)

18–25 C max. (Fancol OA 70)

19 C max. (Adol 320)

22 C max. (Adol 330)

28 C max. (Adol 340)

30–40 C max. (Fancol OA 50)

Acid No.:

0.1 max. (Novol)

0.2 max. (Fancol OA 50, OA 70, OA 80, OA 90, OA 95)

0.5 (Adol 90 NF)

1.0 max. (Adol 80, 85 NF, 320, 330, 340)

Iodine No.:

45–55 (Fancol OA 50)

60–70 (Adol 340)

70–80 (Fancol OA 70)

72–82 (Adol 330)

75–85 (Adol 320)

80–85 (Fancol OA 80)

82–92 (Adol 80)
85–95 (Adol 85 NF; Fancol OA 90)
88–98 (Novol)
90 (Adol 90 NF)
90–96 (Fancol OA 95)

Saponification No.:
0.3 max. (Novol)
1.0 max. (Fancol OA 50, OA 70, OA 80, OA 90, OA 95)
1.5 (Adol 90 NF)
2.0 max. (Adol 85 NF)
3.0 max. (Adol 80, 320, 330, 340)

Hydroxyl No.:
195–210 (Novol)
200–212 (Fancol OA 95)
205–215 (Adol 85 NF; Fancol OA 90)
205–220 (Fancol OA 70, OA 80)
205–222 (Adol 80)
205–225 (Adol 320, 330, 340)
210 (Adol 90 NF)
210–220 (Fancol OA 50)

Stability:
Not prone to rancidity; resistant to oxidation on aging (Novol)

STD. PKGS.:
55-gal steel drums, tankcars, tankwagons (Adol 90 NF)

Oleyl amide

SYNONYMS:
9-Octadecenamide
Oleamide (CTFA)

EMPIRICAL FORMULA:
$C_{18}H_{35}NO$

STRUCTURE:

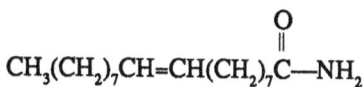

$$CH_3(CH_2)_7CH=CH(CH_2)_7 \overset{\overset{\textstyle O}{\|}}{C}-NH_2$$

CAS No.:
301-02-0

TRADENAME EQUIVALENTS:
Armid O [Akzo/Armak]
Armoslip CPM [Akzo/Noury]

Oleyl amide (cont'd.)

Crodamide O, OR [Croda Ltd.]
Kemamide O, U [Humko/Witco]
Petrac Slip-Eze [Synthetic Products]

CATEGORY:

Lubricant, release agent, slip agent, penetrant, stabilizer, dispersant, foam stabilizer, antiblock agent, foaming agent, water repellent, intermediate, dispersant, thickener, corrosion inhibitor, antitack agent

APPLICATIONS:

Cosmetic industry preparations: (Armid O)

Food applications: food packaging (Petrac Slip-Eze)

Household detergents: (Armid O; Kemamide O, U)

Industrial applications: (Kemamide O, U); asphalts (Kemamide O, U); coatings (Armid O); dyes and pigments (Armid O; Kemamide O, U); metalworking (Kemamide O, U); paint mfg. (Armid O; Kemamide O, U); paper mfg. (Armid O); petroleum industry (Kemamide O, U); plastics (Armid O; Armoslip CPM; Crodamide O, OR; Kemamide O, U; Petrac Slip-Eze); printing inks (Armid O; Kemamide O, U); resins (Kemamide O, U); textile/leather processing (Armid O; Kemamide O, U); water repellents (Kemamide O, U); waxes (Armid O; Kemamide O, U)

PROPERTIES:

Form:

Solid (Kemamide O)

Flake (Armid O; Armoslip CPM; Petrac Slip-Eze)

Pellets (Armid O; Armoslip CPM; Kemamide U; Petrac Slip-Eze)

Powder (Kemamide U; Petrac Slip-Eze)

Wax (Crodamide O)

Waxy solid (Kemamide U)

Color:

APHA 10 (Petrac Slip-Eze)

Gardner 5 max. (Kemamide U)

Gardner 7 (Armid O)

Odor:

Bland (Armid O)

Characteristic mild bland (Petrac Slip-Eze)

Composition:

90% active (Armid O)

95% amide min. (Kemamide U)

99.5% total amide (Petrac Slip-Eze)

100% active (Kemamide O)

Solubility:

Sol. 15 g/100 g in acetone with heating (Armid O)

Sol. 30 g/100 g in 95% ethanol with heating (Armid O)

Sol. 11 g/100 g in ethyl acetate with heating (Armid O)

Sol. 59 g/100 g 95% isopropanol with heating (Armid O); sol. > 30 g/100 g in isopropanol @ 30 C (Kemamide U)

Sol. > 20 g/100 g in MEK @ 30 C (Kemamide U)

Sol. 28 g/100 g in methanol @ 30 C (Kemamide U)

Sol. 11 g/100 g in MIBK with heating (Armid O)

Slightly sol. in organic solvents at room temp. (Kemamide U)

Sol. 25 g/100 g in toluene @ 30 C (Kemamide U)

Sol. 15 g/100 g in trichlorethylene with heating (Armid O)

Insol. in water (Armid O)

M.W.:

275 (Kemamide U)

Sp.gr.:

0.830 (100 C) (Armid O)

Density:

0.823 g/ml (130 C) (Kemamide U)

Visc.:

5.5 cP (130 C) (Kemamide U)

25 cps (Armid O)

M.P.:

68 C (Armid O)

68–78 C (Kemamide U)

73 C (Petrac Slip-Eze)

Flash Pt.:

207 C (Armid O)

245 C (COC) (Kemamide U)

Fire Pt.:

228 C (Armid O)

Acid No.:

4 max. (Kemamide U)

Iodine No.:

72–90 (Kemamide U)

pH:

Neutral (Kemamide U)

STD. PKGS.:

Cardboard cartons, 55-gal steel drums (Armid O)

Oleyl betaine (CTFA)

SYNONYMS:

N-(Carboxymethyl)-N,N-dimethyl-9-octadecen-1-aminium hydroxide, inner salt

9-Octadecen-1-aminium, N-(carboxymethyl-N,N-dimethyl-, hydroxide, inner salt

Oleyl dimethyl glycine

Oleyl betaine (cont'd.)

EMPIRICAL FORMULA:

$C_{22}H_{43}NO_2$

STRUCTURE:

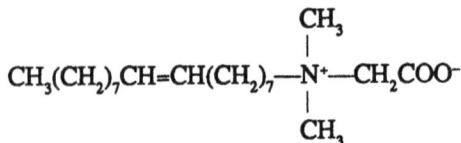

$$CH_3(CH_2)_7CH{=}CH(CH_2)_7{-}\overset{\overset{\displaystyle CH_3}{|}}{\underset{\underset{\displaystyle CH_3}{|}}{N^+}}{-}CH_2COO^-$$

CAS No.:

871-37-4

TRADENAME EQUIVALENTS:

Alkateric OB [Alkaril]

Mackam OB, OB-30 [McIntyre]

Mirataine ODMB-35 [Miranol]

Standapol OLB-30, OLB-50 [Henkel]

Velvetex OLB-30, OLB-50 [Henkel; Henkel Canada]

CATEGORY:

Conditioner, emollient, viscosity builder, foaming agent, wetting agent, detergent, processing aid, surfactant

APPLICATIONS:

Bath products: (Velvetex OLB-30, OLB-50)

Cleansers: (Velvetex OLB-30, OLB-50)

Cosmetic industry preparations: hair preparations (Mackam OB; Standapol OLB-30, OLB-50; Velvetex OLB-30, OLB-50); personal care products (Mirataine ODMB-35; Standapol OLB-30, OLB-50; Velvetex OLB-30, OLB-50); shampoos (Mirataine ODMB-35; Velvetex OLB-30, OLB-50); skin preparations (Mirataine ODMB-35; Standapol OLB-30, OLB-50; Velvetex OLB-30, OLB-50)

Household detergents: (Mackam OB-30)

Industrial applications: lubricating/cutting oils (Velvetex OLB-30, OLB-50); textile/leather processing (Velvetex OLB-30, OLB-50)

PROPERTIES:

Form:

Liquid (Mackam OB, OB-30)

Viscous liquid (Standapol OLB-30)

Gel (Alkateric OB; Mirataine ODMB-35; Standapol OLB-50; Velvetex OLB-50)

Color:

Light amber (Standapol OLB-50)

Translucent amber (Velvetex OLB-50)

Amber (Mackam OB-30; Standapol OLB-30)

Pale yellow (Mirataine ODMB-35)

Composition:

30% active (Mackam OB-30; Standapol OLB-30; Velvetex OLB-30)

30% active, 35% solids (Mirataine ODMB-35)

48–52% solids (Velvetex OLB-50)
50% active (Alkateric OB; Mackam OB; Standapol OLB-50)
Solubility:
Sol. in water (Velvetex OLB-50)
Ionic Nature:
Amphoteric (Alkateric OB; Mackam OB, OB-30; Mirataine ODMB-35; Standapol
OLB-30, OLB-50; Velvetex OLB-30, OLB-50)
Sp.gr.:
0.953 (60 C) (Velvetex OLB-50)
Density:
8.4 lb/gal (Mirataine ODMB-35)
pH:
6.0–8.0 (10%) (Velvetex OLB-50)
7.0 (10% aq. sol'n.) (Mirataine ODMB-35)

Oleyl dimethyl amine oxide

SYNONYMS:
N,N-Dimethyl-9-octadecen-1-amine-N-oxide
9-Octadecen-1-amine, N,N-dimethyl-, N-oxide
Oleamine oxide (CTFA)
Oleylamine oxide
EMPIRICAL FORMULA:
$C_{20}H_{41}NO$
STRUCTURE:

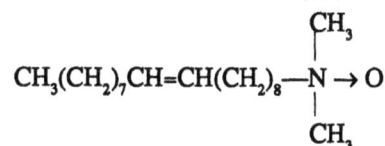

$$CH_3(CH_2)_7CH=CH(CH_2)_8 - \overset{\overset{\displaystyle CH_3}{|}}{\underset{\underset{\displaystyle CH_3}{|}}{N}} \rightarrow O$$

CAS No.:
14351-50-9; 61792-38-9; RD No. 977057-48-9
TRADENAME EQUIVALENTS:
Alkamox ODM [Alkaril]
Conco XA-O [Continental Chem.]
Incromine Oxide OD-50 [Croda]
Jordamox ODA [PPG-Mazer]
Mackamine O2 [McIntyre]
Ninox O [Stepan Europe]
Standamox O1 [Henkel; Henkel Canada]

Oleyl dimethyl amine oxide *(cont'd.)*

CATEGORY:
Conditioner, softener, lubricant, foaming agent, foam stabilizer, thickener, viscosity builder, wetting agent, detergency enhancer

APPLICATIONS:
Cosmetic industry preparations: (Incromine Oxide OD-50); conditioners (Jordamox ODA); hair preparations (Jordamox ODA)

Household detergents: (Incromine Oxide OD-50); liquid detergents (Jordamox ODA)

Industrial applications: (Standamox O1)

Industrial cleaners: janitorial cleaners (Incromine Oxide OD-50)

PROPERTIES:

Form:
Liquid (Alkamox ODM; Conco XA-O; Incromine Oxide OD-50; Jordamox ODA; Mackamine O2; Ninox O)

Clear liquid (Standamox O1)

Color:
Amber (Standamox O1)

Gardner 1 max. (Incromine Oxide OD-50)

Odor:
Mild, low (Jordamox ODA)

Composition:
30% active (Conco XA-O; Mackamine O2)

50% active (Alkamox ODM)

50–52% active (Jordamox ODA)

50–52% amine oxide, 26.5–29% ethanol, 17.5–20.5% water (Incromine Oxide OD-50)

55% active (Standamox O1)

Ionic Nature:
Cationic (Conco XA-O)

Nonionic (Incromine Oxide OD-50; Mackamine O2; Standamox O1)

Nonionic/cationic (Alkamox ODM)

pH:
7.0–8.0 (10%) (Incromine Oxide OD-50)

TOXICITY/HANDLING:
Avoid prolonged contact with skin and eyes (Incromine Oxide OD-50)

STORAGE/HANDLING:
Store in a cool, dry place (Incromine Oxide OD-50)

STD. PKGS.:
55-gal (425 lb net) lined drums (Incromine Oxide OD-50)

Oleyl dimethyl benzyl ammonium chloride

SYNONYMS:
Benzenemethanaminium, N,N-dimethyl-N-9-octadecenyl-, chloride
N,N-Dimethyl-N-9-octadecenylbenzenemethanaminium chloride
Olealkonium chloride (CTFA)

EMPIRICAL FORMULA:
$C_{27}H_{48}N \cdot Cl$

STRUCTURE:

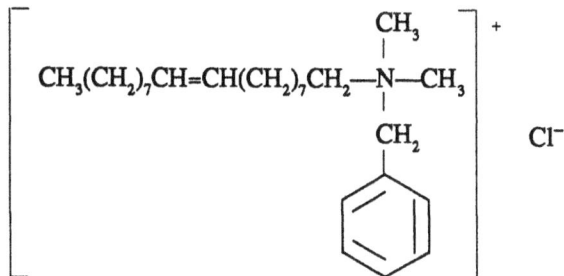

CAS No.:
37139-99-4

TRADENAME EQUIVALENTS:
Alacsan 7LUF [Alcolac]
Ammonyx KP [Millmaster-Onyx]
Empigen BCJ-50 [Albright & Wilson/Australia]
Jordaquat JO-50 [PPG-Mazer]

CATEGORY:
Conditioner, detergent, antistat, solubilizer

APPLICATIONS:
Cosmetic industry preparations: conditioners (Empigen BCJ-50); hair preparations (Ammonyx KP; Empigen BCJ-50; Jordaquat JO-50); shaving preparations (Alacsan 7LUF); toiletries (Alacsan 7LUF)
Household detergents: (Alacsan 7LUF)
Industrial applications: plastics (Alacsan 7LUF); textile/leather processing (Alacsan 7LUF)
Pharmaceutical applications: antiperspirant/deodorant (Alacsan 7LUF); medicated soaps (Alacsan 7LUF); protective hand creams (Alacsan 7LUF); vaginal preparations (Alacsan 7LUF)

PROPERTIES:
Form:
Liquid (Ammonyx KP)
Clear liquid (Empigen BCJ-50; Jordaquat JO-50)
Viscous liquid (Alacsan 7LUF)

213

Oleyl dimethyl benzyl ammonium chloride *(cont'd.)*

Color:
Yellow (Empigen BCJ-50; Jordaquat JO-50)
Gardner 5 max. (Alacsan 7LUF)
Composition:
50% active (Alacsan 7LUF; Jordaquat JO-50)
50% quaternary min. (Ammonyx KP)
$50 \pm 3\%$ quaternary active in water (Empigen BCJ-50)
Solubility:
Excellent sol. in water (Empigen BCJ-50)
Ionic Nature:
Cationic (Alacsan 7LUF; Ammonyx KP)
Sp.gr.:
0.977 (20 C) (Empigen BCJ-50)
0.98 (25/20 C) (Ammonyx KP)
Visc.:
450 cP (Brookfield, #2 spindle) (Empigen BCJ-50)
Flash Pt.:
> 200 F (Ammonyx KP)
Cloud Pt.:
< 0 C (Quick Cool) (Empigen BCJ-50)
pH:
4.0 ± 1.0 (1% aq. sol'n.) (Empigen BCJ-50)
TOXICITY/HANDLING:
May be irritating to skin and eyes on prolonged contact; spillages are slippery (Empigen BCJ-50)
STD. PKGS.:
200-kg net lined closed-head mild-steel drums (Empigen BCJ-50)

Oleyl erucate (CTFA)

SYNONYMS:
13-Docosenoic acid, 9-octadecenyl ester
Erucic acid, oleyl ester
9-Octadecenyl 13-docosenoate
EMPIRICAL FORMULA:
$C_{40}H_{76}O_2$
STRUCTURE:

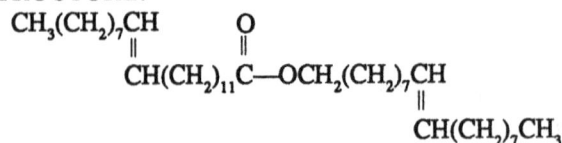

214

CAS No.:
17673-56-2
TRADENAME EQUIVALENTS:
Cetiol J 600 [Henkel Canada; Henkel KGaA]
Dynacerin 660 [Huls AG]
CATEGORY:
Emollient, fatting agent
APPLICATIONS:
Cosmetic industry preparations: (Dynacerin 660); creams and lotions (Dynacerin 660); makeup (Cetiol J 600; Dynacerin 660); personal care products (Cetiol J 600); skin preparations (Cetiol J 600; Dynacerin 660)
PROPERTIES:
Form:
Liquid wax (Cetiol J 600; Dynacerin 660)
Composition:
100% active (Cetiol J 600)
Solubility:
Sol. in fat solvents (Dynacerin 660)
Insol. in water (Dynacerin 660)

Oleyl monooleate

SYNONYMS:
9-Octadecenoic acid, 9-octadecenyl ester
Oleic acid, oleyl ester
Oleyl oleate (CTFA)
EMPIRICAL FORMULA:
$C_{36}H_{68}O_2$
STRUCTURE:

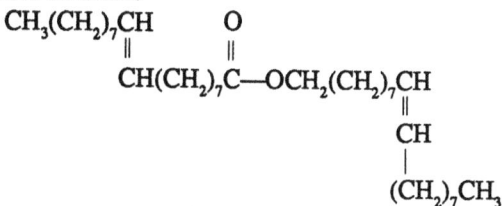

CAS No.:
3687-45-4
TRADENAME EQUIVALENTS:
Cetiol [Henkel Canada; Henkel KGaA]
Rilanit OLO [Henkel KGaA]

Oleyl monooleate *(cont'd.)*

Schercemol OLO [Scher]
Starfol OO [Sherex]
Wickenol 143 [CasChem]
CATEGORY:
Emollient, lubricant, fatting agent, cosolvent, solubilizer
APPLICATIONS:
Bath products: bath oils (Wickenol 143)
Cosmetic industry preparations: (Schercemol OLO; Wickenol 143); creams and lotions (Schercemol OLO; Starfol OO; Wickenol 143); hair preparations (Schercemol OLO); makeup (Wickenol 143); perfumery (Schercemol OLO); skin preparations (Schercemol OLO); toiletries (Wickenol 143)
Industrial applications: (Wickenol 143); ceramics (Rilanit OLO); lubricating/cutting oils (Rilanit OLO); metalworking (Rilanit OLO; Schercemol OLO); textile/leather processing (Rilanit OLO; Starfol OO)
Pharmaceutical applications: (Cetiol)
PROPERTIES:
Form:
Liquid (Cetiol; Rilanit OLO; Starfol OO)
Clear liquid (Schercemol OLO)
Color:
Amber (Schercemol OLO)
Odor:
Mild oleic (Schercemol OLO)
Solubility:
Sol. in alcohols (Schercemol OLO; Wickenol 143)
Sol. in animal oils (Wickenol 143)
Sol. in aromatic hydrocarbons (Schercemol OLO)
Sol. in ethers (Schercemol OLO)
Insol. in glycerin (Wickenol 143)
Sol. in high glycols (Schercemol OLO)
Sol. in ketones (Schercemol OLO)
Sol. in min. oil (Starfol OO)
Insol. in propylene glycol (Wickenol 143)
Sol. in veg. oil (Wickenol 143)
Insol. in water (Wickenol 143)
Sol. in white oils (Wickenol 143)

M.W.:
332 (Schercemol OLO)

Sp.gr.:
0.860 ± 0.01 (Schercemol OLO)

Density:
7.2 lb/gal (Schercemol OLO)

Flash Pt.:
  > 180 C (OC) (Schercemol OLO)
Cloud Pt.:
  13 C (Schercemol OLO)
Acid No.:
  1.0 max. (Schercemol OLO)
Saponification No.:
  105 ± 5 (Schercemol OLO)
Ref. Index:
  1.4630 ± 0.001 (Schercemol OLO)

Oleyl palmitamide (CTFA)

SYNONYMS:
  Hexadecanamide, N-9-octadecenyl-
  N-9-Octadecenyl hexadecanamide
EMPIRICAL FORMULA:
  $C_{34}H_{67}NO$
STRUCTURE:

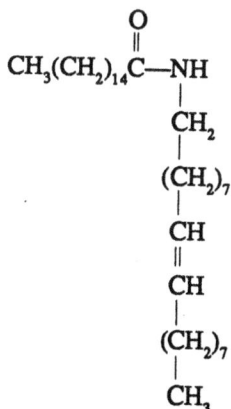

CAS No.:
  16260-09-6
TRADENAME EQUIVALENTS:
  HTSA #1 [Hexcel]
  Kemamide P-181 [Humko/Witco]
CATEGORY:
  Lubricant, release agent, slip agent, antiblocking agent, defoamer, water repellent,
    intermediate

Oleyl palmitamide *(cont'd.)*

APPLICATIONS:
Household products: (Kemamide P-181); detergents (Kemamide P-181)
Industrial applications: (Kemamide P-181); asphalts (Kemamide P-181); dyes and pigments (Kemamide P-181); lubricating/cutting oils (Kemamide P-181); metal-working (Kemamide P-181); paint mfg. (Kemamide P-181); petroleum products (Kemamide P-181); plastics (HTSA #1; Kemamide P-181); polishes and waxes (Kemamide P-181); printing inks (Kemamide P-181); textile/leather processing (Kemamide P-181); waxy products (Kemamide P-181)

PROPERTIES:
Form:
Free-flowing powder (Kemamide P-181)
Color:
White (Kemamide P-181)
Solubility:
Sol. < 50 g/100 g @ 50 C in dichlorethane (Kemamide P-181)
Sol. < 50 g/100 g @ 50 C in isopropanol (Kemamide P-181)
Sol. 50 g/100 g @ 50 C in MEK (Kemamide P-181)
Sol. 5 g/100 g @ 50 C in methanol (Kemamide P-181)
Sol. < 50 g/100 g @ 50 C in VM&P naphtha (Kemamide P-181)
Sol. < 50 g/100 g @ 50 C in toluene (Kemamide P-181)
Density:
0.8076 g/ml (110 C) (Kemamide P-181)
M.P.:
66–80 C (HTSA #1)
69–72 C (Kemamide P-181)
Flash Pt.:
262 C (CC) (Kemamide P-181)
Acid No.:
10 max. (Kemamide P-181)
Iodine No.:
43 (Kemamide P-181)
Amine No.:
10 max. (Kemamide P-181)
Sp. Heat:
0.45 Cal/g (55 C) (Kemamide P-181)
TOXICITY/HANDLING:
Nontoxic; handle powders to prevent eye irritation; wear protective goggles and gloves when handling (Kemamide P-181)
STORAGE/HANDLING:
Handle powders to prevent dust explosions; not classified as flammable, but will burn if ignited; keep away from strong oxidizing agents; should not remain at temps. higher than 175–185 C for long periods (Kemamide P-181)

Palmitamidopropyl dimethylamine (CTFA)

SYNONYMS:
N-[3-(Dimethylamino) propyl] hexadecanamide
Dimethylaminopropyl palmitamide
Hexadecanamide, N-[3-(dimethylamino) propyl]-

EMPIRICAL FORMULA:
$C_{21}H_{44}N_2O$

STRUCTURE:

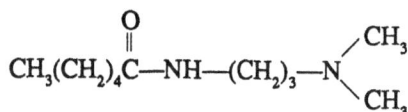

CAS No.:
39669-97-1

TRADENAME EQUIVALENTS:
Incromine PB [Croda]
Lexamine P-13 [Inolex]
Schercodine P [Scher]

CATEGORY:
Conditioner, emulsifier, intermediate, antistat

APPLICATIONS:
Cosmetic industry preparations: conditioners (Incromine PB; Lexamine P-13; Schercodine P); creams and lotions (Incromine PB; Lexamine P-13; Schercodine P);
shampoos (Incromine PB; Lexamine P-13)

PROPERTIES:
Form:
Flakes (Incromine PB)
Hard wax (Schercodine P)
Color:
Off-white (Incromine PB)
Tan (Schercodine P)
Composition:
98% amide min. (Schercodine P)
100% active (Incromine PB)
Solubility:
Sol. in acidic media (Lexamine P-13)
Sol. in organic solvents (Schercodine P)

Palmitamidopropyl dimethylamine *(cont'd.)*

Ionic Nature:
Nonionic/cationic (Incromine PB)
M.W.:
340 (Schercodine P)
340–350 (Incromine PB)
M.P.:
55–60 C (Schercodine P)
56–57 C (Incromine PB)
Acid No.:
4.0 max. (Incromine PB)
Alkali No.:
160–170 (Schercodine P)
TOXICITY/HANDLING:
Avoid prolonged contact with skin and eyes (Incromine PB)
STORAGE/HANDLING:
Store in a cool, dry place (Incromine PB)
STD. PKGS.:
44-gal (150 lb net) Leverpak with polyliner (Incromine PB)

Palmitic acid (CTFA)

SYNONYMS:
Cetylic acid
n-Hexadecanoic acid
EMPIRICAL FORMULA:
$C_{16}H_{32}O_2$
STRUCTURE:
$CH_3(CH_2)_{14}COOH$
CAS No.:
57-10-3
TRADENAME EQUIVALENTS:
Emersol 143 [Henkel/Emery]
Emersol Palmitic Acid [Henkel/Emery]
Hydrofol Acid 1690 [Sherex]
Hystrene 4516, 8016, 9016 [Humko/Witco]
Industrene 4516 [Humko/Witco]
CATEGORY:
Lubricant, opacifier, defoamer, emulsifier, plasticizer, intermediate
APPLICATIONS:
Cleansers: soaps (Emersol 143; Industrene 4516)

Palmitic acid (cont'd.)

Cosmetic industry preparations: (Emersol 143; Hystrene 4516, 8016, 9016); emulsions (Hydrofol Acid 1690); shampoos (Hystrene 4516, 8016, 9016)

Industrial applications: chemical specialties (Emersol 143); enamels (Hystrene 4516, 8016, 9016); lubricating oils/greases/cutting oils (Hystrene 4516, 8016, 9016; Industrene 4516); metallic soaps (Hystrene 4516, 8016, 9016); paper mfg. (Emersol Palmitic Acid); polishes and waxes (Industrene 4516); resins (Industrene 4516); rubber (Industrene 4516); textile/leather processing (Hystrene 4516, 8016, 9016); water repellents (Industrene 4516)

Pharmaceutical applications: (Hystrene 4516, 8016, 9016)

PROPERTIES:

Form:

Solid (Hystrene 4516, 8016, 9016; Industrene 4516)

Color:

Lovibond 2Y 0.5R max. (Hydrofol Acid 1690)

Composition:

45% palmitic acid; 100% conc. (Hystrene 4516; Industrene 4516)

80% palmitic acid; 100% conc. (Hystrene 8016)

90% palmitic acid (Hydrofol Acid 1690)

90% palmitic acid; 100% conc. (Hystrene 9016)

Acid No.:

206–211 (Hystrene 4516)

213–217 (Hystrene 8016)

216–220 (Hystrene 9016)

Saponification No.:

206–212 (Industrene 4516)

207–212 (Hystrene 4516)

214–218 (Hystrene 8016)

216–222 (Hydrofol Acid 1690)

217–221 (Hystrene 9016)

Pentaerythritol tetrabehenate (CTFA)

SYNONYMS:

2,2-Bis [[(1-oxodocosyl) oxy] methyl]-1,3-propanediyl docosanoate

Docosanoic acid, 2,2-bis [[(1-oxodocosyl) oxy] methyl]-1,3-propanediyl ester

Pentaerythrityl tetrabehenate

EMPIRICAL FORMULA:

$C_{93}H_{180}O_8$

Pentaerythritol tetrabehenate *(cont'd.)*

STRUCTURE:

$$
\begin{array}{c}
\quad\quad\quad\quad \overset{\displaystyle O}{\overset{\|}{O\!-\!CR}} \\
\quad\quad\quad\quad | \\
\overset{\displaystyle O}{\overset{\|}{}}\quad\quad\quad CH_2 \quad\quad \overset{\displaystyle O}{\overset{\|}{}} \\
RC\!-\!OCH_2\!-\!C\!-\!CH_2O\!-\!CR \\
\quad\quad\quad\quad | \\
\quad\quad\quad\quad CH_2 \quad O \\
\quad\quad\quad\quad | \quad\quad\;\; \| \\
\quad\quad\quad\quad O\!\!-\!\!-\!\!-\!CR
\end{array}
$$

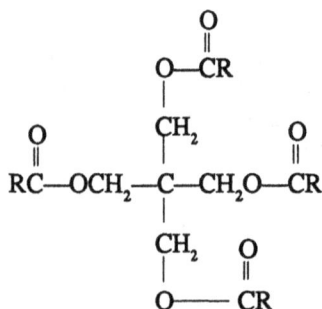

where RCO⁻ represents the behenic acid radical

CAS No.:
61682-73-3

TRADENAME EQUIVALENTS:
Liponate PB-4 [Lipo]
Radia 7514 [Synfina-Oleofin]

CATEGORY:
Emollient, lubricant, thickener, viscosity control agent, chemical intermediate

APPLICATIONS:
Bath products: (Liponate PB-4)
Cosmetic industry preparations: creams and lotions (Liponate PB-4); makeup (Liponate PB-4)
Industrial applications: chemical synthesis (Radia 7514); corrosion inhibitors (Radia 7514); lamination (Radia 7514); lubricating/cutting oils (Radia 7514); textile/leather processing (Radia 7514)

PROPERTIES:

Form:
Flakes (Liponate PB-4)

Color:
Off-white (Liponate PB-4)

Solubility:
Sol. cloudy @ 75 C and 10% in min. oil (Radia 7514)
Miscible in oils (Liponate PB-4)
Sol. cloudy @ 75 C and 10% in trichlorethylene (Radia 7514)
Insol. in water (Liponate PB-4)

M.W.:
1300 (avg.) (Radia 7514)

Sp.gr.:
0.860 (98.9 C) (Radia 7514)

Visc.:
17.20 cps (98.9 C) (Radia 7514)

M.P.:
 73.5 C (Radia 7514)
Flash Pt.:
 289 C (COC) (Radia 7514)
Acid No.:
 10 max. (Liponate PB-4)
Saponification No.:
 159–169 (Liponate PB-4)

Pentaerythritol tetraoleate (CTFA)

SYNONYMS:
 9-Octadecenoic acid, 2,2-bis [[(1-oxo-9-octadecenyl) oxy] methyl]-1,3-propanediyl
 ester
 Pentaerythrityl tetraoleate
EMPIRICAL FORMULA:
 $C_{77}H_{140}O_8$
STRUCTURE:

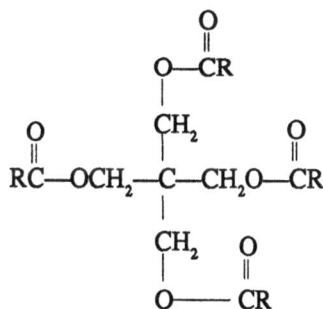

 where R represents the oleic acid radical

CAS No.:
 19321-40-5
TRADENAME EQUIVALENTS:
 Cyclochem PETO [Alcolac]
 Liponate PO-4 [Lipo]
 Radia 7171 [Synfina-Oleofin]
CATEGORY:
 Lubricant, thickener, viscosity controller, emollient, chemical intermediate, gloss
 agent

Pentaerythritol tetraoleate *(cont'd.)*

APPLICATIONS:

Cosmetic industry preparations: makeup (Cyclochem PETO); personal care products (Liponate PO-4)

Industrial applications: chemical synthesis (Radia 7171); corrosion inhibitors (Radia 7171); lamination (Radia 7171); lubricating/cutting oils (Radia 7171); textile/leather processing (Radia 7171)

PROPERTIES:

Form:

Liquid (Cyclochem PETO; Liponate PO-4)

Color:

Yellow (Cyclochem PETO; Liponate PO-4)

Solubility:

Sol. cloudy in benzene @ 10% (Radia 7171)

Sol. cloudy in hexane @ 10% (Radia 7171)

Sol. cloudy in isopropanol @ 10% (Radia 7171)

Sol. in min. oil @ 10% (Radia 7171)

Sol. in trichlorethylene @ 10% (Radia 7171)

Sol. cloudy in veg. oil @ 10% (Radia 7171)

M.W.:

1160 (avg.) (Radia 7171)

Sp.gr.:

0.913 (37.8 C) (Radia 7171)

Visc.:

70.80 cps (37.8 C) (Radia 7171)

M.P.:

< 0 C (Cyclochem PETO)

Flash Pt.:

302 C (COC) (Radia 7171)

Cloud Pt.:

–9.5 C (Radia 7171)

Acid No.:

5.0 max. (Cyclochem PETO)

10 max. (Liponate PO-4)

Saponification No.:

185–195 (Liponate PO-4)

187 (Cyclochem PETO)

Pentaerythritol tetrapelargonate

CAS No.:
14450-05-6
TRADENAME EQUIVALENTS:
Emerest 2485, 2486 [Henkel/Emery]
CATEGORY:
Lubricant, emollient, modifier
APPLICATIONS:
Cosmetic industry preparations: creams and lotions (Emerest 2486); makeup (Emerest 2486); shaving preparations (Emerest 2486); skin preparations (Emerest 2486)
Industrial applications: lubricating/cutting oils (Emerest 2485); metalworking (Emerest 2485); textile/leather processing (Emerest 2485)
PROPERTIES:
Form:
Liquid (Emerest 2485, 2486)
Color:
Gardner 1 (Emerest 2485)
Gardner 2 (Emerest 2486)
Composition:
100% active (Emerest 2486)
Solubility:
Disp. in glycerin @ 5% (Emerest 2486)
Sol. in isopropanol @ 5% (Emerest 2485, 2486)
Sol. in isopropyl myristate @ 5% (Emerest 2486)
Sol. in min. oil @ 5% (Emerest 2485, 2486)
Sol. in toluene @ 5% (Emerest 2485)
Sol. in triolein @ 5% (Emerest 2486)
Insol. in water @ 5% (Emerest 2485, 2486)
Sol. in xylene @ 5% (Emerest 2485)
Density:
8.0 lb/gal (Emerest 2485, 2486)
Visc.:
34 cSt (100 F) (Emerest 2486)
35 cSt (100 F) (Emerest 2485)
Pour Pt.:
10 C (Emerest 2485, 2486)
Flash Pt.:
550 F (Emerest 2486)
555 F (Emerest 2485)
HLB:
3.3 (Emerest 2486)

Pentaerythritol tetrastearate (CTFA)

SYNONYMS:

Octadecanoic acid, 2,2-bis [[(1-oxooctadecyl) oxy] methyl]-1,2-propanediyl ester

Pentaerythrityl tetrastearate

EMPIRICAL FORMULA:

$C_{77}H_{148}O_8$

STRUCTURE:

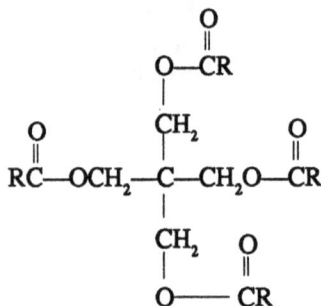

where R represents the stearic acid radical

CAS No.:

115-83-3

TRADENAME EQUIVALENTS:

Cyclochem PETS [Alcolac]

Liponate PS-4 [Lipo]

Radia 7176 [Synfina-Oleofin]

CATEGORY:

Emollient, lubricant, thickener, viscosity control agent, chemical intermediate, heat
stabilizer

APPLICATIONS:

Cosmetic industry preparations: (Cyclochem PETS); personal care products (Liponate
PS-4)

Industrial applications: adhesives (Cyclochem PETS); chemical synthesis (Radia
7176); coatings (Cyclochem PETS); corrosion inhibitors (Radia 7176); lubricating/
cutting oils (Cyclochem PETS; Radia 7176); polishes and waxes (Cyclochem
PETS); textile/leather processing (Radia 7176)

PROPERTIES:

Form:

Flake (Cyclochem PETS; Liponate PS-4)

Color:

White (Liponate PS-4)

Off-white (Cyclochem PETS)

Solubility:

Sol. in trichlorethylene @ 10% (Radia 7176)

Pentaerythritol tetrastearate *(cont'd.)*

Ionic Nature:
  Nonionic (Cyclochem PETS)
M.W.:
  1140 (avg.) (Radia 7176)
Sp.gr.:
  0.868 (98.9 C) (Radia 7176)
Visc.:
  14.50 cps (98.9 C) (Radia 7176)
M.P.:
  57 C (Cyclochem PETS)
  64.5 C (Radia 7176)
Flash Pt.:
  290 C (Radia 7176)
Acid No.:
  9.0 max. (Cyclochem PETS)
  10 max. (Liponate PS-4)
Saponification No.:
  183–198 (Liponate PS-4)
  190 (Cyclochem PETS)

Petrolatum *(CTFA)*

SYNONYMS:
  Petrolatum amber
  Petrolatum white
  Petroleum jelly
CAS No.:
  8009-03-8 (NF); 8027-32-5 (USP)
TRADENAME EQUIVALENTS:
  Chem I Sorb CSD 1035 [Polychem Dispersions]
  Dark Green No. 2 [Witco] (tech.)
  Fonoline White, Yellow [Witco] (USP)
  Mineral Jelly No. 5, No. 10, No. 15, No. 20, No. 25 [Penreco]
  Ointment Base No. 3, 4, 6 [Penreco] (white USP)
  Penreco Amber, Blond, Royal [Penreco] (USP)
  Penreco Cream, Frost, Lily, Regent, Snow, Super, Ultima [Penreco] (white USP)
  Penreco Green, Red [Penreco] (tech.)
  Perfecta [Witco]
  Perfecta USP [Witco] (cosmetic/pharmaceutical grade)
  Petrolatum DLC (also avail. tech. grade) [Harwick; Natrochem]

Petrolatum (cont'd.)

Petrolatum Dry Liquid Conc. [Polymerics]
Petrolatum RPB [Witco] (tech.)
Petrolatum SR-172 [C.P. Hall]
Polytrap 210 (polymer) [Dow Corning/Wickhen]
Protopet Alba, White 1S, White 2L, White 3L, Yellow 1E, Yellow 2A [Witco] (USP)
Sonojell No. 4, No. 9 [Witco]
Tech Pet F, M [Witco] (tech.)

CATEGORY:

Emollient, lubricant, protective coating, binder, conditioner, base, carrier, water repellent, moisture barrier, waterproofing agent, plasticizer, protective agent, softener, solvent, dispersant aid, release aid, flow aid, detackifier

APPLICATIONS:

Cleansers: cleansing/moisturizing creams (Penreco Snow); cold creams (Penreco Frost)

Cosmetic industry preparations: (Mineral Jelly No. 5, No. 10, No. 15, No. 20, No. 25; Penreco Ambe, Blond, Cream, Lily, Regent, Royal, Snow, Super, Ultima; Perfecta USP; Protopet Alba, White 1S, White 2L, White 3L, Yellow 1E, Yellow 2A); creams and lotions (Penreco Amber, Blond, Cream, Frost, Lily, Regent, Royal, Snow, Super, Ultima); hair preparations (Penreco Amber, Blond, Cream, Frost, Lily, Regent, Royal, Snow, Super, Ultima); makeup (Penreco Snow)

Food applications: (Penreco Amber, Blond, Cream, Frost, Lily, Regent, Royal, Snow, Super, Ultima); food production machinery (Penreco Snow)

Industrial applications: (Fonoline White, Yellow; Penreco Green; Protopet Alba, White 1S, White 2L, White 3L, Yellow 1E, Yellow 2A); adhesives (Penreco Snow); flux, soldering paste (Penreco Amber); paper mfg. (Penreco Amber, Blond, Cream, Frost, Lily, Regent, Royal, Snow, Super, Ultima); polishes and waxes (Penreco Amber, Blond, Red); printing inks (Penreco Amber); putty (Penreco Regent); rubber (Petrolatum DLC; Petrolatum Dry Liquid Conc.; Petrolatum SR-172); textile/leather processing (Penreco Amber, Blond, Cream, Frost, Lily, Regent, Royal, Snow, Super, Ultima)

Pharmaceutical applications: (Fonoline White, Yellow; Penreco Amber, Blond, Cream, Frost, Lily, Regent, Royal, Snow, Super, Ultima; Perfecta USP; Protopet Alba, White 1S, White 2L, White 3L, Yellow 1E, Yellow 2A); medications (Ointment Base No. 3, 4, 6); ointments (Fonoline White, Yellow; Ointment Base No. 3, 4, 6; Penreco Amber, Blond, Cream, Frost, Lily, Regent, Royal, Snow, Super, Ultima; Protopet Alba, White 1S, White 2L, White 3L, Yellow 1E, Yellow 2A); sunscreens (Penreco Snow)

PROPERTIES:

Form:

Powder (Polytrap 210)
Dustless powder (Petrolatum Dry Liquid Conc.)
Free-flowing powder (Chem I Sorb CSD 1035; Petrolatum DLC)
Soft solid (Petrolatum SR-172; Protopet Yellow 1E)

Color:
Off-white (Petrolatum DLC; Petrolatum Dry Liquid Conc.)
Dark green (Dark Green No. 2; Penreco Green)
Red (Penreco Red)
Dark brown (Petrolatum RPB; Tech Pet F, M)
Lovibond 0.3Y (Perfecta USP)
Lovibond 0.5Y (Sonojell No. 4, No. 9); (2 in. cell) (Penreco Super, Ultima)
Lovibond 1.0Y (Protopet Alba)
Lovibond 1.5Y (Protopet White 1S); (2 in. cell) (Penreco Frost)
Lovibond 1.7Y (Fonoline White)
Lovibond 2Y (2 in. cell) (Mineral Jelly No. 5; Penreco Regent, Snow)
Lovibond 8Y 0.5R (2 in. cell) (Penreco Lily)
Lovibond 8Y 0.6R (Protopet White 2L)
Lovibond 18Y 0.5R (2 in. cell) (Penreco Cream)
Lovibond 25Y 1.0R (Protopet White 3C)
Lovibond 30Y 2.5R (Fonoline Yellow; Protopet Yellow 2A)
Lovibond 35Y 3R (2 in. cell) (Penreco Blond)
Lovibond 35Y 7R (2 in. cell) (Penreco Amber, Royal)
Odor:
Odorless (Mineral Jelly No. 5)
Mild (Petrolatum SR-172)
Taste:
Tasteless (Mineral Jelly No. 5)
Composition:
72% petrolatum on inert silicate carrier (Petrolatum DLC)
72% petrolatum on Microcel E (Chem I Sorb CSD 1035)
72% petrolatum in a microcel binder (Petrolatum Dry Liquid Conc.)
Solubility:
Miscible with most animal fats, oils, and waxes (Perfecta)
Miscible with cosmetic ingredients used in oil-based formulations (Mineral Jelly No.
    5, No. 10, No. 15, No. 20, No. 25)
Miscible with many essential oils (Perfecta)
Miscible with all petroleum products (Perfecta)
Miscible with most vegetable fats, oils, and waxes (Perfecta)
Sp.gr.:
1.04 (Petrolatum DLC; Petrolatum Dry Liquid Conc.)
1.08 (Chem I Sorb CSD 1035)
Visc.:
2.6–5.7 cSt (100 C) (Sonojell No. 4, No. 9)
9–14 cSt (100 C) (Fonoline White, Yellow; Perfecta USP)
10–16 cSt (100 C) (Protopet Alba, White 1S, White 2L, White 3C, Yellow 2A)
37–40 SUS (210 F) (Mineral Jelly No. 20)
38–40 SUS (210 F) (Mineral Jelly No. 25)

Petrolatum *(cont'd.)*

38–43 SUS (210 F) (Mineral Jelly No. 5)
40–43 SUS (210 F) (Mineral Jelly No. 10)
40–44 SUS (210 F) (Mineral Jelly No. 15)
55–65 SUS (210 F) (Ointment Base No. 3)
55–75 SUS (210 F) (Perfecta)
57–70 SUS (210 F) (Penreco Regent, Royal)
60–70 SUS (210 F) (Ointment Base No. 4, 6; Penreco Ultima)
60–75 SUS (210 F) (Penreco Frost, Super)
60–125 SUS (210 F) (Penreco Green)
64-75 SUS (210 F) (Penreco Cream, Lily, Snow)
68–82 SUS (210 F) (Penreco Amber, Blond)
70–82 SUS (210 F) (Penreco Red)

M.P.:

38–52 C (Sonojell No. 4)
42–49 C (Sonojell No. 9)
51–65 C (Tech Pet M)
53–58 C (Fonoline White, Yellow)
54–60 C (Protopet Alba, White 1S, White 2L, White 3C, Yellow 2A)
57–60 C (Perfecta USP)
57–66 C (Tech Pet F)
71–77 C (Petrolatum RPB)
60–68 C (Dark Green No. 2)
97–105 F (Saybolt) (Mineral Jelly No. 10)
97–108 F (Saybolt) (Mineral Jelly No. 15)
103–108 F (Saybolt) (Mineral Jelly No. 25)
111–116 F (Saybolt) (Mineral Jelly No. 20)
115–140 F (Penreco Green)
118–125 F (Ointment Base No. 3, 4)
118–130 F (Penreco Regent, Royal)
120–135 F (Penreco Red)
122–133 F (Ointment Base No. 6)
122–135 F (Penreco Amber, Blond, Cream, Lily, Snow, Super)
125–135 F (Penreco Frost)
130–140 F (Penreco Ultima; Protopet Yellow 1E)
135–140 F (Perfecta)

Pour Pt.:

20 F (Fonoline White, Yellow)
75–85 F (Mineral Jelly No. 5)
95–105 F (Mineral Jelly No. 10, 15)
100–110 F (Mineral Jelly No. 25)
110–120 F (Mineral Jelly No. 20)

Congeal Pt.:

104–115 F (Ointment Base No. 3)

230

109–119 F (Ointment Base No. 4)
118 F (Penreco Royal)
120 F (Penreco Regent)
120–130 F (Ointment Base No. 6)
123 F (Penreco Amber, Blond, Snow)
124 F (Penreco Frost, Lily)
125 F (Penreco Cream, Super)
130 F (Penreco Ultima)

Solidification Pt.:
105-115 F (Ointment Base No. 4)
114-124 F (Ointment Base No. 6)
115 F (Penreco Royal)
119 F (Penreco Regent)
121 F (Penreco Snow)
122 F (Penreco Amber, Blond, Cream)
123 F (Penreco Frost, Lily)
124 F (Penreco Super)
128 F (Penreco Ultima)

STD. PKGS.:
Bags (Petrolatum DLC)
Drums (Petrolatum SR-172)

POE (2) castor oil

SYNONYMS:
PEG-2 castor oil (CTFA)
PEG 100 castor oil

CAS No.:
61791-12-6

TRADENAME EQUIVALENTS:
Hetoxide C2 [Heterene]

CATEGORY:
Emollient, emulsifier, solubilizer, viscosity control agent, lubricant, dispersant, scouring agent, dyeing assistant, carrier, intermediate

APPLICATIONS:
Cosmetic industry preparations: (Hetoxide C2); creams and lotions (Hetoxide C2); perfumery (Hetoxide C2)
Household detergents: (Hetoxide C2)
Industrial applications: dyes and pigments (Hetoxide C2); metal treating/plating (Hetoxide C2)

POE (2) castor oil (cont'd.)

Industrial cleaners: textile scouring (Hetoxide C2)
PROPERTIES:
Form:
Liquid @ 30 C (Hetoxide C2)
Solubility:
Sol. in isopropanol (Hetoxide C2)
Sol. in min. oil (Hetoxide C2)
Ionic Nature:
Nonionic (Hetoxide C2)
HLB:
4.3 (Hetoxide C2)
Acid No.:
1.0 (Hetoxide C2)
Saponification No.:
155–170 (Hetoxide C2)

POE (20) castor oil

SYNONYMS:
PEG-20 castor oil (CTFA)
PEG 1000 castor oil
CAS No.:
61791-12-6 (generic)
TRADENAME EQUIVALENTS:
Alkasurf CO-20 [Alkaril]
Etocas 20 [Croda Ltd.]
Merpoxen RO200 [Kempen]
Nikkol CO-20TX [Nikko Ltd.]
Nopalcol 10-CO [Henkel/Process]
CATEGORY:
Lubricant, emulsifier, hydrotrope, degreaser, antistat, pigment dispersant, wetting agent, surfactant
APPLICATIONS:
Cosmetic industry preparations: (Nikkol CO-20TX)
Industrial applications: dyes and pigments (Alkasurf CO-20); metalworking (Etocas 20); textile/leather processing (Alkasurf CO-20; Etocas 20; Merpoxen RO200); waxes (Alkasurf CO-20)
Pharmaceutical applications: (Nikkol CO-20TX)

PROPERTIES:
Form:
   Liquid (Alkasurf CO-20; Etocas 20; Nikkol CO-20TX; Nopalcol 10-CO)
   Viscous liquid (Merpoxen RO200)
Color:
   Yellow (Merpoxen RO200)
   Gardner 4 (Alkasurf CO-20)
Composition:
   97% active (Etocas 20)
   99% active (Nopalcol 10-CO)
   100% active (Merpoxen RO200; Nikkol CO-20TX)
Solubility:
   Sol. in aromatic hydrocarbons @ 10% (Alkasurf CO-20)
   Sol. in perchloroethylene @ 10% (Alkasurf CO-20)
   Sol. in water @ 10% (Alkasurf CO-20)
Ionic Nature:
   Nonionic (Alkasurf CO-20; Etocas 20; Merpoxen RO200; Nikkol CO-20TX; Nopalcol 10-CO)
Density:
   1.02 g/ml (Alkasurf CO-20)
Cloud Pt.:
   65–69 C (10% in 25% butyl Carbitol) (Alkasurf CO-20)
HLB:
   10.3 (Alkasurf CO-20; Nopalcol 10-CO)
   101.5 (Nikkol CO-20TX)
Saponification No.:
   91–96 (Alkasurf CO-20)

POE (60) castor oil

SYNONYMS:
   PEG-60 castor oil (CTFA)
   PEG (60) castor oil
CAS No.:
   61791-12-6 (generic)
   RD No.: 977065-79-4
TRADENAME EQUIVALENTS:
   Etocas 60 [Croda Ltd.]
   Incrocas 60 [Croda]
   Nikkol CO-60TX [Nikko]

POE (60) castor oil (cont'd.)

CATEGORY:
  Emollient, emulsifier, solubilizer, hydrotrope, superfatting agent, lubricant, antistat, softener, detergent, leveling agent

APPLICATIONS:
  Bath products: bubble bath (Etocas 60)
  Cosmetic industry preparations: (Etocas 60; Nikkol CO-60TX); perfumery (Etocas 60); personal care products (Incrocas 60); shampoos (Etocas 60); skin preparations (Etocas 60)
  Farm products: herbicides (Etocas 60; Incrocas 60); insecticides/pesticides (Etocas 60; Incrocas 60)
  Household products: (Etocas 60; Incrocas 60)
  Industrial applications: dyes and pigments (Etocas 60); metalworking (Etocas 60; Incrocas 60); polymers/polymerization (Etocas 60); textile/leather processing (Etocas 60; Incrocas 60)
  Pharmaceutical applications: (Nikkol CO-60TX)

PROPERTIES:

Form:
  Soft paste (Etocas 60; Incrocas 60)
  Paste (Nikkol CO-60TX)

Color:
  Pale yellow (Etocas 60; Incrocas 60)

Composition:
  100% active (Incrocas 60; Nikkol CO-60TX)

Solubility:
  Partly sol. in butyl stearate (Etocas 60)
  Sol. in ethanol (Etocas 60; Incrocas 60)
  Sol. in MEK (Etocas 60; Incrocas 60)
  Sol. in naphtha (Etocas 60; Incrocas 60)
  Sol. in oleic acid (Etocas 60; Incrocas 60)
  Sol. in oleyl alcohol (Incrocas 60); partly sol. (Etocas 60)
  Sol. in trichlorethylene (Etocas 60; Incrocas 60)
  Sol. in water (Etocas 60; Incrocas 60)

Ionic Nature:
  Nonionic (Etocas 60; Incrocas 60; Nikkol CO-60TX)

Cloud Pt.:
  60 C (Etocas 60); (1% in brine) (Incrocas 60)

HLB:
  14.0 (Nikkol CO-60TX)
  14.7 (Etocas 60; Incrocas 60)

Acid No.:
  1.0 max. (Etocas 60)

Iodine No.:
  22–26 (Etocas 60)

POE (60) castor oil (cont'd.)

Saponification No.:
45–50 (Etocas 60; Incrocas 60)
Hydroxyl No.:
50–60 (Etocas 60)
pH:
6.0–7.5 (Etocas 60); (3% aq.) (Incrocas 60)
Surface Tension:
43.2 dynes/cm (0.1% deionized water) (Etocas 60; Incrocas 60)

POE (200) castor oil

SYNONYMS:
PEG-200 castor oil (CTFA)
PEG (200) castor oil
CAS No.:
61791-12-6 (generic)
RD No.: 977063-54-9
TRADENAME EQUIVALENTS:
Alkasurf CO-200 [Alkaril]
Chemax CO-200/50 [Chemax]
Etocas 200 [Croda Ltd.]
Hetoxide C-200, C-200-50% [Heterene]
Industrol CO-200, CO-200-50% [BASF]
Trylox 5918, CO-200, CO-200/50 [Henkel/Emery]
CATEGORY:
Lubricant, emollient, emulsifier, antistat, humectant, dispersant, solubilizer, viscosity
control agent, dyeing assistant, carrier, intermediate, surfactant
APPLICATIONS:
Cosmetic industry preparations: (Hetoxide C-200, C-200-50%); perfumery (Hetoxide
C-200, C-200-50%)
Household detergents: (Hetoxide C-200, C-200-50%)
Industrial applications: chemical synthesis (Hetoxide C-200, C-200-50%); dyes and
pigments (Chemax CO-200/50; Hetoxide C-200, C-200-50%); metalworking
(Hetoxide C-200, C-200-50%); textile/leather processing (Alkasurf CO-200;
Chemax CO-200/50; Etocas 200; Hetoxide C-200, C-200-50%; Trylox 5918, CO-
200, CO-200/50)
Industrial cleaners: textile scouring (Hetoxide C-200, C-200-50%)
PROPERTIES:
Form:
Liquid (Chemax CO-200/50; Hetoxide C-200-50%; Industrol CO-200-50%; Trylox
5918, CO-200)

POE (200) castor oil (cont'd.)

Solid (Alkasurf CO-200; Etocas 200; Trylox CO-200); @ 30 C (Hetoxide C-200)
Wax (Industrol CO-200)

Color:
White (Industrol CO-200)
APHA 200 max. (Industrol CO-200-50%)
Gardner < 1 (Trylox CO-200/50)
Gardner 1 (Trylox 5918; Trylox CO-200)
Gardner 4 (Alkasurf CO-200)
Gardner 4 max. (Hetoxide C-200-50%)

Composition:
50% active (Chemax CO-200/50; Hetoxide C-200-50%; Industrol CO-200-50%;
Trylox 5918, CO-200/50)
97% active (Etocas 200)
100% active (Industrol CO-200; Trylox CO-200)

Solubility:
Insol. in aromatic hydrocarbons (Alkasurf CO-200)
Insol. in butyl stearate (Trylox CO-200, CO-200/50)
Insol. in glycerol trioleate (Trylox CO-200, CO-200/50)
Sol. in isopropanol (Hetoxide C-200, C-200-50%)
Insol. in min. oil (Trylox CO-200, CO-200/50)
Insol. in Stoddard solvent (Trylox CO-200, CO-200/50)
Sol. in water (Hetoxide C-200, C-200-50%; Industrol CO-200, CO-200-50%; Trylox
5918); sol. @ 10% (Alkasurf CO-200); sol. @ 5% (Trylox CO-200, CO-200/50)
Insol. in xylene (Trylox CO-200, CO-200/50)

Ionic Nature:
Nonionic (Alkasurf CO-200; Chemax CO-200/50; Etocas 200; Hetoxide C-200, C-
200-50%; Industrol CO-200, CO-200-50%; Trylox CO-200, CO-200/50)

M.W.:
9800 (theoret.) (Industrol CO-200)

Sp.gr.:
1.04 (Industrol CO-200-50%)
1.08 (Industrol CO-200)

Density:
8.5 lb/gal (Trylox 5918)
8.6 lb/gal (Trylox CO-200, CO-200/50)

Visc.:
1015 cSt (100 F) (Trylox 5918, CO-200/50)
2000 cps (Industrol CO-200-50%)

M.P.:
40 C (Industrol CO-200; Trylox CO-200, CO-200/50)

Pour Pt.:
7 C (Industrol CO-200-50%; Trylox 5918, CO-200/50)

236

Flash Pt.:
470 F (Trylox CO-200)
Cloud Pt.:
78–84 C (1% in 5% sodium chloride) (Alkasurf CO-200)
80 C (5% saline) (Trylox 5918, CO-200, CO-200/50)
HLB:
18.0 (Etocas 200)
18.1 (Alkasurf CO-200; Chemax CO-200/50; Industrol CO-200, CO-200-50%; Trylox 5918, CO-200, CO-200/50)
Acid No.:
1.0 (Hetoxide C-200)
Saponification No.:
7–10 (Hetoxide C-200-50%)

POE (10) dinonyl phenyl ether phosphate

SYNONYMS:
Nonyl nonoxynol-10 phosphate (CTFA)
PEG-10 dinonyl phenyl ether phosphate
PEG 500 dinonyl phenyl ether phosphate
TRADENAME EQUIVALENTS:
Gafac RM-510 [GAF]
Monafax L10 [Mona]
CATEGORY:
Lubricant, antistat, corrosion inhibitor, detergent, emulsifier, solubilizer
APPLICATIONS:
Cosmetic industry preparations: (Gafac RM-510)
Farm products: insecticides/pesticides (Gafac RM-510; Monafax L10)
Industrial cleaners: drycleaning compositions (Monafax L10)
PROPERTIES:
Form:
Hazy, viscous liquid (Gafac RM-510)
Slightly hazy, viscous liquid (Monafax L10)
Composition:
100% active (Gafac RM-510; Monafax L10)
Solubility:
Sol. in cottonseed oil (Gafac RM-510: Monafax L10)
Sol. in ethanol (Gafac RM-510; Monafax L10)
Sol. in ethylene glycol monobutyl ether (Gafac RM-510; Monafax L10)
Sol. in kerosene (Gafac RM-510; Monafax L10)

POE (10) dinonyl phenyl ether phosphate *(cont'd.)*

Disp. in min. oil (Gafac RM-510; Monafax L10)
Sol. in perchloroethylene (Gafac RM-510; Monafax L10)
Sol. in Stoddard solvent (Gafac RM-510; Monafax L10)
Disp in water (Gafac RM-510; Monafax L10)
Sol. in xylene (Gafac RM-510; Monafax L10)
Ionic Nature:
Anionic (Gafac RM-510; Monafax L10)
Sp.gr.:
1.05–1.07 (Gafac RM-510)
1.06 (Monafax L10)
Density:
8.8 lb/gal (Gafac RM-510; Monafax L10)
Pour Pt.:
5 C (Gafac RM-510)
Acid No.:
45–55 (Gafac RM-510)
48 ± 3 @ pH 5.0–5.5 (Monafax L10)
Stability:
Stable to alkaline and neutral media (Gafac RM-510)
pH:
< 2.5 (10% sol'n.) (Gafac RM-510; Monafax L10)
TOXICITY/HANDLING:
Protect skin and eyes from contact (Gafac RM-510)

POE (7) glyceryl ether

SYNONYMS:
Glycereth-7 (CTFA)
PEG-7 glyceryl ether
PEG (7) glyceryl ether
CAS No.:
31694-55-0
TRADENAME EQUIVALENTS:
Hetoxide G7 [Heterene]
Liponic EG-7 [Lipo]
CATEGORY:
Emollient, conditioner, lubricant, humectant, emulsifier, solubilizer, viscosity control
agent, lubricant, dispersant, scouring agent, dyeing assistant, carrier, intermediate,
foam modifier

POE (7) glyceryl ether *(cont'd.)*

APPLICATIONS:

Cleansers: soaps (Liponic EG-7)

Cosmetic industry preparations: (Hetoxide G7; Liponic EG-7); creams and lotions (Liponic EG-7); hair preparations (Liponic EG-7); perfumery (Hetoxide G7); personal care products; shampoos (Liponic EG-7); shaving preparations (Liponic EG-7); skin preparations (Liponic EG-7); toiletries (Liponic EG-7)

Household detergents: (Hetoxide G7)

Industrial applications: chemical synthesis (Hetoxide G7); dyes and pigments (Hetoxide G7); metalworking (Hetoxide G7); textile/leather processing (Hetoxide G7)

Industrial cleaners: textile scouring (Hetoxide G7)

PROPERTIES:

Form:

Liquid (Liponic EG-7); @ 30 C (Hetoxide G7)

Color:

Colorless to pale yellow (Liponic EG-7)

Odor:

Practically odorless (Liponic EG-7)

Composition:

99% active (Liponic EG-7)

Solubility:

Sol. in alcohols (Liponic EG-7)

Sol. in isopropanol (Hetoxide G7)

Insol. in min. oil (Liponic EG-7)

Insol. in veg. oil (Liponic EG-7)

Sol. in water (Hetoxide G7; Liponic EG-7)

Ionic Nature:

Nonionic (Hetoxide G7)

Acid No.:

1.0 (Hetoxide G7)

Hydroxyl No.:

400–428 (Hetoxide G7)

412–428 (Liponic EG-7)

Stability:

Stable over pH range normally encountered in cosmetics (Liponic EG-7)

POE (26) glyceryl ether

SYNONYMS:

Glycereth-26 (CTFA)

PEG-26 glyceryl ether

PEG (26) glyceryl ether

POE (26) glyceryl ether (cont'd.)

CAS No.:
31694-55-0 (generic)
RD No.: 977054-72-0

TRADENAME EQUIVALENTS:
Acconon ETG [Capital City]
Ethosperse G-26 [Lonza]
Hetoxide G-26 [Heterene]
Liponic EG-1 [Lipo]

CATEGORY:
Lubricant, humectant, dispersing medium, plasticizer, foam modifier

APPLICATIONS:
Cleansers: soaps (Liponic EG-1)
Cosmetic industry preparations: (Ethosperse G-26; Liponic EG-1); creams and lotions (Acconon ETG; Liponic EG-1); hair preparations (Liponic EG-1); makeup (Liponic EG-1); shaving preparations (Liponic EG-1); skin preparations (Acconon ETG; Liponic EG-1); toiletries (Liponic EG-1)
Industrial applications: (Ethosperse G-26); adhesives (Hetoxide G-26)

PROPERTIES:

Form:
Liquid (Hetoxide G-26)
Clear to slightly hazy viscous liquid (Liponic EG-1)

Color:
Colorless (Liponic EG-1)
Gardner 2 max. (Hetoxide G-26)

Odor:
Practically odorless (Liponic EG-1)

Composition:
99% active (Liponic EG-1)
99.5% active (Hetoxide G-26)

Solubility:
Sol. in acetone (Liponic EG-1)
Sol. in alcohols (Acconon ETG; Liponic EG-1)
Sol. in esters (Acconon ETG)
Sol. in ethyl acetate (Liponic EG-1)
Sol. in isopropanol (Hetoxide G-26)
Sol. in ketones (Acconon ETG)
Insol. in min. oil (Hetoxide G-26; Liponic EG-1)
Insol. in veg. oil (Liponic EG-1)
Sol. in water (Acconon ETG; Hetoxide G-26; Liponic EG-1)

Acid No.:
0.5 max. (Liponic EG-1)
2.0 max. (Hetoxide G-26)

Hydroxyl No.:
127–137 (Hetoxide G-26)
128–138 (Liponic EG-1)
Stability:
Stable over the pH range normally encountered in cosmetics and toiletries (Liponic EG-1)
STD. PKGS.:
500 lb net closed-head steel drums (Liponic EG-1)

POE (7) glyceryl monococoate

SYNONYMS:
PEG-7 glyceryl cocoate (CTFA)
PEG (7) glyceryl monococoate
STRUCTURE:

$$\underset{\displaystyle \underset{OH}{|}}{RC}{-}OCH_2CHCH_2(OCH_2CH_2)_nOH$$

with O double-bonded to the first carbon (RC)

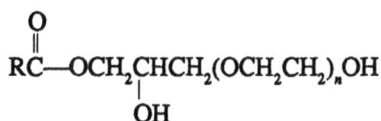

where   RCO⁻ represents the coconut radical and
avg. $n = 7$

CAS No.:
RD No. 977064-68-8
TRADENAME EQUIVALENTS:
Cetiol HE [Henkel; Henkel Canada; Henkel KGaA] (self-emulsifying)
Mazol 159 [PPG-Mazer]
Standamul HE [Henkel] (self-emulsifying)
CATEGORY:
Emollient, emulsifier, plasticizer, refatting agent, solubilizer, coupling agent
APPLICATIONS:
Bath products: bubble bath (Cetiol HE; Standamul HE)
Cleansers: skin cleanser (Standamul HE)
Cosmetic industry preparations: (Mazol 159; Standamul HE); creams and lotions (Standamul HE); perfumery (Standamul HE); shampoos (Cetiol HE; Standamul HE); toiletries (Mazol 159)
Food applications: (Mazol 159)
Industrial applications: lubricating/cutting oils (Mazol 159); mold release (Mazol 159); plastics (Mazol 159); textile/leather processing (Mazol 159)
Pharmaceutical applications: (Mazol 159; Standamul HE); sunscreens (Cetiol HE); topical preparations (Standamul HE)

241

POE (7) glyceryl monococoate *(cont'd.)*

PROPERTIES:
Form:
  Clear low-viscosity oil (Cetiol HE; Standamul HE)
  Clear liquid (Mazol 159)
Color:
  Gardner 1 (Mazol 159)
Odor:
  Faint, characteristic (Standamul HE)
Composition:
  100% active (Standamul HE)
Solubility:
  Sol. in castor oil (Standamul HE); sol. @ 10% (Cetiol HE)
  Sol. in anhydrous ethanol (Standamul HE); sol. in 95% SD-40 ethanol @ 10% (Cetiol HE); sol. in 70% aq. 3A-ethanol (Cetiol HE)
  Sparingly sol. in isopropyl myristate (Standamul HE)
  Sol. in oleyl alcohol (Standamul HE); sol. @ 10% (Cetiol HE)
  Sparingly sol. in most polyols (Standamul HE)
  Disp. in propylene glycol @ 10% (Cetiol HE)
  Disp. in sorbo @ 10% (Cetiol HE)
  Sol. in water (Standamul HE); sol. @ 10% (Cetiol HE)
Ionic Nature:
  Nonionic (Standamul HE)
Sp.gr.:
  1.050 (Standamul HE)
Density:
  1.050 g/ml (Cetiol HE)
Visc.:
  200 cps (Standamul HE)
Solidification Pt.:
  < 0 C (Cetiol HE)
Cloud Pt.:
  < 0 C (Cetiol HE)
  ≈ 0 C (Standamul HE)
HLB:
  13.0 (Mazol 159)
  16.0 (Standamul HE)
Acid No.:
  5 max. (Cetiol HE; Standamul HE)
Iodine No.:
  5 max. (Cetiol HE; Standamul HE)
Saponification No.:
  90 (Mazol 159)
  90–100 (Cetiol HE; Standamul HE)

POE (7) glyceryl monococoate (cont' d.)

Hydroxyl No.:
180–195 (Standamul HE)
Ref. Index:
1.460 (Cetiol HE)
STD. PKGS.:
385 lb net closed-head steel drums or bulk (Standamul HE)

POE (16) hydrogenated castor oil

SYNONYMS:
PEG-16 hydrogenated castor oil (CTFA)
PEG (16) hydrogenated castor oil
CAS No.:
61788-85-0 (generic)
RD No.: 977065-36-3
TRADENAME EQUIVALENTS:
Chemax HCO-16 [Chemax]
Hetoxide HC-16 [Heterene]
Trylox 5921, HCO-16 [Henkel/Emery]
CATEGORY:
Lubricant, softener, emulsifier, dispersant
APPLICATIONS:
Cosmetic industry preparations: (Hetoxide HC-16); perfumery (Hetoxide HC-16)
Household products: fabric softeners (Trylox 5921, HCO-16)
Industrial applications: chemical synthesis (Hetoxide HC-16); dyes and pigments (Hetoxide HC-16); metalworking (Hetoxide HC-16); textile/leather processing (Hetoxide HC-16)
Industrial cleaners: textile scouring (Hetoxide HC-16)
PROPERTIES:
Form:
Liquid (Chemax HCO-16; Trylox 5921, HCO-16); @ 30C (Hetoxide HC-16)
Color:
Gardner 1 (Trylox 5921, HCO-16)
Composition:
100% active (Chemax HCO-16)
Solubility:
Disp. in butyl stearate @ 5% (Trylox HCO-16)
Sol. in glycerol trioleate (Trylox 5921); sol. @ 5% (Trylox HCO-16)
Sol. in isopropanol (Hetoxide HC-16)
Sol. in min. oil (Hetoxide HC-16) sol. @ 5% (Trylox HCO-16); disp. (Trylox 5921)
Disp. in Stoddard solvent @ 5% (Trylox HCO-16)

POE (16) hydrogenated castor oil *(cont'd.)*

Disp. in water (Trylox 5921); disp. @ 5% (Trylox HCO-16)
Sol. in xylene (Trylox 5921); sol. @ 5% (Trylox HCO-16)
Ionic Nature:
Nonionic (Chemax HCO-16; Hetoxide HC-16; Trylox HCO-16)
Density:
8.4 lb/gal (Trylox 5921, HCO-16)
Visc.:
569 cSt (100 F) (Trylox 5921, HCO-16)
Pour Pt.:
7 C (Trylox 5921, HCO-16)
Flash Pt.:
565 F (Trylox 5921, HCO-16)
Cloud Pt.:
< 25 C (Trylox 5921, HCO-16)
HLB:
8.4 (Chemax HCO-16)
8.6 (Trylox 5921, HCO-16)
11.3 (Hetoxide HC-16)
Acid No.:
1.0 (Hetoxide HC-16)
Saponification No.:
85–95 (Hetoxide HC-16)

POE (30) lanolin

SYNONYMS:
PEG-30 lanolin (CTFA)
PEG (30) lanolin
CAS No.:
61790-81-6 (generic)
RD No.: 977065-51-2
TRADENAME EQUIVALENTS:
Aqualose L30 [Westbrook Lanolin]
Nikkol TW-30 [Nikko]
CATEGORY:
Emollient, emulsifier, plasticizer, solubilizer, bodying agent
APPLICATIONS:
Cleansers: soaps (Nikkol TW-30)
Cosmetic industry preparations: shampoos (Nikkol TW-30)

244

PROPERTIES:
Form:
Paste (Nikkol TW-30)
Wax (Aqualose L30)
Composition:
100% active (Aqualose L30; Nikkol TW-30)
Ionic Nature:
Nonionic (Aqualose L30; Nikkol TW-30)
HLB:
14.0 (Aqualose L30)
15.0 (Nikkol TW-30)

POE (60) lanolin

SYNONYMS:
PEG-60 lanolin (CTFA)
PEG (60) lanolin
CAS No.:
61790-81-6 (generic)
RD No.: 977059-06-5
TRADENAME EQUIVALENTS:
Laneto 60 [RITA]
Solan, 50 [Croda; Croda Ltd.]
CATEGORY:
Lubricant, emollient, conditioner, superfatting agent, solubilizer, moisturizer, surfactant, resin modifier, glossing agent, foam stabilizer, plasticizer, humectant, emulsifier
APPLICATIONS:
Cleansers: skin cleaners (Solan, 50); soaps (Laneto 60; Solan, 50)
Cosmetic industry preparations: aerosols (Solan, 50); hair preparations (Laneto 60; Solan, 50); perfumery (Solan, 50); personal care products (Laneto 60); shampoos (Solan, 50)
Household detergents: (Laneto 60)
Pharmaceutical applications: antiperspirant/deodorant (Solan, 50)
PROPERTIES:
Form:
Liquid (Solan 50)
Wax (Solan)
Color:
Gardner 11 max. (Solan, 50)

POE (60) lanolin *(cont'd.)*

Composition:
    50% active (Solan 50)
    100% active (Solan)
Solubility:
    Sol. in alcohols (Laneto 60)
    Sol. in most nonpolar solvents (Solan)
    Sol. in most polar solvents (Solan)
    Sol. in water/alcohol (Solan)
Ionic Nature:
    Nonionic (Solan, 50)
M.P.:
    46–54 C (Solan)
    48–52 C (Laneto 60)
Acid No.:
    2.0 max. (Solan, 50)
Iodine No.:
    6 max. (Solan 50)
    10 max. (Solan)
Saponification No.:
    8 max. (Solan 50)
    8–16 (Solan)
    10–16 (Laneto 60)
Stability:
    Unaffected by hard water, high electrolyte concentrations, or variations in pH (Solan, 50)
pH:
    5.5–7.0 (1% aq. sol'n.) (Solan, 50)

POE (75) lanolin

SYNONYMS:
    PEG-75 lanolin (CTFA)
    PEG 4000 lanolin
CAS No.:
    8039-09-6; 61790-81-6 (generic)
TRADENAME EQUIVALENTS:
    Aqualose L75, L75/50 [Westbrook Lanolin] (USP)
    Cralane LR-10 [Pulcra SA]
    Ethoxylan 1685, 1686 [Henkel/Emery]
    Lan-Aqua-Sol xtra-Hydrophilic 50, 100 [Fanning]
    Laneto-50, -100 [RITA]

Lanogel 41 [Amerchol]
Solangel 401 [Croda]
Solulan 75, L-575 [Amerchol]

CATEGORY:

Emollient, emulsifier, plasticizer, solubilizer, humectant, conditioner, lubricant, superfatting agent, wetting agent, detergent, dispersant, foam stabilizer, moisturizer, surfactant, resin modifier, glossing agent, surfactant

APPLICATIONS:

Bath products: bubble bath (Solulan 75, L-575)

Cleansers: hand cleanser (Lan-Aqua-Sol xtra-Hydrophilic 50, 100; Laneto-50, -100); soaps (Aqualose L75, L75/50; Lan-Aqua-Sol xtra-Hydrophilic 50, 100; Laneto-50, -100; Lanogel 41; Solangel 401; Solulan 75)

Cosmetic industry preparations: (Ethoxylan 1685, 1686; Laneto-50, -100; Solulan 75); conditioners (Lan-Aqua-Sol xtra-Hydrophilic 50, 100; Lanogel 41); creams and lotions (Lan-Aqua-Sol xtra-Hydrophilic 50, 100; Solulan 75, L-575); hair preparations (Aqualose L75, L75/50; Laneto-50, -100; Solulan 75); makeup (Solulan 75, L-575); perfumery (Aqualose L75, L75/50; Lan-Aqua-Sol xtra-Hydrophilic 50, 100; Laneto-50; Solulan 75); personal care products (Aqualose L75, L75/50; Cralane LR-10; Laneto-50, -100); shampoos (Aqualose L75, L75/50; Lan-Aqua-Sol xtra-Hydrophilic 50, 100; Laneto-50, -100; Lanogel 41; Solulan 75, L-575); shaving preparations (Lan-Aqua-Sol xtra-Hydrophilic 50, 100; Solulan 75); toiletries (Solulan 75)

Household detergents: (Laneto-50, -100; Solulan 75, L-575); dishwashing (Lan-Aqua-Sol xtra-Hydrophilic 50, 100)

Pharmaceutical applications: (Ethoxylan 1685, 1686; Laneto-50, -100; Solulan 75); antiperspirant/deodorant (Lan-Aqua-Sol xtra-Hydrophilic 50, 100; Solulan 75); germicides (Aqualose L75, L75/50; Cralane LR-10)

PROPERTIES:

Form:

Liquid (Ethoxylan 1686; Laneto-50; Solangel 401)
Clear viscous liquid (Solulan L-575)
Viscous liquid (Cralane LR-10)
Gel (Aqualose L75/50; Lanogel 41)
Solid (Ethoxylan 1685)
Soft, waxy solid (Solulan 75)
Wax (Aqualose L75; Laneto-100)

Color:

Light yellow-amber (Solulan 75, L-575)
Yellow-amber (Lanogel 41)
Yellow (Laneto-100)
Gardner < 10 (Ethoxylan 1686)
Gardner 10 max. (Lan-Aqua-Sol xtra-Hydrophilic 50)
Gardner < 11 (Ethoxylan 1685)

Gardner 11 max. (Lan-Aqua-Sol xtra-Hydrophilic 100)
Gardner 12 max. (Laneto-50)

Odor:

Faint (Lan-Aqua-Sol xtra-Hydrophilic 50, 100)
Characteristic (Laneto-50)
Faint, pleasant (Lanogel 41; Solulan 75, L-575)

Composition:

50% active (Aqualose L75/50; Cralane LR-10; Lan-Aqua-Sol xtra-Hydrophilic 50; Solangel 401)
50% active in water (Ethoxylan 1686; Lanogel 41; Solulan L-575)
50 ± 1% active in water (Laneto-50)
100% active (Aqualose L75; Ethoxylan 1685; Lan-Aqua-Sol xtra-Hydrophilic 100; Laneto-100)

Solubility:

Sol. in isopropanol @ 5% (Ethoxylan 1685, 1686)
Sol. in water (Lan-Aqua-Sol xtra-Hydrophilic 50, 100; Lanogel 41; Solulan 75, L-575); sol. @ 5% (Ethoxylan 1685, 1686; Laneto-50, -100)

Ionic Nature:

Nonionic (Ethoxylan 1685, 1686; Lan-Aqua-Sol xtra-Hydrophilic 50, 100; Laneto-50, -100; Lanogel 41; Solangel 401)

Sp.gr.:

1.00–1.05 (50/4 C) (Lan-Aqua-Sol xtra-Hydrophilic 50)
1.00–1.10 (Laneto-50)
1.03–1.07 (50/4 C) (Lan-Aqua-Sol xtra-Hydrophilic 100)

Density:

8.9 lb/gal (Ethoxylan 1686)
9.6 lb/gal (Ethoxylan 1685)

Visc.:

1767 cSt (100 F) (Ethoxylan 1686)

M.P.:

39 C (Ethoxylan 1685)
46–52 C (Lan-Aqua-Sol xtra-Hydrophilic 100)
48–52 C (Laneto-100)

Pour Pt.:

1 C (Ethoxylan 1686)

Flash Pt.:

530 F (Ethoxylan 1685)

Cloud Pt.:

80–87 C (1% sol'n. in 5% sodium chloride aq.) (Solulan 75)
80–87 C (1% active) (Solulan L-575)
85 C (Ethoxylan 1685)
86 C (Ethoxylan 1686)

HLB:
14–16 (Aqualose L75)
16.0 (Solangel 401)
16.9 (Ethoxylan 1685, 1686)
Acid No.:
1.0 max. (Solulan L-575)
3.0 max. (Lan-Aqua-Sol xtra-Hydrophilic 50, 100; Solulan 75)
Iodine No.:
2.0–7.5 (Hanus) (Lan-Aqua-Sol xtra-Hydrophilic 50)
3–12 (Hanus) (Lan-Aqua-Sol xtra-Hydrophilic 100)
8 max. (Lanogel 41)
Saponification No.:
1.0-1.10 (Laneto-50)
6.0–15.0 (Lanogel 41)
6.5–15.0 (Lan-Aqua-Sol xtra-Hydrophilic 50)
10 max. (Solulan L-575)
10–20 (Solulan 75)
14–22 (Lan-Aqua-Sol xtra-Hydrophilic 100)
pH:
4.0–6.0 (4% in dist. water) (Lan-Aqua-Sol xtra-Hydrophilic 50, 100)
4.5–7.0 (10% aq. sol'n.) (Solulan 75, L-575)

POE (20) lanolin ether

SYNONYMS:
Laneth-20 (CTFA)
PEG-20 lanolin ether
PEG 1000 lanolin ether
CAS No.:
61791-20-6 (generic)
RD No. 977054-79-7
TRADENAME EQUIVALENTS:
Aqualose W20, W20/50 [Westbrook Lanolin]
Polychol 20, 20-40 [Croda Ltd.]
CATEGORY:
Emollient, plasticizer, solubilizer, emulsifier
APPLICATIONS:
Cosmetic industry preparations: (Polychol 20-40); personal care products (Polychol 20)

POE (20) lanolin ether (cont'd.)

PROPERTIES:
Form:
Gel (Aqualose W20/50)
Wax (Aqualose W20; Polychol 20-40)
Color:
Gardner 8–10 (Polychol 20)
Composition:
50% active (Aqualose W20/50)
Solubility:
Sol. in water (Polychol 20, 20-40)
Ionic Nature:
Nonionic (Polychol 20)
Acid No.:
5.0 max. (Polychol 20)
pH:
3.5–5.5 (10% aq.) (Polychol 20, 20-40)

POE (3) myristyl ether myristate

SYNONYMS:
Myreth-3 myristate (CTFA)
PEG-3 myristyl ether myristate
PEG (3) myristyl ether myristate
STRUCTURE:

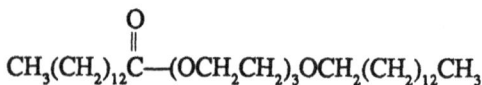

$$CH_3(CH_2)_{12}\overset{\overset{\displaystyle O}{\|}}{C}\text{—}(OCH_2CH_2)_3OCH_2(CH_2)_{12}CH_3$$

CAS No.:
RD No.: 977054-88-8
TRADENAME EQUIVALENTS:
Cetiol 1414E [Henkel Canada]
Liponate 143M [Lipo]
Schercemol MEM-3 [Scher]
Standamul 1414-E [Henkel]
CATEGORY:
Emollient, thickener, viscosity control agent, penetrant, coupling agent, emulsifier, solubilizer, carrier
APPLICATIONS:
Bath products: (Liponate 143M); bath oils (Cetiol 1414E; Schercemol MEM-3; Standamul 1414-E)

Cleansers: skin cleanser (Standamul 1414-E)
Cosmetic industry preparations: (Cetiol 1414E; Schercemol MEM-3); creams and lotions (Cetiol 1414E; Liponate 143M; Schercemol MEM-3; Standamul 1414-E); makeup (Cetiol 1414E; Schercemol MEM-3); personal care products (Schercemol MEM-3); shampoos (Standamul 1414-E)
Pharmaceutical applications: antiperspirant/deodorant (Standamul 1414-E)

PROPERTIES:

Form:
Clear liquid (Schercemol MEM-3)
Clear low viscosity oily liquid (Standamul 1414-E)
Liquid/paste (Liponate 143M)

Color:
White to light straw (Schercemol MEM-3)
White/yellow (Liponate 143M)

Odor:
Slight, typical (Schercemol MEM-3)

Composition:
98% ester content (Schercemol MEM-3)

Solubility:
Sol. in alcohols (Cetiol 1414E)
Sol. in aromatic hydrocarbons (Cetiol 1414E; Schercemol MEM-3)
Sol. in castor oil (Standamul 1414-E)
Sol. in cosmetic oils (Cetiol 1414E)
Sol. in esters (Cetiol 1414E; Schercemol MEM-3)
Sol. in anhyd. ethanol (Standamul 1414-E)
Sol. in glycols (Cetiol 1414E); disp. (Schercemol MEM-3)
Sol. in isopropyl myristate (Standamul 1414-E)
Sol. in ketones (Cetiol 1414E)
Sol. in min. oil (Schercemol MEM-3; Standamul 1414-E)
Sol. in min. spirits (Schercemol MEM-3)
Sol. in oleyl alcohol (Standamul 1414-E)
Disp. in polyols (Schercemol MEM-3)
Disp. in triols (Schercemol MEM-3)
Sol. in veg. oil (Schercemol MEM-3)
Sparingly sol. in water (Standamul 1414-E); disp. in water (Cetiol 1414E; Schercemol MEM-3)

M.W.:
556 (theoret.) (Schercemol MEM-3)

Sp.gr.:
0.901 (Schercemol MEM-3)

Flash Pt.:
160 C (OC) (Schercemol MEM-3)

POE (3) myristyl ether myristate *(cont'd.)*

Cloud Pt.:
25 C max. (Standamul 1414-E)
HLB:
12.0 (Standamul 1414-E)
Acid No.:
3.0 max. (Schercemol MEM-3)
6.0 max. (Liponate 143M)
Saponification No.:
90–100 (Liponate 143M; Standamul 1414-E)
95–110 (Schercemol MEM-3)
Hydroxyl No.:
10–20 (Standamul 1414-E)
Ref. Index:
1.4525 (Schercemol MEM-3)
STD. PKGS.:
450 lb net closed-head steel drums or bulk (Standamul 1414-E)

POE (3) oleyl ether

SYNONYMS:
Oleth-3 (CTFA)
PEG-3 oleyl ether
PEG (3) oleyl ether
STRUCTURE:

$CH_3(CH_2)_7CH=CH(CH_2)_7CH_2(OCH_2CH_2)_nOH$
where avg. $n = 3$
CAS No.:
9004-98-2 (generic); 25190-05-0 (generic)
RD No. 977057-50-3
TRADENAME EQUIVALENTS:
Hetoxol OA-3 Special [Heterene]
Volpo 3, O3 [Croda Ltd.]
Volpo N3 [Croda Ltd.] (dist.)
CATEGORY:
Emollient, lubricant, superfatting agent, emulsifier, pigment dispersant, solubilizer, spreading agent, wetting agent, scouring agent, gelling agent, leveling agent, antistat, plasticizer
APPLICATIONS:
Bath products: bubble bath (Volpo 3); bath oils (Volpo 3)
Cosmetic industry preparations: (Hetoxol OA-3 Special; Volpo 3); astringent creams and lotions (Volpo 3); cold waves/hair straighteners (Volpo 3); makeup (Volpo 3);

personal care products (Volpo O3); shampoos (Hetoxol OA-3 Special; Volpo 3)
Farm applications: (Volpo O3)
Household detergents: hard surface cleaners (Volpo O3)
Industrial applications: metalworking (Volpo O3); paper mfg. (Volpo O3); textile/
   leather processing (Volpo O3)
Pharmaceutical applications: (Volpo O3); depilatories (Volpo 3)
PROPERTIES:
Form:
   Liquid (Hetoxol OA-3 Special; Volpo N3, O3)
   Hazy liquid (Volpo 3)
Color:
   Pale straw (Volpo O3)
   Pale yellow (Hetoxol OA-3 Special)
Composition:
   97% active (Volpo N3)
Solubility:
   Sol. in alcohols (Volpo 3)
   Sol. in aromatic solvents (Volpo 3)
   Sol. in chlorinated solvents (Volpo 3)
   Sol. in ethanol (Volpo O3)
   Sol. in glycols (Volpo 3)
   Sol. in isopropanol (Hetoxol OA-3 Special)
   Sol. in kerosene (Volpo O3)
   Sol. in ketones (Volpo 3)
   Sol. in min. oil (Hetoxol OA-3 Special; Volpo 3, O3)
   Sol. in nonpolar oils (Volpo 3)
   Sol. in oleic acid (Volpo O3)
   Sol. in oleyl alcohol (Volpo O3)
   Sol. in trichloroethylene (Volpo O3)
   Insol. in water (Hetoxol OA-3 Special; Volpo 3)
Ionic Nature:
   Nonionic (Volpo 3, N3, O3)
HLB:
   6.4 (Hetoxol OA-3 Special)
   6.6 (Volpo 3)
   6.7 (Volpo O3)
Acid No.:
   1.0 max. (Volpo O3)
   2.0 max. (Volpo 3)
Iodine No.:
   57–62 (Wijs) (Volpo 3)
Hydroxyl No.:
   135–150 (Volpo 3)

POE (3) oleyl ether (cont'd.)

Stability:
Stable in sol'ns. of metallic ions and to many acids and alkalis (Volpo 3)
pH:
5.0–7.0 (3% aq.) (Volpo 3)
6.0–7.5 (3%) (Volpo O3)

POE (3) oleyl ether phosphate

SYNONYMS:
Oleth-3 phoshate (CTFA)
Oleyl triethoxy monodiphosphate
PEG-3 oleyl ether phosphate
PEG (3) oleyl ether phosphate
CAS No.:
39464-69-2 (generic)
RD No.: 977060-93-7
TRADENAME EQUIVALENTS:
Crodafos N3 Acid [Croda; Croda Ltd.]
Cyclophos PO3 [Alcolac]
CATEGORY:
Surfactant, conditioner, antistat, emulsifier, gelling agent, antigelling agent, corrosion
inhibitor
APPLICATIONS:
Cleansers: skin cleanser (Crodafos N3 Acid)
Cosmetic industry preparations: (Crodafos N3 Acid); conditioners (Crodafos N3 Acid;
Cyclophos PO3); hair preparations (Crodafos N3 Acid); shampoos (Crodafos N3
Acid; Cyclophos PO3); toiletries (Crodafos N3 Acid)
Pharmaceutical applications: (Crodafos N3 Acid); antiperspirant/deodorant (Crodafos
N3 Acid); depilatories (Crodafos N3 Acid)
PROPERTIES:
Form:
Liquid (Cyclophos PO3)
Color:
Gardner 11 max. (Crodafos N3 Acid)
Composition:
100% active (Crodafos N3 Acid; Cyclophos PO3)
Ionic Nature:
Anionic (Crodafos N3 Acid; Cyclophos PO3)
Acid No.:
120–135 (Crodafos N3 Acid)

POE (3) oleyl ether phosphate *(cont'd.)*

Iodine No.:
45–58 (Wijs) (Crodafos N3 Acid)
Saponification No.:
125–145 (Crodafos N3 Acid)
Stability:
Stable over wide pH range (Cyclophos PO3)
pH:
1–3 (2% aq. sol'n.) (Crodafos N3 Acid)

POE (10) oleyl ether phosphate

SYNONYMS:
Oleth-10 phosphate (CTFA)
PEG-10 oleyl ether phosphate
PEG 500 oleyl ether phosphate
CAS No.:
39464-69-2 (generic)
RD No.: 977060-95-9
TRADENAME EQUIVALENTS:
Crodafos N10 Acid [Croda; Croda Ltd.]
CATEGORY:
Surfactant, conditioner, antistat, emulsifier, gelling agent, antigelling agent, corrosion
inhibitor
APPLICATIONS:
Cleansers: skin cleanser (Crodafos N10 Acid)
Cosmetic industry preparations: (Crodafos N10 Acid); conditioners (Crodafos N10
Acid); hair preparations (Crodafos N10 Acid); shampoos (Crodafos N10 Acid);
toiletries (Crodafos N10 Acid)
Pharmaceutical applications: (Crodafos N10 Acid); antiperspirant/deodorant (Cro-
dafos N10 Acid); depilatories (Crodafos N10 Acid)
PROPERTIES:
Color:
Gardner 11 max. (Crodafos N10 Acid)
Composition:
100% active (Crodafos N10 Acid)
Ionic Nature:
Anionic (Crodafos N10 Acid)
Acid No.:
70-100 (Crodafos N10 Acid)
Iodine No.:
25–35 (Wijs) (Crodafos N10 Acid)

POE (10) oleyl ether phosphate *(cont'd.)*

Saponification No.:
 75–115 (Crodafos N10 Acid)
pH:
 1–3 (2% aq. sol'n.) (Crodafos N10 Acid)

POE (10) sorbitan monolaurate

SYNONYMS:
 PEG-10 sorbitan laurate (CTFA)
 PEG 500 sorbitan monolaurate
CAS No.:
 9005-64-5 (generic)
 RD No.: 977055-35-8
TRADENAME EQUIVALENTS:
 Atlas G-7596-J [ICI Americas]
 Hetsorb L-10 [Heterene]
 Liposorb L-10 [Lipo]
CATEGORY:
 Lubricant, emulsifier, detergent, antistat, viscosity control agent, surfactant, solubil-
 izer
APPLICATIONS:
 Cosmetic industry preparations: (Hetsorb L-10)
PROPERTIES:
Form:
 Liquid (Atlas G-7596-J; Hetsorb L-10; Liposorb L-10)
Color:
 Yellow (Atlas G-7596-J; Liposorb L-10)
Composition:
 100% active (Liposorb L-10)
Solubility:
 Sol. in acetone (Atlas G-7596-J)
 Sol. in lower alcohols (Atlas G-7596-J)
 Sol. in aniline (Atlas G-7596-J)
 Sol. in Cellosolve (Atlas G-7596-J)
 Sol. in dioxane (Atlas G-7596-J)
 Sol. in ethyl acetate (Atlas G-7596-J)
 Sol. in isopropanol (Hetsorb L-10)
 Sol. in water (Atlas G-7596-J; Hetsorb L-10)
Ionic Nature:
 Nonionic (Atlas G-7596-J; Liposorb L-10)

Sp.gr.:
   1.1 (Atlas G-7596-J)
Visc.:
   ≈ 400 cps (Atlas G-7596-J)
Flash Pt.:
   > 300 F (Atlas G-7596-J)
Fire Pt.:
   > 300 F (Atlas G-7596-J)
HLB:
   8.4 (Hetsorb L-10)
   14.9 (Atlas G-7596-J)
   14.9 ± 1 (Liposorb L-10)
Saponification No.:
   66–76 (Hetsorb L-10; Liposorb L-10)
Hydroxyl No.:
   150–170 (Hetsorb L-10; Liposorb L-10

POE (2) soya amine

SYNONYMS:
   PEG-2 soyamine (CTFA)
   PEG 100 soya amine
STRUCTURE:

$$R-N \begin{cases} (CH_2CH_2O)_xH \\ (CH_2CH_2O)_yH \end{cases}$$

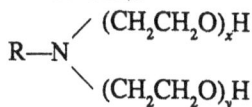

   where   R represents the soya radical and
           avg. $(x + y) = 2$
CAS No.:
   61791-24-0 (generic)
   RD No.: 977063-44-7
TRADENAME EQUIVALENTS:
   Accomeen S2 [Capital City]
   Chemeen S-2 [Chemax]
   Ethomeen S/12 [Akzo/Armak]
   Hetoxamine S-2 [Heterene]
   Mazeen S2 [PPG-Mazer]
   Teric 16M2 [ICI Australia]
   Varonic L202 [Sherex]

POE (2) soya amine (cont'd.)

CATEGORY:
Lubricant, softener, emulsifier, antistat, surfactant, dispersant, desizing agent, wetting/ rewetting agent

APPLICATIONS:
Cosmetic industry preparations: (Mazeen S2)

Farm products: (Hetoxamine S-2); herbicides (Mazeen S2); insecticides/pesticides (Mazeen S2)

Industrial applications: construction (Teric 16M2); dyes and pigments (Varonic L202); latex (Varonic L202); printing inks (Mazeen S2); textile/leather processing (Ethomeen S/12; Hetoxamine S-2; Mazeen S2; Teric 16M2); waxes, fats, greases, and oils (Hetoxamine S-2; Mazeen S2; Teric 16M2; Varonic L202)

Industrial cleaners: metal processing surfactants (Hetoxamine S-2; Teric 16M2)

PROPERTIES:

Form:
Liquid (Accomeen S2; Chemeen S-2; Hetoxamine S-2; Mazeen S2; Teric 16M2)
Clear heavy liquid (Ethomeen S/12)

Color:
Gardner 5–10 (Accomeen S2)
Gardner 14 (Mazeen S2)
Gardner 14 max. (Ethomeen S/12)

Odor:
Amine (Accomeen S2)

Composition:
95% tert. amine (Ethomeen S/12)
99% active (Accomeen S2)
100% active (Chemeen S-2; Mazeen S2; Teric 16M2; Varonic L202)

Solubility:
Sol. in acetone (Ethomeen S/12; Mazeen S2)
Sol. in benzene (Ethomeen S/12; Mazeen S2; Teric 16M2)
Sol. in carbon tetrachloride (Ethomeen S/12)
Sol. in ethanol (Teric 16M2)
Sol. in ethyl acetate (Teric 16M2)
Sol. in ethyl Icinol (Teric 16M2)
Sol. in isopropanol (Ethomeen S/12; Hetoxamine S-2; Mazeen S2)
Sol. in kerosene (Teric 16M2)
Sol. in min. oil (Hetoxamine S-2; Mazeen S2; Teric 16M2)
Sol. in olein (Teric 16M2)
Sol. in organic solvents (Accomeen S2)
Sol. in paraffin oil (Teric 16M2)
Sol. in perchloroethylene (Teric 16M2)
Sol. in Stoddard solvent (Ethomeen S/12)
Sol. in veg. oil (Teric 16M2)
Insol. in water (Ethomeen S/12)

Ionic Nature:
Cationic (Ethomeen S/12; Hetoxamine S-2; Mazeen S2)
Nonionic (Teric 16M2; Varonic L202)
M.W.:
350 (Hetoxamine S-2; Mazeen S2)
Sp.gr.:
0.906 (Teric 16M2)
0.91 (Accomeen S2; Ethomeen S/12)
0.911 (Mazeen S2)
Density:
7.6 lb/gal (Accomeen S2)
Visc.:
191 cps (Teric 16M2)
M.P.:
-15 ± 2 C (Teric 16M2)
Flash Pt.:
440 F (TOC) (Ethomeen S/12)
HLB:
4.7 (Varonic L202)
9.5 (Teric 16M2)
pH:
8.0–10.0 (1% aq.) (Teric 16M2)
Surface Tension:
26 dynes/cm (0.1% sol'n.) (Mazeen S2)
26.6 dynes/cm (Teric 16M2)
31.3 dynes/cm (0.1%) (Accomeen S2)
31.4 dynes/cm (0.1%) (Ethomeen S/12)

POE (5) soya sterol

SYNONYMS:
PEG-5 soya sterol (CTFA)
PEG (5) soya sterol
TRADENAME EQUIVALENTS:
Generol 122E5 [Henkel]
CATEGORY:
Emollient, solubilizer, emulsifier, viscosity modifier, stabilizer
APPLICATIONS:
Cosmetic industry preparations: (Generol 122E5); hair preparations (Generol 122E5);
    shampoos (Generol 122E5)

POE (5) soya sterol *(cont'd.)*

Industrial applications: emulsions (Generol 122E5)
PROPERTIES:
Form:
Soft, waxy amorphous solid (Generol 122E5)
Composition:
100% active (Generol 122E5)
Solubility:
Sol. in ethanol (Generol 122E5)
Sol. hot in isopropyl esters (Generol 122E5)
Disp. in water (Generol 122E5)
Ionic Nature:
Nonionic (Generol 122E5)
HLB:
5.0 (Generol 122E5)

POE (10) soya sterol

SYNONYMS:
PEG-10 soya sterol (CTFA)
PEG 500 soya sterol
CAS No.:
RD No.: 977065-31-8
TRADENAME EQUIVALENTS:
Generol 122E10 [Henkel]
CATEGORY:
Emollient, solubilizer, emulsifier, gloss enhancer, viscosity modifier, stabilizer
APPLICATIONS:
Cosmetic industry preparations: (Generol 122E10); hair preparations (Generol 122E10); shampoos (Generol 122E10)
Industrial applications: emulsions (Generol 122E10)
PROPERTIES:
Form:
Soft wax (Generol 122E10)
Color:
Light amber (Generol 122E10)
Composition:
100% active (Generol 122E10)
Solubility:
Sol. in ethanol (Generol 122E10)
Sol. hot in isopropyl esters (Generol 122E10)
Disp. in water (Generol 122E10)

POE (10) soya sterol (cont'd.)

Ionic Nature:
Nonionic (Generol 122E10)
HLB:
12.0 (Generol 122E10)

POE (25) soya sterol

SYNONYMS:
PEG-25 soya sterol (CTFA)
PEG (25) soya sterol
CAS No.:
RD No.: 977065-46-5
TRADENAME EQUIVALENTS:
Generol 122E25 [Henkel]
CATEGORY:
Emollient, solubilizer, wetting agent, deflocculating agent
APPLICATIONS:
Industrial applications: pigments (Generol 122E25)
PROPERTIES:
Form:
Hard wax (Generol 122E25)
Color:
Ivory (Generol 122E25)
Composition:
100% active (Generol 122E25)
Ionic Nature:
Nonionic (Generol 122E25)
HLB:
17.0 (Generol 122E25)

POE (20) tallow amine

SYNONYMS:
PEG-20 tallow amine (CTFA)
PEG 1000 tallow amine

261

POE (20) tallow amine (cont'd.)

STRUCTURE:

$$R-N \begin{cases} (CH_2CH_2O)_xH \\ (CH_2CH_2O)_yH \end{cases}$$

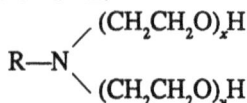

where R represents the tallow radical, and
avg. $(x + y) = 20$

CAS No.:

61791-26-2 (generic)
RD No.: 977063-52-7

TRADENAME EQUIVALENTS:

Atlas G-3780A [Atlas Refinery]
Chemcol T-20 [Chemform]
Chemeen T-20 [Chemax]
DeSomeen TA-20 [DeSoto]
Hetoxamine T20 [Heterene]
Icomeen T-20 [BASF]
Katapol PN-810 [GAF]
Polyfac TA-20 [Stepan]
Serdox NJAD20 [Servo]
Trymeen 6607, TAM-20 [Henkel/Emery]

CATEGORY:

Antistat, conditioner, lubricant, softener, surfactant, emulsifier, coemulsifier, dispersant, antiprecipitant, leveling agent, migrating agent, scouring agent, dyeing assistant, desizing agent, water repellent, wetting agent, penetrant, stabilizer

APPLICATIONS:

Farm products: (Hetoxamine T20)
Household detergents: carpet & upholstery shampoos (Chemeen T-20)
Industrial applications:; dyes and pigments (Chemeen T-20; DeSomeen TA-20; Trymeen 6607, TAM-20); textile/leather processing (Atlas G-3780A; Chemcol T-20; Chemeen T-20; DeSomeen TA-20; Hetoxamine T20; Katapol PN-810; Polyfac TA-20; Trymeen 6607, TAM-20); waxes and oils (Hetoxamine T20; Serdox NJAD20)
Industrial cleaners: metal processing surfactants (Hetoxamine T20; Serdox NJAD20)

PROPERTIES:

Form:

Liquid (Chemcol T-20; Chemeen T-20; DeSomeen TA-20; Icomeen T-20; Polyfac TA-20; Trymeen 6607, TAM-20)
Liquid (may become hazy) (Atlas G-3780A)
Clear liquid (Katapol PN-810)
Liquid/paste (Hetoxamine T20)
Solid (Serdox NJAD20)

Color:

Amber (Atlas G-3780A; Chemcol T-20; Katapol PN-810; Polyfac TA-20)

Gardner 6 (Trymeen 6607, TAM-20)
Gardner 8 max. (Icomeen T-20)
Composition:
99% active (Katapol PN-810)
100% active (Atlas G-3780A; Chemcol T-20; Chemeen T-20; DeSomeen TA-20; Icomeen T-20; Serdox NJAD20; Trymeen TAM-20)
Solubility:
Sol. in acetone (Atlas G-3780A)
Sol. in lower alcohols (Atlas G-3780A)
Sol. in ethyl acetate (Atlas G-3780A)
Sol. in ethylene glycol (Atlas G-3780A)
Sol. in isopropanol (Hetoxamine T20)
Sol. in water (Atlas G-3780A; Hetoxamine T20; Icomeen T-20; Katapol PN-810; Trymeen 6607)
Ionic Nature:
Nonionic (Atlas G-3780A; Serdox NJAD20)
Cationic (Chemcol T-20; Hetoxamine T20; Katapol PN-810; Trymeen TAM-20)
Cationic/nonionic (Icomeen T-20)
M.W.:
1120 (Chemeen T-20)
1145 (Icomeen T-20)
1150 (Hetoxamine T20)
Sp.gr.:
1.04 (Atlas G-3780A; Icomeen T-20)
1.05 (Polyfac TA-20)
Density:
8.7 lb/gal (Chemcol T-20; Trymeen 6607, TAM-20)
8.74 lb/gal (Polyfac TA-20)
Visc.:
119 cSt (100 F) (Trymeen 6607)
200 cps (Polyfac TA-20)
240 cs (Trymeen TAM-20)
≈ 250 cps (Atlas G-3780A)
Pour Pt.:
–2 C (Trymeen 6607)
10 C (Chemcol T-20)
Flash Pt.:
> 300 F (Atlas G-3780A); (PMCC) (Polyfac TA-20)
550 F (Trymeen 6607)
Cloud Pt.:
87 C (10% saline) (Trymeen 6607)
> 100 C (Chemcol T-20; Trymeen TAM-20)
179–181 F (10% NaCl) (DeSomeen TA-20)

POE (20) tallow amine *(cont'd.)*

HLB:
15.4 (Icomeen T-20; Trymeen 6607, TAM-20)
15.5 (Atlas G-3780A)
pH:
9 (5% aq.) (Icomeen T-20)
10.0 (1% aq.) (Polyfac TA-20)
11.59 (Chemcol T-20)

POP (30) lanolin ether

SYNONYMS:
PPG-30 lanolin ether (CTFA)
PPG (30) lanolin ether
CAS No.:
68439-53-2 (generic)
RD No.: 977055-71-2
TRADENAME EQUIVALENTS:
Hetoxol PLA [Heterene]
Wickenol 727 [CasChem]
CATEGORY:
Emollient, conditioner
APPLICATIONS:
Cosmetic industry preparations: (Hetoxol PLA; Wickenol 727)
PROPERTIES:
Form:
Liquid (Hetoxol PLA; Wickenol 727)
Odor:
Low (Wickenol 727)
Composition:
99% active (Hetoxol PLA)
Solubility:
Sol. in animal oils (Wickenol 727)
Sol. in ethanol (Wickenol 727)
Insol. in glycerin (Wickenol 727)
Sol. in isopropanol (Hetoxol PLA)
Sol. in min. oil (Hetoxol PLA)
Insol. in propylene glycol (Wickenol 727)
Sol. in veg. oils (Wickenol 727)
Insol. in water (Hetoxol PLA; Wickenol 727)
Acid No.:
1.0 max. (Hetoxol PLA)

SYNONYMS:
   PPG-10 methyl glucose ether (CTFA)
   PPG (10) methyl glucose ether
STRUCTURE:

$$CH_3(C_6H_{10}O_5)(OCHCH_2)_nOH$$
$$|$$
$$CH_3$$

   where avg. $n = 10$
CAS No.:
   RD No.: 977067-01-8
TRADENAME EQUIVALENTS:
   Glucam P-10 [Amerchol]
CATEGORY:
   Emollient, cosolvent, humectant, spreading aid, freezing point depressant, conditioner
APPLICATIONS:
   Bath products: afterbath friction lotion (Glucam P-10)
   Cleansers: personal cleansers (Glucam P-10)
   Cosmetic industry preparations: (Glucam P-10); astringents (Glucam P-10); hair preparations (Glucam P-10); makeup (Glucam P-10); shampoos (Glucam P-10); shaving preparations (Glucam P-10)
   Pharmaceutical applications: (Glucam P-10); topical preparations (Glucam P-10)
PROPERTIES:
Form:
   Viscous syrup (Glucam P-10)
Color:
   Pale yellow (Glucam P-10)
Odor:
   Practically odorless (Glucam P-10)
Solubility:
   Freely sol. in alcohols (Glucam P-10)
   Sol. in castor oil (Glucam P-10)
   Sol. in esters (Glucam P-10)
   Sol. in ethanol (Glucam P-10)
   Freely sol. in hydroalcoholic systems (Glucam P-10)
   Sol. in isopropyl myristate to 1% (Glucam P-10)
   Sol. in isopropyl palmitate to 1% (Glucam P-10)
   Freely sol. in water (Glucam P-10)
Visc.:
   8500 cps (Glucam P-10)
Acid No.:
   1.0 max. (Glucam P-10)
Iodine No.:
   0.5 max. (Glucam P-10)

POP (10) methyl glucose ether (cont'd.)

Saponification No.:
   1.0 max. (Glucam P-10)
Hydroxyl No.:
   302 (Glucam P-10)

POP (20) methyl glucose ether

SYNONYMS:
   PPG-20 methyl glucose ether (CTFA)
   PPG (20) methyl glucose ether
STRUCTURE:

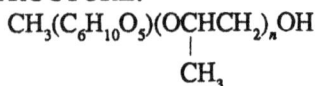

$$CH_3(C_6H_{10}O_5)(OCHCH_2)_nOH$$
$$\underset{CH_3}{|}$$

   where avg. $n = 20$
CAS No.:
   RD No.: 977067-02-9
TRADENAME EQUIVALENTS:
   Glucam P-20 [Amerchol]
CATEGORY:
   Emollient, cosolvent, humectant, spreading aid, freezing point depressant, conditioner
APPLICATIONS:
   Bath products: afterbath friction lotion (Glucam P-20)
   Cleansers: personal cleansers (Glucam P-20)
   Cosmetic industry preparations: (Glucam P-20); astringents (Glucam P-20); hair preparations (Glucam P-20); makeup (Glucam P-20); shampoos (Glucam P-20); shaving preparations (Glucam P-20)
   Pharmaceutical applications: (Glucam P-20); topical preparations (Glucam P-20)
PROPERTIES:
Form:
   Viscous syrup (Glucam P-20)
Color:
   Pale yellow (Glucam P-20)
Odor:
   Practically odorless (Glucam P-20)
Solubility:
   Freely sol. in alcohols (Glucam P-20)
   Sol. in castor oil (Glucam P-20)
   Sol. in esters (Glucam P-20)
   Sol. in ethanol (Glucam P-20)
   Freely sol. in hydroalcoholic systems (Glucam P-20)

POP (20) methyl glucose ether *(cont'd.)*

Sol. in isopropyl myristate to 10% (Glucam P-20)
Sol. in isopropyl palmitate to 5% (Glucam P-20)
Freely sol. in water (Glucam P-20)
Visc.:
1700 cps (Glucam P-20)
Acid No.:
1.0 max. (Glucam P-20)
Iodine No.:
0.5 max. (Glucam P-20)
Saponification No.:
1.0 max. (Glucam P-20)
Hydroxyl No.:
176 (Glucam P-20)

POP (2) myristyl ether propionate

SYNONYMS:
PPG-2 myristyl ether propionate (CTFA)
PPG (2) myristyl ether propionate
STRUCTURE:

$$CH_3(CH_2)_{12}CH_2(OCHCH_2)_2O—CCH_2CH_3$$
$$\underset{CH_3}{} \qquad \overset{O}{\underset{\parallel}{}}$$

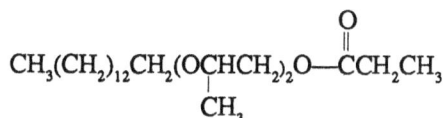

TRADENAME EQUIVALENTS:
Crodamol PMP [Croda; Croda Ltd.]
CATEGORY:
Emollient, solvent, stabilizer
APPLICATIONS:
Cosmetic industry preparations: (Crodamol PMP)
Pharmaceutical applications: sunscreens (Crodamol PMP)
PROPERTIES:
Form:
Liquid (Crodamol PMP)
Color:
Colorless (Crodamol PMP)
Solubility:
Sol. in alcohol (Crodamol PMP)
Sol. in aq. alcohol (Crodamol PMP)
Sol. in min. oil (Crodamol PMP)

POP (2) myristyl ether propionate *(cont'd.)*

Sol. in oils (Crodamol PMP)
Sol. in volatile silicone (Crodamol PMP)
Insol. in water (Crodamol PMP)
Cloud Pt.:
−5 C (Crodamol PMP)

POP (15) stearyl ether

SYNONYMS:
PPG-15 stearyl ether (CTFA)
PPG (15) stearyl ether
STRUCTURE:

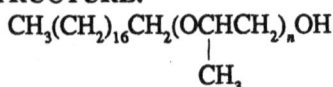

where avg. $n = 15$
CAS No.:
25231-21-4 (generic)
RD No.: 977061-23-6
TRADENAME EQUIVALENTS:
Arlamol E [ICI Americas]
Hetoxol SP-15 [Heterene]
Prostearyl 15 [Croda; Croda Ltd.]
CATEGORY:
Emollient, solvent, coupler
APPLICATIONS:
Bath products: bath oils (Prostearyl 15)
Cosmetic industry preparations: (Hetoxol SP-15; Prostearyl 15); hair preparations (Prostearyl 15); hand and body lotions (Prostearyl 15); personal care products (Arlamol E)
Pharmaceutical applications: aerosol antiperspirant/deodorant (Prostearyl 15); sunscreens (Prostearyl 15)
PROPERTIES:
Form:
Oily liquid (Arlamol E; Hetoxol SP-15)
Clear liquid (Prostearyl 15)
Color:
ASTM 100 max. (Hetoxol SP-15)
Composition:
100% active (Arlamol E)

Solubility:
Sol. in alcohols (Prostearyl 15)
Sol. in cottonseed oil (Arlamol E)
Sol. in ethanol (Arlamol E)
Sol. in hexadecyl alcohol (Arlamol E)
Sol. in isopropanol (Arlamol E; Hetoxol SP-15)
Sol. in isopropyl esters (Arlamol E)
Sol. in min. oil (Arlamol E; Hetoxol SP-15)
Insol. in water (Hetoxol SP-15; Prostearyl 15)
Ionic Nature:
Nonionic (Arlamol E)
Sp.gr.:
0.95 (Arlamol E)
Visc.:
90 cps (Arlamol E)
Pour Pt.:
< 0 C (Arlamol E)
Acid No.:
1.5 max. (Hetoxol SP-15)
Saponification No.:
2.0 max. (Hetoxol SP-15)
Hydroxyl No.:
62–70 (Hetoxol SP-15)

Potassium oleate *(CTFA)*

SYNONYMS:
9-Octadecenoic acid, potassium salt
Potassium 9-octadecenoate
EMPIRICAL FORMULA:
$C_{18}H_{34}O_2 \cdot K$
STRUCTURE:
$CH_3(CH_2)_7CH=CH(CH_2)_7COOK$
CAS No.:
143-18-0
TRADENAME EQUIVALENTS:
Atlas KO [Guelph Soap]
Emkapol PO-18 [Emkay]
Norfox KO [Norman, Fox]

Potassium oleate (cont'd.)

CATEGORY:
Lubricant, detergent, stabilizer, emulsifier, corrosion inhibitor
APPLICATIONS:
Cleansers: hand cleanser (Norfox KO)
Industrial applications: (Atlas KO); latex (Emkapol PO-18); lubricating/cutting oils
(Norfox KO); paint strippers (Norfox KO); textile/leather processing (Atlas KO)
PROPERTIES:
Form:
Liquid (Emkapol PO-18; Norfox KO)
Viscous paste (Atlas KO)
Composition:
80% active (Norfox KO)
Solubility:
Sol. in water (Atlas KO)
Ionic Nature:
Anionic (Emkapol PO-18; Norfox KO)
HLB:
20.0 (Norfox KO)

Propylene glycol dipelargonate (CTFA)

SYNONYMS:
Nonanoic acid, 1-methyl-1,2-ethanediyl ester
Propylene glycol dinonanoate
EMPIRICAL FORMULA:
$C_{21}H_{40}O_4$
STRUCTURE:

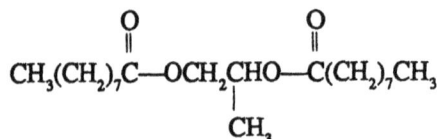

CAS No.:
41395-83-9
TRADENAME EQUIVALENTS:
DPPG [Gattefosse Ets.]
Emerest 2388 [Henkel/Emery]
Lexol PG 900 [Inolex]
Schercemol PGDP [Scher]

CATEGORY:
Emollient, lubricant, cosolvent, carrier
APPLICATIONS:
Bath products: bath oils (Emerest 2388; Lexol PG 900; Schercemol PGDP)
Cosmetic industry preparations: (DPPG); aerosols (Lexol PG 900); creams and lotions (Emerest 2388; Schercemol PGDP); makeup (Lexol PG 900); perfumery (Lexol PG 900; Schercemol PGDP); shaving preparations (Emerest 2388; Lexol PG 900)
Pharmaceutical applications: (DPPG)
PROPERTIES:
Form:
Liquid (Lexol PG 900)
Oily liquid (DPPG)
Clear liquid (Schercemol PGDP)
Color:
Colorless (Emerest 2388; Lexol PG 900)
Gardner 4 max. (Schercemol PGDP)
Odor:
Odorless (Emerest 2388; Lexol PG 900)
Slight, typical (Schercemol PGDP)
Composition:
100% active (Schercemol PGDP)
Solubility:
Sol. in alcohols (Schercemol PGDP)
Sol. in aliphatic hydrocarbons (Schercemol PGDP)
Sol. in aromatic hydrocarbons (Schercemol PGDP)
Sol. in castor oil (Emerest 2388)
Sol. in chlorinated hydrocarbons (Schercemol PGDP)
Sol. in esters (Schercemol PGDP)
Sol. in ethanol (Emerest 2388)
Insol. in glycerol (Emerest 2388)
Sol. in glycol ethers (Schercemol PGDP)
Disp. in glycols (Schercemol PGDP); insol. (Emerest 2388)
Sol. in isopropanol (Emerest 2388)
Sol. in ketones (Schercemol PGDP)
Sol. in lanolin (Emerest 2388)
Sol. in min. oil (Emerest 2388; Schercemol PGDP)
Disp. in polyols (Schercemol PGDP)
Sol. in silicone oil (Emerest 2388)
Disp. in triols (Schercemol PGDP)
Sol. in veg. oil (Schercemol PGDP)
Insol. in water (Emerest 2388; Schercemol PGDP)
Ionic Nature:
Nonionic (Schercemol PGDP)

Propylene glycol dipelargonate *(cont'd.)*

M.W.:
  360 (theoret.) (Schercemol PGDP)
Sp.gr.:
  0.917 (Schercemol PGDP)
Density:
  7.6 lb/gal (Schercemol PGDP)
F.P.:
  −25 C (Schercemol PGDP)
Flash Pt.:
  > 170 C (OC) (Schercemol PGDP)
Acid No.:
  5.0 max. (Schercemol PGDP)
Iodine No.:
  Nil (Schercemol PGDP)
Saponification No.:
  290–310 (Schercemol PGDP)
Ref. Index:
  1.440 (Schercemol PGDP)

Propylene glycol myristyl ether acetate (CTFA)

SYNONYMS:
  PPG-1 myristyl ether acetate
TRADENAME EQUIVALENTS:
  Hetester PMA [Heterene]
CATEGORY:
  Emollient, solvent, plasticizer
APPLICATIONS:
  Cosmetic industry preparations: emulsions (Hetester PMA); oil systems (Hetester PMA)
PROPERTIES:
Form:
  Clear liquid (Hetester PMA)
Color:
  Colorless (Hetester PMA)
Composition:
  100% active (Hetester PMA)
Solubility:
  Sol. in most cosmetic oils @ 5% (Hetester PMA)
  Insol. in 70% ethanol @ 5% (Hetester PMA)

Propylene glycol myristyl ether acetate (cont'd.)

Insol. in glycols @ 5% (Hetester PMA)

Insol. in water @ 5% (Hetester PMA)

Cloud Pt.:

0 C (neat) (Hetester PMA)

Acid No.:

0.2 max. (Hetester PMA)

Saponification No.:

140–160 (Hetester PMA)

Hydroxyl No.:

5.0 max. (Hetester PMA)

Stability:

If partially or fully solidified, product returns to complete liquidity and homogeneity
when allowed to warm to 25 C (Hetester PMA)

pH:

6.0–7.0 (5% in 50/50 IPA/water) (Hetester PMA)

Quaternium-27 (CTFA)

SYNONYMS:
Methyl-1-tallow amido ethyl-2-tallow imidazolinium-methyl sulfate

STRUCTURE:

$$\left[\begin{array}{c} \overset{\displaystyle O}{\underset{\displaystyle \|}{}} \\ RC-NH-CH_2 \\ | \\ CH_2 \\ | \\ R \underset{N}{\overset{}{=\!=}} N-CH_3 \end{array} \right]^{+} \quad CH_3OSO_3^{-}$$

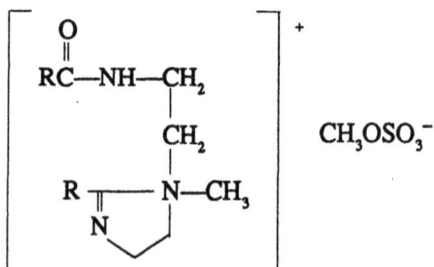

where R is derived from the tallow acid radical

CAS No.:
RD No.: 977065-92-1

TRADENAME EQUIVALENTS:
Carsosoft S-75, S-90, S-90M [Lonza]
Incroquat S-75CG [Croda]
Rewoquat W-7500, W-7500/H [Rewo; Rewo GmbH]
Varisoft 475 [Sherex]

CATEGORY:
Softener, conditioner, lubricant, antistat, rewetting agent

APPLICATIONS:
Cosmetic industry preparations: conditioners (Rewoquat W-7500/H); hair preparations (Incroquat S-75CG; Rewoquat W-7500/H)
Household products: fabric softener (Carsosoft S-75, S-90, S-90M; Rewoquat W-7500)
Industrial applications: commercial laundry softeners (Varisoft 475); textile/leather processing (Varisoft 475)

PROPERTIES:
Form:
Liquid (Carsosoft S-75, S-90, S-90M; Incroquat S-75CG; Varisoft 475)
Viscous liquid (Rewoquat W-7500)
Soft paste (Rewoquat W-7500/H)
Color:
Yellow (Incroquat S-75CG)

Gardner 4 (Carsosoft S-75, S-90M)
Gardner 5 max. (Varisoft 475)
Composition:
   75% active (Incroquat S-75CG; Rewoquat W-7500, W-7500/H)
   75% active in isopropanol (Carsosoft S-75)
   75–77% solids (Varisoft 475)
   88–91% active in isopropanol (Carsosoft S-90M)
   90% active (Carsosoft S-90)
Solubility:
   Readily disp. in warm and cold water (Carsosoft S-90M)
Ionic Nature:
   Cationic (Carsosoft S-75, S-90, S-90M; Incroquat S-75CG; Rewoquat W-7500, W-7500/H)
M.W.:
   720 (Carsosoft S-75)
Flash Pt.:
   72 F (PM) (Varisoft 475)
Stability:
   Excellent thermal stability (Carsosoft S-75, S-90M)
   Good hard water stability (Carsosoft S-90M)
pH:
   5.0–7.0 (5% sol'n.) (Carsosoft S-75)

Quaternium-52 (CTFA)

STRUCTURE:

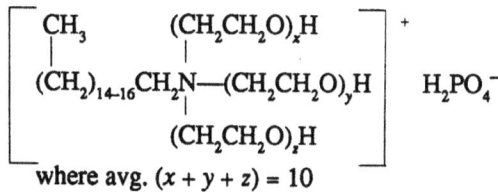

where avg. $(x + y + z) = 10$

CAS No.:
   58069-11-7
TRADENAME EQUIVALENTS:
   Dehyquart SP [Henkel]
CATEGORY:
   Surfactant, conditioner, softener, germicide, antistat, wetting agent, emulsifier, corrosion inhibitor

Quaternium-52 *(cont'd.)*

APPLICATIONS:
    Cosmetic industry preparations: (Dehyquart SP); conditioners (Dehyquart SP); hair preparations (Dehyquart SP)
    Industrial applications: metal corrosion protection (Dehyquart SP)

PROPERTIES:

Form:
    Clear viscous liquid (Dehyquart SP)

Color:
    Light yellow (Dehyquart SP)

Composition:
    49–51% solids (Dehyquart SP)

Ionic Nature:
    Cationic (Dehyquart SP)

pH:
    6.8–7.2 (10% sol'n.) (Dehyquart SP)

TOXICITY/HANDLING:
    Irritating to skin and eyes in conc. form (Dehyquart SP)

STORAGE/HANDLING:
    Store in sealed containers below 30 C and protected against frost to obtain 3 yr shelf life (Dehyquart SP)

STD. PKGS.:
    220 lb net plastic containers (Dehyquart SP)

Ricinoleamidopropyl betaine (CTFA)

SYNONYMS:

N-(Carboxymethyl)-N,N-dimethyl-3-[(1-oxoricinoleyl) amino]-1-propanaminium hydroxide, inner salt

1-Propanaminium, N-(carboxymethyl)-N,N-dimethyl-3-[(1-oxoricinoleyl) amino]-, hydroxide, inner salt

Ricinoleamidopropyl dimethyl glycine

EMPIRICAL FORMULA:

$C_{25}H_{48}N_2O_4$

STRUCTURE:

```
    OH
    |
CH₂CH(CH₂)₅CH₃
    |
    CH
    ||
    CH    O
    |     ||
   (CH₂)₇C
```

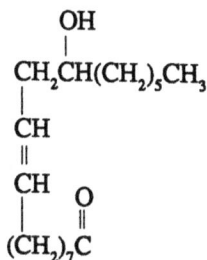

TRADENAME EQUIVALENTS:

Mackam RA [McIntyre]

Rewoteric AM-R40 [Rewo GmbH]

Varion AM-R40 [Sherex]

CATEGORY:

Conditioner

APPLICATIONS:

Cleansers: mild cleansers (Rewoteric AM-R40)

Cosmetic industry preparations: baby and child care formulations (Rewoteric AM-R40; Varion AM-R40); shampoos (Rewoteric AM-R40)

Pharmaceutical applications: medicated soaps (Rewoteric AM-R40)

PROPERTIES:

Form:

Liquid (Mackam RA; Rewoteric AM-R40; Varion AM-R40)

Color:

Amber (Mackam RA)

Composition:

40% active (Rewoteric AM-R40; Varion AM-R40)

Ricinoleamidopropyl betaine (cont'd.)

Ionic Nature:
Amphoteric (Rewoteric AM-R40; Varion AM-R40)

Ricinoleamidopropyl ethyldimonium ethosulfate (CTFA)

SYNONYMS:
N-Ethyl-N-N-dimethyl-3-[(1-oxoricinoleyl) amino]-1-propanaminium ethosulfate
1-Propanaminium, N-ethyl-N-N-dimethyl-3-[(1-oxoricinoleyl) amino]-ethosulfate

EMPIRICAL FORMULA:
$C_{25}H_{51}N_2O_2 \cdot C_2H_5O_4S$

STRUCTURE:

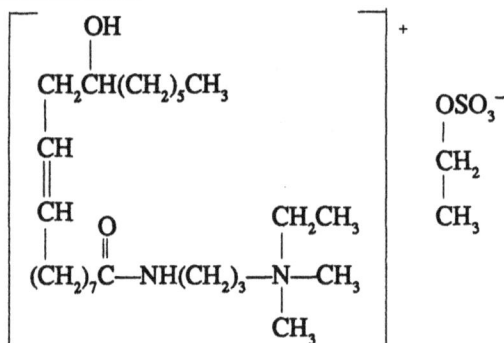

CAS No.:
112324-16-0

TRADENAME EQUIVALENTS:
Jordaquat JN [PPG-Mazer]
Lipoquat R [Lipo]

CATEGORY:
Conditioner, glosser, antistat, emollient, softener, substantivity agent

APPLICATIONS:
Cosmetic industry preparations: conditioners (Jordaquat JN); hair preparations (Jordaquat JN; Lipoquat R); skin preparations (Lipoquat R)

PROPERTIES:

Form:
Liquid (Jordaquat JN)
Viscous liquid (Lipoquat R)

Color:
Amber (Lipoquat R)

Odor:
Characteristic (Lipoquat R)

278

Ricinoleamidopropyl ethyldimonium ethosulfate (cont'd.)

Composition:
100% active (Jordaquat JN; Lipoquat R)
Solubility:
Sol. in water (Jordaquat JN; Lipoquat R)
Ionic Nature:
Cationic (Jordaquat JN; Lipoquat R)
pH:
6.5–7.5 (3% aq. sol'n.) (Lipoquat R)

Ricinoleic diethanolamide

SYNONYMS:
N,N-Bis(2-hydroxyethyl) ricinoleamide
Diethanolamine ricinoleic acid amide
12-Hydroxy-N,N-bis (2-hydroxyethyl)-9-octadecenamide
9-Octadecenamide, 12-hydroxy-N,N-bis (2-hydroxyethyl)-
Ricinoleamide DEA (CTFA)
Ricinoleoyl diethanolamide
EMPIRICAL FORMULA:
$C_{22}H_{43}NO_4$
STRUCTURE:

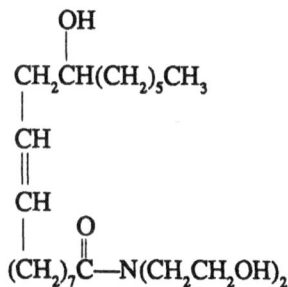

CAS No.:
40716-42-5
TRADENAME EQUIVALENTS:
Alkamide RDO [Alkaril; Alkaril Canada]
Aminol CA-2 [Finetex]
Mackamide R [McIntyre]
Rodea [Alcolac] (2:1)
CATEGORY:
Lubricant, emulsifier, softener, antiblocking agent, release agent, rust inhibitor
APPLICATIONS:
Cosmetic industry preparations: emulsions (Aminol CA-2)

Ricinoleic diethanolamide *(cont'd.)*

Household detergents: (Alkamide RDO)
Industrial applications: lubricating/cutting oils (Alkamide RDO; Rodea)
PROPERTIES:
Form:
Liquid (Alkamide RDO; Aminol CA-2; Mackamide R)
Viscous liquid (Rodea)
Color:
Amber (Alkamide RDO)
Dark amber (Rodea)
Composition:
100% active (Alkamide RDO; Aminol CA-2; Mackamide R)
Solubility:
Sol. in alcohols (Rodea)
Sol. in aliphatic hydrocarbons (Rodea)
Sol. in aromatic hydrocarbons (Rodea); disp. @ 1% (Alkamide RDO)
Sol. in esters (Rodea)
Sol. in glycol ethers (Rodea)
Sol. in glycols (Rodea)
Sol. in ketones (Rodea)
Disp. in min. oil @ 10% (Alkamide RDO)
Disp. in min. spirits @ 1% (Alkamide RDO)
Disp. in perchloroethylene @ 1% (Alkamide RDO)
Disp. in water (Rodea); disp. @ 10% (Alkamide RDO)
Ionic Nature:
Nonionic (Alkamide RDO; Aminol CA-2; Mackamide R)
Sp.gr.:
1.016 (Rodea)
Density:
1.0 g/ml (Alkamide RDO)
Congeal Pt.:
−20 to −10 C (Rodea)
pH:
8.0–11.0 (1% DW) (Alkamide RDO)
9.0–10.5 (1% disp.) (Rodea)

Sodium isostearoyl lactylate (CTFA)

SYNONYMS:
Sodium isostearoyl-2-lactylate

EMPIRICAL FORMULA:
$C_{24}H_{44}O_6 \cdot Na$

STRUCTURE:

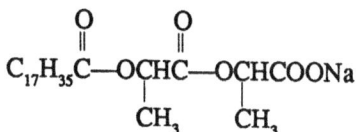

CAS No.:
66988-04-3

TRADENAME EQUIVALENTS:
Crodactil SISL [Croda]
Pationic ISL [RITA]

CATEGORY:
Emulsifier, emollient, solubilizer

APPLICATIONS:
Cosmetic industry preparations: (Pationic ISL); perfumery (Pationic ISL); skin preparations (Crodactil SISL; Pationic ISL)

PROPERTIES:
Form:
Liquid (Crodactil SISL)
Clear viscous liquid (Pationic ISL)

Color:
Straw-honey (Pationic ISL)

Composition:
100% active (Pationic ISL)

Solubility:
Sol. in isopropanol (Pationic ISL)
Sol. in isopropyl myristate (Pationic ISL)
Sol. in min. oil (Pationic ISL)
Sol. in oils (Pationic ISL)
Sol. in propylene glycol (Pationic ISL)
Disp. in water (Pationic ISL)

Sodium isostearoyl lactylate *(cont'd.)*

Ionic Nature:
Anionic (Crodactil SISL; Pationic ISL)
HLB:
5.9 (Pationic ISL)
Saponification No.:
205–225 (Pationic ISL)
pH:
6.30 (2% aq.) (Pationic ISL)
Surface Tension:
26.28 dynes/cm (Pationic ISL)

Sodium lactate (CTFA)

SYNONYMS:
2-Hydroxypropanoic acid, monosodium salt
Propanoic acid, 2-hydroxy-, monosodium salt
EMPIRICAL FORMULA:
$C_3H_6O_3 \cdot Na$
STRUCTURE:

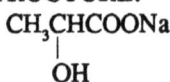

CAS No.:
72-17-3
TRADENAME EQUIVALENTS:
Patlac NAL [RITA]
Generically sold by:
[Croda Ltd.; Patco; Pfanstiehl; RITA]
CATEGORY:
Humectant, moisturizer, stabilizer, buffering agent, emollient, lubricant, filler, solvent, corrosion inhibitor
APPLICATIONS:
Cosmetic industry preparations: (generic—Croda, Pfanstiehl); creams and lotions (generic—Pfanstiehl, RITA)
Industrial applications: (generic—Pfanstiehl); antifreezes (generic—Pfanstiehl)
Pharmaceutical applications: (generic—RITA); hygiene products (generic—Pfanstiehl); ointments (generic—Pfanstiehl)
PROPERTIES:
Form:
Liquid (generic—Croda)

Sodium lactate *(cont'd.)*

Clear liquid (generic—Patco, RITA; Patlac NAL)
Syrupy liquid (generic—Pfanstiehl)
Color:
Colorless (generic—Pfanstiehl)
APHA 30 max. (generic—Patco)
Odor:
Essentially odorless (generic—Patco)
Odorless (generic—Pfanstiehl)
Composition:
60% active (generic—Patco)
60% solids in water (generic—Pfanstiehl)
Solubility:
Sol. in alcohols (generic—Pfanstiehl)
Sp.gr.:
1.31–1.34 (generic—Patco)
pH:
8.0–9.0 (generic—Patco)
STD. PKGS.:
600 lb net open-head steel drum with polyethylene liner; 50 lb net polyethylene
container-in-a-carton (generic—Patco)

Sodium lauroyl sarcosinate (CTFA)

SYNONYMS:
Glycine, N-methyl-N-(1-oxododecyl)-, sodium salt
N-Methyl-N-(1-oxododecyl) glycine, sodium salt
EMPIRICAL FORMULA:
$C_{15}H_{29}NO_3 \cdot$ Na
STRUCTURE:

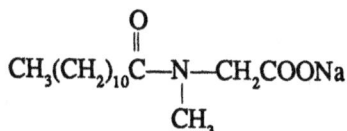

CAS No.:
137-16-6
TRADENAME EQUIVALENTS:
Crodasinic LS30, LS35 [Croda; Croda Ltd.]
Hamposyl L-30, L-95 [W.R. Grace]
Maprosyl 30 [Millmaster-Onyx]

Sodium lauroyl sarcosinate (cont'd.)

Nikkol Sarcosinate LN [Nikko]
Sarkosyl NL-30 [Ciba-Geigy]
Secosyl [Stepan Europe]

CATEGORY:
Surfactant, foaming agent, foam stabilizer, wetting agent, detergent, lubricant, conditioner, corrosion inhibitor, bacteriostat, antistat, penetrant, base, emulsifier

APPLICATIONS:
Cleansers: hand/skin cleanser (Hamposyl L-30, L-95); soaps (Crodasinic LS35)

Cosmetic industry preparations: (Hamposyl L-30, L-95); personal care products (Crodasinic LS30; Maprosyl 30; Secosyl); shampoos (Crodasinic LS35; Hamposyl L-30, L-95; Nikkol Sarcosinate LN; Sarkosyl NL-30); shaving preparations (Crodasinic LS35)

Food applications: food packaging (Crodasinic LS35)

Household detergents: (Crodasinic LS30, LS35; Maprosyl 30); carpet & upholstery shampoos (Crodasinic LS35; Hamposyl L-30, L-95; Sarkosyl NL-30; Secosyl); laundry detergent (Sarkosyl NL-30); window cleaners (Sarkosyl NL-30)

Industrial applications: (Crodasinic LS30); lubricating/cutting oils (Crodasinic LS35); polymers/polymerization (Crodasinic LS35)

Industrial cleaners: (Crodasinic LS35); textile scouring (Crodasinic LS35)

Pharmaceutical applications: (Crodasinic LS30, LS35; Sarkosyl NL-30); antiseptic soaps (Crodasinic LS35); dental preparations (Crodasinic LS30, LS35; Hamposyl L-95; Nikkol Sarcosinate LN; Sarkosyl NL-30); depilatories (Crodasinic LS35); mouthwash (Crodasinic LS35)

PROPERTIES:
Form:
Liquid (Hamposyl L-30; Maprosyl 30; Sarkosyl NL-30; Secosyl)
Clear liquid (Crodasinic LS30, LS35)
Powder (Hamposyl L-95; Nikkol Sarcosinate LN)

Color:
Colorless (Sarkosyl NL-30)
White (Hamposyl L-95)
APHA 60 (Hamposyl L-30)

Composition:
29.5–30.5% active (Maprosyl 30)
30% active (Crodasinic LS30; Hamposyl L-30; Sarkosyl NL-30; Secosyl)
35% active (Crodasinic LS35)
94% active (Hamposyl L-95)
100% active (Nikkol Sarcosinate LN)

Solubility:
Sol. in water (Crodasinic LS30, LS35; Hamposyl L-95); miscible (Hamposyl L-30)

Ionic Nature:
Anionic (Crodasinic LS30, LS35; Hamposyl L-30, L-95; Maprosyl 30; Nikkol Sarcosinate LN; Sarkosyl NL-30; Secosyl)

Sodium lauroyl sarcosinate *(cont'd.)*

Sp.gr.:
1.02–1.03 (Hamposyl L-30)
1.03 (25/20 C) (Maprosyl 30)
Density:
25 lb/ft³ (Hamposyl L-95)
Visc.:
30 cps (Hamposyl L-30)
F.P.:
−1 C (Hamposyl L-30)
Stability:
Stable to alkaline hydrolysis, under moderately acid conditions at normal temps.
(Crodasinic LS35)
pH:
7.5–8.5 (1% sol'n.) (Sarkosyl NL-30)
Surface Tension:
30 dynes/cm (Hamposyl L-30)
Biodegradable: (Crodasinic LS35; Hamposyl L-30, L-95)
STD. PKGS.:
55-gal drums (Hamposyl L-30, L-95)

Sodium POE (4) lauryl ether phosphate

SYNONYMS:
Sodium laureth-4 phosphate (CTFA)
Sodium PEG 200 lauryl ether phosphate
CAS No.:
42612-52-2 (generic)
RD No.: 977067-47-2
TRADENAME EQUIVALENTS:
Gafac MC-470F [GAF]
CATEGORY:
Lubricant, softener, antistat, detergent, emulsifier
APPLICATIONS:
Industrial applications: textile/leather processing (Gafac MC-470F)
PROPERTIES:
Form:
Clear viscous liquid (Gafac MC-470F)
Composition:
95% active (Gafac MC-470F)
Solubility:
Disp. in butyl Cellosolve (Gafac MC-470F)

Sodium POE (4) lauryl ether phosphate *(cont'd.)*

Disp. in ethanol (Gafac MC-470F)
Sol. in kerosene (Gafac MC-470F)
Sol. in min. oil (Gafac MC-470F)
Sol. in perchloroethylene (Gafac MC-470F)
Disp. in water (Gafac MC-470F)
Sol. in xylene (Gafac MC-470F)
Ionic Nature:
Anionic (Gafac MC-470F)
Sp.gr.:
1.02–1.04 (Gafac MC-470F)
Density:
8.6 lb/gal (Gafac MC-470F)
Pour Pt.:
0 C (Gafac MC-470F)
pH:
5.0–6.5 (10% sol'n.) (Gafac MC-470F)

Sodium stearate (CTFA)

SYNONYMS:
Octadecanoic acid, sodium salt
Sodium octadecanoate
EMPIRICAL FORMULA:
$C_{18}H_{36}O_2 \cdot Na$
STRUCTURE:
$CH_3(CH_2)_{16}COONa$
CAS No.:
822-16-2
TRADENAME EQUIVALENTS:
Norfox B [Norman, Fox]
Witco Sodium Stearate C-1, C-7, Heat Stable, Polymer Grade, T-1 [Witco]
Witco EA [Witco] (food grade)
CATEGORY:
Lubricant, gelling agent, thickener, stabilizer, emulsifier, opacifier, anticaking agent
APPLICATIONS:
Cosmetic industry preparations: (Norfox B; Witco Sodium Stearate C-1, C-7); creams
and lotions (Witco Sodium Stearate C-1, C-7); makeup (Witco Sodium Stearate C-1, C-7); perfumery (Witco Sodium Stearate C-1, C-7); shampoos (Witco Sodium Stearate C-1, C-7); toiletries (Witco Sodium Stearate C-1, C-7)
Food applications: (Witco Sodium Stearate EA)

Industrial applications: plastics (Witco Sodium Stearate Heat-Stable, Polymer Grade)
Pharmaceutical applications: antiperspirant/deodorant (Witco Sodium Stearate C-1, C-7)

PROPERTIES:
Form:
Powder (Witco Sodium Stearate C-1, C-7, EA, Heat-Stable, Polymer Grade, T-1)
Fineness:
35% through 100 mesh sieve (Witco Sodium Stearate C-1)
65% through 100 mesh sieve (Witco Sodium Stearate C-7)
92% through 100 mesh sieve (Witco Sodium Stearate EA)
96% through 100 mesh sieve (Witco Sodium Stearate Heat-Stable, Polymer Grade)
Color:
White (Witco Sodium Stearate C-1, C-7, EA, Heat-Stable, Polymer Grade, T-1)
Composition:
96% active (Norfox B)
98.5% active (Witco Sodium Stearate C-1, C-7, EA, Heat-Stable, Polymer Grade)
Solubility:
Sol. in hot alcohols with gelation on cooling (Witco Sodium Stearate C-1, C-7)
Insol. in benzene (Witco Sodium Stearate EA, Heat-Stable, Polymer Grade)
Insol. in carbon tetrachloride (Witco Sodium Stearate EA, Heat-Stable, Polymer Grade)
Insol. in esters (Witco Sodium Stearate EA, Heat-Stable, Polymer Grade)
Sol. in ethanol (Witco Sodium Stearate EA, Heat-Stable, Polymer Grade)
Sol. in hot glycols with gelation on cooling (Witco Sodium Stearate C-1, C-7)
Insol. in ketones (Witco Sodium Stearate EA, Heat-Stable, Polymer Grade)
Sol. in methanol (Witco Sodium Stearate EA, Heat-Stable, Polymer Grade)
Insol. in min. oils (Witco Sodium Stearate EA, Heat-Stable, Polymer Grade)
Insol. in oleic acid (Witco Sodium Stearate EA, Heat-Stable, Polymer Grade)
Insol. in toluene (Witco Sodium Stearate EA, Heat-Stable, Polymer Grade)
Insol. in turpentine (Witco Sodium Stearate EA, Heat-Stable, Polymer Grade)
Insol. in veg. oil (Witco Sodium Stearate EA, Heat-Stable, Polymer Grade)
Sol. in hot water (Witco Sodium Stearate EA, Heat-Stable, Polymer Grade); sol. in hot water with gelation on cooling (Witco Sodium Stearate C-1, C-7)
Insol. in waxes (Witco Sodium Stearate EA, Heat-Stable, Polymer Grade)
Insol. in xylene (Witco Sodium Stearate EA, Heat-Stable, Polymer Grade)
Ionic Nature:
Anionic (Norfox B)
Sp.gr.:
1.02 (Witco Sodium Stearate C-1, C-7, EA, Heat-Stable, Polymer Grade, T-1)
Softening Pt.:
205 C (Witco Sodium Stearate C-1, C-7, EA, Heat-Stable, Polymer Grade, T-1)

Soluble collagen (CTFA)

SYNONYMS:
Soluble animal collagen
TRADENAME EQUIVALENTS:
Collagen CLR [Henkel Canada]
Collasol [Croda; Croda Ltd.]
Pancogene S [Gattefosse Ets.]
Sollagen [Hormel]
CATEGORY:
Conditioner, humectant, film former
APPLICATIONS:
Cosmetic industry preparations: (Pancogene S; Sollagen); hair preparations (Sollagen); skin preparations (Collagen CLR; Collasol; Sollagen)
PROPERTIES:
Form:
Liquid (Collagen CLR)
Aq. dispersion (Collasol)
Aq. sol'n. (Pancogene S)
Composition:
1% sol'n.; 7.5–9.5% total solids (Sollagen)
Stability:
Not fully stable to freezing (Sollagen)
pH:
3.6–3.9 (Sollagen)
STORAGE/HANDLING:
Store at refrigerated temps. (35–55 F); protect from freezing; temps. over 105 F will allow protein degradation to occur (Sollagen)

Sorbitan monooleate

SYNONYMS:
1,4-Anhydro-D-glucitol, 6-(9-octadecenoate)
Anhydrosorbitol monooleate
D-Glucitol, 1,4-anhydro-, 6-(9-octadecenoate)
SMO
Sorbitan mono-9-octadecenoate
Sorbitan oleate (CTFA)
EMPIRICAL FORMULA:
$C_{24}H_{44}O_6$

STRUCTURE:

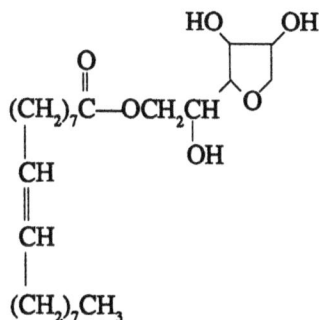

CAS No.:
1338-43-8; 5938-38-5

TRADENAME EQUIVALENTS:
Ablunol S-80 [Taiwan]
Ahco 832, 944 [ICI Americas]
Alkamuls SMO [Alkaril]
Anfomul S4, S50 [Croda Ltd.]
Arlacel 80 [ICI Americas]
Armotan MO [Akzo BV; Akzo Italia]
Atlas G-4884 [Atlas Refinery] (tech.)
Atpet 80, 100, 200 [ICI Americas; ICI Specialty]
Capmul O [Capital City]
Crill 4, 50 [Croda] (tech.)
Dehymuls SMO [Henkel KGaA]
DeSonic SMO [DeSoto]
DeSotan SMO [DeSoto]
Drewmulse SMO [Stepan/PVO]
Durtan 80 [Durkee]
Emasol O-10 [Kao]
Emsorb 2500 [Henkel/Emery]
Ethylan GO-80 [Harcros UK]
Glycomul O [Lonza]
Hodag SMO [Hodag]
Ionet S-80 [Sanyo]
Kuplur SMO [BASF]
Liposorb O [Lipo]
Lonzest SMO [Lonza]
MO-33-F [Hefti Ltd.]
Montane 80 [Seppic]
Newcol 3-80, 3-85, 80 [Nippon Nyukazai]
Nikkol SO-10 [Nikko]
Nissan Nonion OP-80R [Nippon Oils & Fats]

Sorbitan monooleate (cont'd.)

Norfox Sorbo S-80 [Norman, Fox]
Radiamuls 155, SORB 2155 [Synfina-Oleofin]
Radiasurf 7155 [Synfina-Oleofin]
Rheodol AO-10, SP-O10 [Kao]
S-Maz 80 [PPG-Mazer]
Soprofor S/80 [Rhone-Poulenc Geronazzo]
Sorbax SMO [Chemax]
Sorbon S-80 [Toho]
Sorgen 40, S-40-H [Dai-ichi Kogyo Seiyaku]
Span 80 [ICI Americas]

CATEGORY:
Emulsifier, coemulsifier, lubricant, softener, solubilizer, corrosion inhibitor, stabilizer, wetting agent, surfactant, detergent, dispersant, coupling agent, antifoam, cosolvent, thickener, viscosity control agent, release agent, antiblock agent, antistat, antistaling agent, spray-drying aid, whipping aid, descouring aid, superfatting agent, bodying aid, antifog, opacifier

APPLICATIONS:
Cosmetic industry preparations: (Alkamuls SMO; Armotan MO; Crill 4, 50; Drewmulse SMO; Emsorb 2500; Glycomul O; Ionet S-80; Nissan Nonion OP-80R; Radiasurf 7155; Rheodol AO-10; S-Maz 80; Sorbax SMO; Sorgen 40, S-40-H; Span 80); creams and lotions (Dehymuls SMO; Drewmulse SMO; Ionet S-80; MO-33-F); makeup (Crill 4); perfumery (Drewmulse SMO); shampoos (Drewmulse SMO; MO-33-F); shaving preparations (Drewmulse SMO)
Farm products: herbicides (Alkamuls SMO; Crill 50); insecticides/pesticides (Alkamuls SMO; Crill 4, 50; Durtan 80; Ionet S-80; Radiasurf 7155)
Food applications: (Alkamuls SMO; Capmul O; Crill 4, 50; Glycomul O; Lonzest SMO; Nikkol SO-10; Nissan Nonion OP-80R; Radiamuls 155, SORB 2155; S-Maz 80; Sorgen 40, S-40-H; Span 80); food packaging (Crill 50)
Household products: (Emsorb 2500; S-Maz 80); detergents (Radiasurf 7155; Span 80)
Industrial applications: (Crill 4; Glycomul O; S-Maz 80); concrete (Radiasurf 7155); dyes and pigments (Alkamuls SMO; Emasol O-10; Ionet S-80; Radiasurf 7155); explosives (Anfomul S4, S50); lubricating/cutting oils (Alkamuls SMO; Armotan MO; Crill 4; Ionet S-80; MO-33-F; Radiasurf 7155); metalworking (Crill 4, 50; Ionet S-80); paint mfg. (Crill 4, 50; MO-33-F; Rheodol SP-O10); paper mfg. (Radiasurf 7155); petroleum industry (Crill 4, 50; Radiasurf 7155); plastics (Crill 50; Durtan 80; Radiasurf 7155); polishes (Crill 4, 50; Durtan 80; MO-33-F); printing inks (Crill 4, 50; Durtan 80; Rheodol SP-O10); resins (Alkamuls SMO); textile/leather processing (Ablunol S-80; Ahco 832, 944; Alkamuls SMO; Crill 4, 50; DeSonic SMO; DeSotan SMO; Emsorb 2500; Ethylan GO-80; Ionet S-80; Lonzest SMO; Radiasurf 7155; Soprofor S/80; Sorbax SMO; Span 80); water treatment (Durtan 80); waxes, fats, and oils (Ablunol S-80; Alkamuls SMO; Emsorb 2500)

Sorbitan monooleate *(cont'd.)*

Industrial cleaners: drycleaning compositions (Crill 4; Radiasurf 7155); metal processing surfactants (Radiasurf 7155)

Pharmaceutical applications: (Armotan MO; Crill 4, 50; Drewmulse SMO; Glycomul O; Nissan Nonion OP-80R; Radiasurf 7155; Rheodol AO-10; Sorgen 40, S-40-H; Span 80); medicinals (Crill 4); ointments (Dehymuls SMO; Drewmulse SMO; Ionet S-80; MO-33-F); topical preparations (Drewmulse SMO); vitamins (Drewmulse SMO; Radiamuls 155, SORB 2155)

PROPERTIES:
Form:

Liquid (Ablunol S-80; Ahco 832, 944; Alkamuls SMO; Anfomul S4, S50; Armotan MO; Atlas G-4884; Atpet 80, 200; Crill 50; Dehymuls SMO; DeSonic SMO; DeSotan SMO; Drewmulse SMO; Emasol O-10; Emsorb 2500; Ethylan GO-80; Glycomul O; Hodag SMO; Ionet S-80; Kuplur SMO; Liposorb O; Lonzest SMO; MO-33-F; Montane 80; Newcol 3-80, 3-85, 80; Nikkol SO-10; Norfox Sorbo S-80; Radiamuls 155, SORB 2155; Radiasurf 7155; Rheodol AO-10, SP-O10; S-Maz 80; Soprofor S/80; Sorbax SMO; Sorbon S-80; Sorgen 40, S-40-H; Span 80)

Viscous liquid (Crill 4)

Viscous liquid—may gel on standing (Atpet 100)

Oily liquid (Arlacel 80; Nissan Nonion OP-80R)

Paste (Crill 50)

Plastic (Capmul O)

Solid (Capmul O)

Color:

Amber (Alkamuls SMO; Atpet 80; Crill 4; Glycomul O; Liposorb O; Radiasurf 7155; S-Maz 80; Span 80)

Yellow (Arlacel 80; Kuplur SMO)

Reddish brown (Atpet 100, 200)

Gardner 8 (Emsorb 2500)

Gardner 8 max. (Armotan MO)

Gardner 9 max. (Nissan Nonion OP-80R)

Odor:

Typical (Alkamuls SMO)

Composition:

97% active (Alkamuls SMO)

98% active (Crill 4, 50)

99% active (Kuplur SMO)

> 99% active (Armotan MO)

100% active (Ablunol S-80; Ahco 832, 944; Anfomul S4, S50; Arlacel 80; Atlas G-4884; Atpet 80; Capmul O; DeSonic SMO; DeSotan SMO; Emasol O-10; Emsorb 2500; Ethylan GO-80; Glycomul O; Hodag SMO; Ionet S-80; Liposorb O; Lonzest SMO; MO-33-F; Montane 80; Newcol 3-80, 3-85, 80; Nikkol SO-10; Nissan Nonion OP-80R; Norfox Sorbo S-80; Rheodol AO-10, SP-O10; S-Maz 80; Soprofor S/80; Sorbon S-80; Sorgen 40, S-40-H; Span 80)

Sorbitan monooleate (cont'd.)

Solubility:
Sol. in lower alcohols (Atpet 80, 100, 200)
Sol. in butyl stearate (Emsorb 2500)
Sol. in chlorinated hydrocarbons (Atpet 80, 100, 200)
Sol. in cottonseed oil (Arlacel 80); sol. @ 1% (Span 80)
Sol. in ethanol (Crill 4); miscible in certain proportions (Glycomul O; S-Maz 80)
Sol. in ethyl acetate (Glycomul O)
Disp. in glycerol trioleate (Emsorb 2500)
Insol. in glycols (Alkamuls SMO)
Sol. in hexane @ 10% (Radiasurf 7155)
Sol. in isopropanol (Arlacel 80); sol. @ 10% (Radiasurf 7155); sol. @ 1% (Span 80)
Sol. in isopropyl myristate (Crill 4)
Miscible with methanol in certain proportions (Glycomul O)
Sol. in min. oils (Arlacel 80; Crill 4; Emsorb 2500; Glycomul O; S-Maz 80); sol. @ 10% (Radiasurf 7155); sol. @ 1% (Span 80); sol. in most min. oils (Alkamuls SMO; Atpet 100, 200)
Miscible with naphtha in certain proportions (Glycomul O; S-Maz 80)
Sol. in oils (DeSonic SMO; DeSotan SMO)
Sol. in oleic acid (Crill 4)
Sol. in oleyl alcohol (Crill 4)
Sol. in olive oil (Crill 4)
Sol. in organic solvents (DeSonic SMO; DeSotan SMO)
Sol. in perchloroethylene (Emsorb 2500); sol. @ 1% (Span 80)
Sol. in most petroleum oils (Atpet 80)
Sol. in Stoddard solvent (Emsorb 2500)
Miscible with toluene in certain proportions (Glycomul O; S-Maz 80)
Sol. in trichlorethylene @ 10% (Radiasurf 7155)
Sol. in veg. oils (Glycomul O; S-Maz 80); sol. in most veg. oils (Alkamuls SMO; Atpet 80, 100, 200); sol. in veg. oils @ 10% (Radiasurf 7155)
Disp. in water (Atpet 80, 100, 200; Glycomul O; Kuplur SMO; S-Maz 80); insol. in water (Alkamuls SMO; DeSonic SMO; DeSotan SMO)
Sol. in xylene @ 1% (Span 80)

Ionic Nature:
Nonionic (Ablunol S-80; Ahco 832, 944; Alkamuls SMO; Anfomul S4, S50; Arlacel 80; Armotan MO; Atlas G-4884; Atpet 80, 100, 200; Capmul O; Dehymuls SMO; DeSonic SMO; Drewmulse SMO; Emasol O-10; Emsorb 2500; Glycomul O; Hodag SMO; Ionet S-80; Kuplur SMO; Liposorb O; Lonzest SMO; MO-33-F; Montane 80; Nikkol SO-10; Nissan Nonion OP-80R; Norfox Sorbo S-80; Radiamuls 155; Radiasurf 7155; Rheodol AO-10, SP-O10; Soprofor S/80; Sorbon S-80; Sorgen 40, S-40-H; Span 80)

M.W.:
510 (avg.) (Radiasurf 7155)

Sp.gr.:
0.987 (37.8 C) (Radiasurf 7155)
1.0 (Alkamuls SMO; Arlacel 80; Atpet 80, 100, 200; Glycomul O; S-Maz 80)
1.01 (Kuplur SMO); (20 C) (Armotan MO)
Density:
8.3 lb/gal (Emsorb 2500)
Visc.:
395.2 cps (37.8 C) (Radiasurf 7155)
≈ 900 cps (Atpet 100)
1000 cps (Emsorb 2500; Glycomul O; Kuplur SMO; S-Maz 80; Span 80)
≈ 1200 cps (Atpet 80, 200)
≈ 1900 cps (Arlacel 80)
9.5–11 poise (Armotan MO)
Pour Pt.:
≤ –12 C (Armotan MO)
≈ 0 F (Atpet 80, 200)
Flash Pt.:
249 C (COC) (Radiasurf 7155)
> 300 F (Arlacel 80; Atpet 80, 100, 200)
Fire Pt.:
> 300 F (Arlacel 80; Atpet 80, 100, 200)
Cloud Pt.:
7 C (Radiasurf 7155)
HLB:
4.3 (Ablunol S-80; Ahco 832, 944; Alkamuls SMO; Anfomul S4, S50; Arlacel 80; Atlas G-4884; Atpet 80, 100, 200; Capmul O; Crill 4, 50; DeSonic SMO; DeSotan SMO; Drewmulse SMO; Durtan 80; Emasol O-10; Glycomul O; Hodag SMO; Ionet S-80; Kuplur SMO; Montane 80; Nissan Nonion OP-80R; Norfox Sorbo S-80; Rheodol AO-10, SP-O10; S-Maz 80; Sorbax SMO; Sorgen 40, S-40-H; Span 80)
4.3 ± 1 (Liposorb O)
4.4 (MO-33-F)
4.5 (Ethylan GO-80)
4.6 (Emsorb 2500)
4.7 (Radiamuls 155, SORB 2155; Radiasurf 7155)
5.0 (Nikkol SO-10)
Acid No.:
7.0 max. (Radiamuls 155; Radiasurf 7155)
7.5 max. (Glycomul O; S-Maz 80)
8.0 max. (Kuplur SMO)
Iodine No.:
1.0 max. (Radiamuls 155)
60–75 (Radiasurf 7155)

Sorbitan monooleate *(cont'd.)*

Saponification No.:
   144–156 (Drewmulse SMO)
   145–160 (Liposorb O)
   145–161 (Radiasurf 7155)
   147–160 (Crill 4)
   148–161 (Durtan 80; Glycomul O)
   149–160 (Kuplur SMO; S-Maz 80)
Hydroxyl No.:
   193–209 (Glycomul O; S-Maz 80)
   193–210 (Liposorb O)
Ref. Index:
   1.4778 (Radiasurf 7155)
STD. PKGS.:
   55-gal steel drums (Arlacel 80)
   55-gal closed-head steel drums (Atpet 200)
   55-gal (450 lb net) steel drums (Kuplur SMO)
   190 kg net bung drums (Radiamuls 155, SORB 2155)
   190 kg net bung drums or bulk (Radiasurf 7155)

Sorbitan sesquioleate *(CTFA)*

SYNONYMS:
   Anhydrohexitol sesquioleate
   Anhydrosorbitol sesquioleate
   Sorbitan, 9-octadecenoate (2:3)
CAS No.:
   8007-43-0
TRADENAME EQUIVALENTS:
   Anfomul S43 [Croda Ltd.]
   Arlacel 83 [ICI Americas; ICI Specialty] (cosmetic/pharmaceutical grade)
   Arlacel C [ICI Americas; ICI Specialty]
   Crill 43 [Croda; Croda Ltd.]
   Dehymuls SSO [Henkel KGaA]
   Emasol O-15 R [Kao]
   Emsorb 2502 [Henkel/Emery]
   Glycomul SOC [Lonza]
   Liposorb SQO [Lipo]
   Montane 83 [Seppic]
   Nikkol SO-15 [Nikko]
   Nissan Nonion OP-83 RAT [Nippon Oils & Fats]

Sorbitan sesquioleate (cont'd.)

QO-33-F [Hefti Ltd.]
Rheodol AO-15 [Kao]
S-Maz 83R [PPG-Mazer]
Sorgen 30, S-30-H [Dai-ichi Kogyo Seiyaku]

CATEGORY:
Emulsifier, coemulsifier, lubricant, softener, wetting agent, dispersant, surfactant, coupling agent, solubilizer, antifoam, thickener, corrosion inhibitor, stabilizer

APPLICATIONS:
Cosmetic industry preparations: (Arlacel 83; Crill 43; Emsorb 2502; Glycomul SOC; Nissan Nonion OP-83 RAT; QO-33-F; Rheodol AO-15; Sorgen 30, S-30-H); makeup (Crill 43)
Farm products: insecticides/pesticides (Crill 43)
Food applications: (Crill 43; Glycomul SOC; Nissan Nonion OP-83 RAT; Sorgen 30, S-30-H); food emulsifying (Nikkol SO-15)
Household products: (Emsorb 2502)
Industrial applications: (Crill 43; Glycomul SOC); explosives (Anfomul S43); lubricating/cutting oils (Crill 43); metalworking (Crill 43); paint mfg. (Crill 43); petroleum industry (Crill 43); polishes and waxes (Crill 43); printing inks (Crill 43); textile/leather processing (Crill 43; Emsorb 2502; Nissan Nonion OP-83 RAT); waxes and oils (Dehymuls SSO; Emsorb 2502)
Industrial cleaners: drycleaning compositions (Crill 43)
Pharmaceutical applications: (Arlacel 83; Crill 43; Glycomul SOC; Nissan Nonion OP-83 RAT; Rheodol AO-15; Sorgen 30, S-30-H); medicinals (Crill 43)

PROPERTIES:
Form:
Liquid (Anfomul S43; Dehymuls SSO; Emasol O-15 R; Emsorb 2502; Glycomul SOC; Liposorb SQO; Montane 83; Nikkol SO-15; QO-33-F; Rheodol AO-15; S-Maz 83R; Sorgen 30, S-30-H)
Viscous liquid (Crill 43)
Oily liquid (Arlacel 83, C; Nissan Nonion OP-83 RAT)
Color:
Amber (Arlacel C; Crill 43; Glycomul SOC)
Yellow (Arlacel 83)
Gardner 6 (Emsorb 2502)
Gardner 9 max. (Nissan Nonion OP-83 RAT)
Composition:
98% active (Crill 43)
100% active (Anfomul S43; Arlacel 83, C; Emasol O-15 R; Emsorb 2502; Glycomul SOC; Liposorb SQO; Montane 83; Nikkol SO-15; Nissan Nonion OP-83 RAT; QO-33-F; Rheodol AO-15; S-Maz 83R; Sorgen 30, S-30-H)
Solubility:
Sol. in acetone (Nissan Nonion OP-83 RAT)
Sol. in butyl stearate (Emsorb 2502)

Sorbitan sesquioleate *(cont'd.)*

Sol. in cottonseed oil (Arlacel 83, C)
Sol. in ethanol (Arlacel 83, C; Glycomul SOC; Nissan Nonion OP-83 RAT)
Sol. in ethyl acetate (Glycomul SOC)
Sol. in ethyl ether (Nissan Nonion OP-83 RAT)
Sol. in glycerol trioleate (Emsorb 2502)
Sol. in isopropanol (Arlacel 83, C)
Sol. in kerosene (Nissan Nonion OP-83 RAT)
Sol. in methanol (Glycomul SOC); slightly sol. (Nissan Nonion OP-83 RAT)
Sol. in min. oil (Arlacel 83, C; Crill 43; Emsorb 2502); miscible in certain proportions
    (Glycomul SOC)
Sol. in oils (S-Maz 83R)
Sol. in oleic acid (Crill 43)
Sol. in oleyl alcohol (Crill 43)
Sol. in olive oil (Crill 43)
Sol. in perchloroethylene (Emsorb 2502)
Sol. in solvents (S-Maz 83R)
Sol. in Stoddard solvent (Emsorb 2502)
Sol. in tetrachloromethan (Nissan Nonion OP-83 RAT)
Disp. in water (S-Maz 83R); disp. warm (Nissan Nonion OP-83 RAT); insol. in water
    (Emsorb 2502)
Sol. in xylene (Nissan Nonion OP-83 RAT)
Ionic Nature:
Nonionic (Arlacel 83, C; Emasol O-15 R; Emsorb 2502; Glycomul SOC; Montane 83;
    Nikkol SO-15; QO-33-F; Rheodol AO-15; S-Maz 83R; Sorgen 30, S-30-H)
Sp.gr.:
1.0 (Arlacel 83, C; Glycomul SOC)
Density:
8.2 lb/gal (Emsorb 2502)
Visc.:
1000 cps (Arlacel C; Glycomul SOC)
1500 cps (Arlacel 83; Emsorb 2502)
Solidification Pt.:
0 C max. 9Nissan Nonion OP-83 RAT)
Flash Pt.:
> 300 F (Arlacel 83, C)
Fire Pt.:
> 300 F (Arlacel 83, C)
HLB:
3.7 (Anfomul S43; Arlacel 83, C; Crill 43; Emasol O-15 R; Emsorb 2502; Liposorb
    SQO; Montane 83; Nissan Nonion OP-83 RAT; Rheodol AO-15; S-Maz 83R;
    Sorgen 30, S-30-H)
4.0 (Glycomul SOC; QO-33-F)
4.5 (Nikkol SO-15)

Acid No.:
  7.0 max. (Nissan Nonion OP-83 RAT)
  7.5–11.5 (Glycomul SOC)
Saponification No.:
  149–160 (Crill 43)
  149–166 (Glycomul SOC)
Hydroxyl No.:
  184–206 (Glycomul SOC)

Soy acid (CTFA)

SYNONYMS:
  Fatty acids, soya
CAS No.:
  68308-53-2
TRADENAME EQUIVALENTS:
  Industrene 225, 226 [Humko] (dist.)
  Industrene 225 FG, 226 FG [Humko]
CATEGORY:
  Lubricant, release agent, binder, defoamer, intermediate
APPLICATIONS:
  Food applications: (Industrene 225 FG, 226 FG); food additives (Industrene 226 FG);
    food emulsifiers (Industrene 225 FG, 226 FG)
  Industrial applications: abrasives (Industrene 225, 226); lubricating/cutting oils (Industrene 225, 226); polishes and waxes (Industrene 225, 226); resins (Industrene 225, 226); rubber (Industrene 225, 226); water repellents (Industrene 225, 226)
  Industrial cleaners: soaps (Industrene 225, 226)
PROPERTIES:
Form:
  Liquid (Industrene 225, 225 FG, 226, 226 FG)
Color:
  Gardner 3–4 (Industrene 225, 226)
  Lovibond 25.0Y/2.5R ($5^{1}/_{4}$ in. cell) (Industrene 226 FG)
Composition:
  100% active (Industrene 225, 225 FG, 226, 226 FG)
Solidification Pt.:
  26 C max. (Industrene 226 FG)
Acid No.:
  195–203 (Industrene 226 FG)
  198–203 (Industrene 226)

Soy acid *(cont'd.)*

Iodine No.:
  125–135 (Industrene 226 FG)
Saponification No.:
  195–204 (Industrene 226 FG)
  197–204 (Industrene 225)
  199–204 (Industrene 226)

Soy sterol (CTFA)

CAS No.:
  RD No.: 977066-08-2
TRADENAME EQUIVALENTS:
  Generol 122 [Henkel]
CATEGORY:
  Emollient, emulsifier, stabilizer, viscosity modifier, solubilizer
APPLICATIONS:
  Cosmetic industry preparations: (Generol 122)
  Industrial applications: (Generol 122)
PROPERTIES:
Form:
  Waxy flake (Generol 122)
Color:
  White (Generol 122)
Composition:
  90% active (Generol 122)
Solubility:
  Sol. hot in ethanol (Generol 122)
  Sol. in isopropyl esters (Generol 122)
  Sol. in isostearic acid (Generol 122)
  Sol. hot in min. oil (Generol 122)
  Sol. hot in veg. oils (Generol 122)
  Insol. in water (Generol 122)
HLB:
  3.0 (Generol 122)

Stearamidoethyl diethylamine (CTFA)

SYNONYMS:

N-[2-(Diethylamino) ethyl] octadecanamide
Diethylaminoethyl stearamide
Octadecanamide, N-[2-(diethylamino) ethyl]-

EMPIRICAL FORMULA:

$C_{24}H_{50}N_2O$

STRUCTURE:

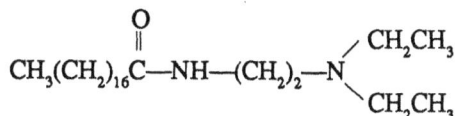

CAS No.:

16889-14-8

TRADENAME EQUIVALENTS:

Chemical Base 6532 [Sandoz]
Lexamine 22 [Inolex]

CATEGORY:

Conditioner, emulsifier, emollient, antistat, substantivity agent, intermediate

APPLICATIONS:

Cosmetic industry preparations: conditioners (Lexamine 22); acid creams and lotions
(Chemical Base 6532; Lexamine 22); hair preparations (Chemical Base 6532);
shampoos (Lexamine 22); skin preparations (Chemical Base 6532)

PROPERTIES:

Form:

Solid (Chemical Base 6532; Lexamine 22)

Color:

Yellowish tan (Chemical Base 6532)

Odor:

Slight (Chemical Base 6532)

Composition:

100% active (Chemical Base 6532; Lexamine 22)

Solubility:

Sol. in acidic media (Lexamine 22)

Sol. in water as the acid salt (Chemical Base 6532)

Sol. in water when neutralized with water-sol. acids such as citric, lactic, etc. and with
weak inorganic acids (Lexamine 22)

Ionic Nature:

Cationic (Chemical Base 6532; Lexamine 22)

M.P.:

35–45 C (Chemical Base 6532)

Stability:

Good (Chemical Base 6532)

Stearamidoethyl diethylamine *(cont' d.)*

STD. PKGS.:
200 lb fiber drums (Chemical Base 6532)

Stearamidoethyl ethanolamine *(CTFA)*

SYNONYMS:
Ethanolaminoethyl stearamide
N-[2-[(2-Hydroxyethyl) amino] ethyl] octadecanamide
Octadecanamide, N-[2-[(2-hydroxyethyl) amino] ethyl]-
EMPIRICAL FORMULA:
$C_{22}H_{46}N_2O_2$
STRUCTURE:

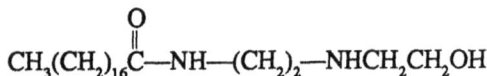

$$CH_3(CH_2)_{16}\overset{\displaystyle O}{\overset{\|}{C}}-NH-(CH_2)_2-NHCH_2CH_2OH$$

CAS No.:
141-21-9
TRADENAME EQUIVALENTS:
Chemical 39 Base [Sandoz]
CATEGORY:
Conditioner, emollient, emulsifier, base, lubricant
APPLICATIONS:
Cosmetic industry preparations: acid creams and lotions (Chemical 39 Base); conditioners (Chemical 39 Base); cosmetic base (Chemical 39 Base); hair preparations (Chemical 39 Base); shaving preparations (Chemical 39 Base); skin preparations (Chemical 39 Base); toiletries (Chemical 39 Base)
PROPERTIES:
Form:
Waxy solid (Chemical 39 Base)
Color:
Gardner 3 (Chemical 39 Base)
Odor:
Very slight amine (Chemical 39 Base)
Solubility:
Insol. in water; disp. as the amine salt (Chemical 39 Base)
Ionic Nature:
Cationic (Chemical 39 Base)

Stearamidopropyl dimethylamine (CTFA)

SYNONYMS:

N-[3-(Dimethylamino) propyl] octadecanamide

Dimethylaminopropyl stearamide

Octadecanamide, N-[3-(dimethylamino) propyl]-

EMPIRICAL FORMULA:

$C_{23}H_{48}N_2O$

STRUCTURE:

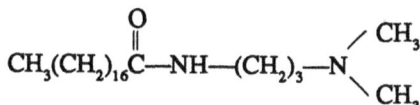

$$CH_3(CH_2)_{16}\overset{\overset{\displaystyle O}{\|}}{C}-NH-(CH_2)_3-N\overset{\diagup CH_3}{\diagdown CH_3}$$

CAS No.:

7651-02-7

TRADENAME EQUIVALENTS:

Cyclomide SODI [Alcolac]

Incromine SB [Croda]

Jordamine DAPSA, S-13 [PPG-Mazer]

Lexamine S13 [Inolex]

Mackine 301 [McIntyre]

Miramine DD [Miranol]

Schercodine S [Scher]

Tegamine 18 [Goldschmidt]

Tego-Amid S18 [Goldschmidt AG]

CATEGORY:

Conditioner, softener, emollient, emulsifier, intermediate

APPLICATIONS:

Bath products: (Tegamine 18)

Cosmetic industry preparations: conditioners (Jordamine DAPSA, S-13; Lexamine S13; Mackine 301); creams and lotions (Incromine SB; Lexamine S13; Tego-Amid S18); emulsions (Cyclomide SODI); hair preparations (Incromine SB; Lexamine S13; Mackine 301; Miramine DD; Schercodine S; Tegamine 18; Tego-Amid S18); personal care products (Incromine SB); shampoos (Jordamine DAPSA, S-13; Lexamine S13); skin preparations (Jordamine DAPSA, S-13; Miramine DD; Schercodine S)

Industrial applications: lubricating/cutting oils (Jordamine DAPSA, S-13)

PROPERTIES:

Form:

Solid (Cyclomide SODI; Jordamine DAPSA, S-13; Mackine 301; Miramine DD)

Flakes (Lexamine S13; Tegamine 18; Tego-Amid S18)

Waxy flakes (Incromine SB)

Hard wax (Schercodine S)

Color:

Off-white (Cyclomide SODI)

Stearamidopropyl dimethylamine *(cont'd.)*

  Cream (Incromine SB)
  Tan (Schercodine S)
Odor:
  Faint characteristic amine (Incromine SB)
Composition:
  98% amide min. (Schercodine S)
  100% active (Cyclomide SODI; Incromine SB; Lexamine S13; Mackine 301; Miramine DD; Tegamine 18)
Solubility:
  Sol. in min. oil (Cyclomide SODI)
  Sol. in organic solvents (Schercodine S)
  Disp. in water (Cyclomide SODI); insol. in water, but forms water-sol. salts with certain acids (Incromine SB)
Ionic Nature:
  Nonionic/cationic (Incromine SB)
  Cationic (Lexamine S13; Mackine 301; Miramine DD; Tegamine 18; Tego-Amid S18)
M.W.:
  368 (Schercodine S)
M.P.:
  63 C (Incromine SB)
  65–70 C (Schercodine S)
HLB:
  7.8 (Lexamine S13)
Acid No.:
  4 max. (Incromine SB)
Alkali No.:
  145–155 (Schercodine S)
Equivalent Wt.:
  369–379 (Incromine SB)

Stearamidopropyl dimethylamine lactate *(CTFA)*

SYNONYMS:
  Propanoic acid, 2-hydroxy-, compd. with N-[3-(dimethylamino) propyl] octadecanamide
STRUCTURE:
  $C_{23}H_{48}N_2O \cdot C_3H_6O_3$
CAS No.:
  55819-53-9

Stearamidopropyl dimethylamine lactate *(cont'd.)*

TRADENAME EQUIVALENTS:
Emcol 3780 [Witco]
Hetamine 5L25 [Heterene]
Incromate SDL [Croda]
Jordamine S-13 Lactate [PPG-Mazer]
Lexamine S-13 Lactate [Inolex]
Mackalene 316 [McIntyre]
Richamate 3780 [Richardson]

CATEGORY:
Conditioner, softener, surfactant, antistat, emollient, foaming agent, substantive agent, solubilizer, emulsifier

APPLICATIONS:
Cosmetic industry preparations: (Richamate 3780); conditioners (Emcol 3780; Jordamine S-13 Lactate; Lexamine S-13 Lactate; Mackalene 316); hair preparations (Hetamine 5L25; Incromate SDL; Jordamine S-13 Lactate; Mackalene 316; Richamate 3780); personal care products (Richamate 3780); shampoos (Emcol 3780; Incromate SDL; Jordamine S-13 Lactate; Lexamine S-13 Lactate); skin preparations (Richamate 3780)
Household products: fabric softener (Incromate SDL)

PROPERTIES:
Form:
Liquid (Emcol 3780; Jordamine S-13 Lactate; Lexamine S-13 Lactate; Mackalene 316)
Stratified liquid to paste (Hetamine 5L25)
Slurry (Incromate SDL; Richamate 3780)
Color:
Gardner 3 (Richamate 3780)
Gardner 3 max. (Incromate SDL)
Gardner 4 max. (Hetamine 5L25)
Composition:
19–21% solids (Hetamine 5L25)
23% active in water (Jordamine S-13 Lactate)
24–26% active (Incromate SDL)
25% active (Emcol 3780; Mackalene 316)
25% solids (Richamate 3780)
Solubility:
Sol. in water (Emcol 3780)
Ionic Nature:
Cationic (Emcol 3780; Lexamine S-13 Lactate; Mackalene 316; Richamate 3780)
Acts like a cationic but compatible with anionics (Hetamine 5L25; Incromate SDL)
Density:
8.3 lb/gal (Richamate 3780)

303

Stearamidopropyl dimethylamine lactate *(cont'd.)*

pH:

4.0–5.0 (Hetamine 5L25; Richamate 3780); (10% sol'n.) (Incromate SDL)

TOXICITY/HANDLING:

Conc. surfactant—avoid prolonged contact with skin and eyes (Incromate SDL)

STORAGE/HANDLING:

Store in cool, dry place (Incromate SDL)

STD. PKGS.:

55-gal (450 lb net) polyethylene-lined leverpak (Incromate SDL)

Stearic acid (CTFA)

SYNONYMS:

n-Octadecanoic acid

EMPIRICAL FORMULA:

$C_{18}H_{36}O_2$

STRUCTURE:

$CH_3(CH_2)_{16}COOH$

CAS No.:

57-11-4

TRADENAME EQUIVALENTS:

Crosterene [Croda Universal]

Dar-Chem 11 [Unichema] (single-pressed)

Dar-Chem 12 [Unichema] (double-pressed)

Dar-Chem 13, 14 [Unichema] (triple-pressed)

Emersol 110, 120 [Henkel/Emery]

Emersol 132 [Henkel/Emery] (triple-pressed)

Emersol Stearic Acid [Henkel/Emery]

Emery 400 [Henkel/Emery]

Groco 55L, 56R Rubber Grade Stearic Acid, 57, 58 [A. Gross]

Harwick F-300, F-1000, F-1500 [Harwick]

Hydrofol Acid 1655-CG-NF, 1855, 1895 [Sherex]

Hystrene 5016 [Humko/Witco] (low IV triple-pressed)

Hystrene 5016 NF FG [Humko/Witco] (triple-pressed food grade)

Hystrene 7018, 7018 FG, 8018, 9018, 9718, 9718 NF FG [Humko/Witco]

Industrene 5016 [Humko/Witco] (double-pressed)

Industrene 5016 FG, 7018, 7018 FG, 8518, 8718 FG, 9018 [Humko/Witco]

Loxiol G20 [Henkel KGaA]

Petrac 250 [Synthetic Products] (rubber grade)

Petrac 270 [Synthetic Products] (tech.)

Radiacid 408, 411, 420, 423 [Synfina-Oleofin]

Stearic Acid-RECCO [R.E. Carroll] (single, double, triple-pressed)

Stearic acid (cont'd.)

Generically sold by:
[C.P. Hall (rubber grade); Housmex (rubber grade); Humko; H.M. Royal; Smith

CATEGORY:
Lubricant, release agent, superfatting agent, opacifier, pearlizing agent, binder, activator, plasticizer, softener, emulsifier, textile auxiliary, intermediate, defoamer, dispersant, thickener, gelling agent, viscosity depressant, stabilizer, processing aid

APPLICATIONS:
Cleansers: soaps (Dar-Chem 11, 12; Emersol 110, 120, 132)
Cosmetic industry preparations: (Dar-Chem 11, 12; Emersol 110, 120, 132; Hystrene 5016, 7018, 8018, 9018, 9718); creams and lotions (Crosterene; Dar-Chem 11, 12; Emersol 132); emulsions (Hydrofol Acid 1655-CG-NF, 1855, 1895); shampoos (Hystrene 5016, 7018, 8018, 9018, 9718)
Food applications: (Hystrene 5016 NF FG, 7018 FG, 9718 NF FG; Industrene 5016 FG)
Household detergents: (Dar-Chem 11, 12); soaps (Industrene 5016, 7018, 8518, 9018)
Industrial applications: abrasives (Industrene 5016, 7018, 8518, 9018); chemical specialties (Emersol 110, 120; Petrac 270); industrial processing (Harwick F-300, F-1000, F-1500); latex (generic—Hall, Humko, Royal, Smith; Harwick F-300, F-1000, F-1500); lubricating/cutting oils (Crosterene; Dar-Chem 13, 14; Emersol 132; Hystrene 5016, 7018, 8018, 9018, 9718; Industrene 5016, 7018, 8518, 9018; Petrac 250, 270); metallic soaps (Hystrene 5016, 7018, 8018, 9018, 9718); paint/enamel mfg. (Hystrene 5016, 7018, 8018, 9018, 9718); plastics (Loxiol G20; Petrac 270; Radiacid 408, 411, 420, 423); polishes and waxes (Crosterene; Dar-Chem 13, 14; Emersol 132; Industrene 5016, 7018, 8518, 9018; Petrac 250); resins (Industrene 5016, 7018, 8518, 9018); rubber (generic—Hall, Housmex, Humko, Royal, Smith; Groco 55L, 56R Rubber Grade Stearic Acid, 57, 58; Harwick F-300, F-1000, F-1500; Hystrene 5016, 7018, 8018, 9018, 9718; Industrene 5016, 7018, 8518, 9018; Petrac 250, 270); textile/leather processing (Dar-Chem 13, 14; Hystrene 5016, 7018, 8018, 9018, 9718); water repellents (Industrene 5016, 7018, 8518, 9018)
Pharmaceutical applications: (Hystrene 5016, 7018, 8018, 9018, 9718)

PROPERTIES:
Form:
Solid (Hystrene 5016, 7018, 8018, 9018, 9718; Industrene 5016, 5016 FG, 7018, 8518, 9018; Loxiol G20)
Wax-like solid (Crosterene)
Beads (Groco 56R Rubber Grade Stearic Acid)
Flakes (generic—Hall, Humko, Housmex, Royal, Smith; Groco 56R Rubber Grade Stearic Acid; Industrene 9018; Radiacid 408, 411, 420, 423; Stearic Acid-RECCO)
Powder (generic—Humko, Royal, Smith; Harwick F-300, F-1000, F-1500; Radiacid 408, 411, 420, 423; Stearic Acid-RECCO)
Slab (generic—Royal, Smith)

Stearic acid *(cont'd.)*

Color:

White (generic—Hall, Housmex, Humko, Royal, Smith; Harwick F-300, F-1000, F-1500; Stearic Acid-RECCO)

FAC 5 (Petrac 250)

Gardner 12 max. (Groco 56R Rubber Grade Stearic Acid)

Lovibond 1.0R ± 5.0Y (5$^1/_4$) (Groco 58)

Lovibond 2.0R ± 1.0Y (5$^1/_4$) (Groco 57)

Lovibond 0.3R ± 1.0Y (5$^1/_4$) (Groco 55L)

Lovibond 1.0Y/0.1R (5$^1/_4$) (Hystrene 5016 NF FG)

Lovibond 2.0Y/0.2R (5$^1/_4$) (Industrene 7018 FG, 8718 FG)

Lovibond 2.0Y/0.5R max. (Hydrofol Acid 1655-CG-NF)

Lovibond 3–5Y/0.3R max. (Hydrofol Acid 1855)

Lovibond 5Y/0.5R max. (Hydrofol Acid 1895)

Lovibond 7Y/0.7R (5$^1/_4$) (Petrac 270)

Lovibond 10.0Y/1.0R (5$^1/_4$) (Industrene 9018)

Odor:

Faint (Harwick F-1500)

Mild fat (generic—Hall, Humko, Royal, Smith)

Composition:

45% C_{18}, 59% C_{16} (Groco 55L)

48% C_{18}, 52% C_{16} (Hydrofol Acid 1655-CG-NF)

55% stearic acid (Hydrofol Acid 1855)

55% C_{18}, 28% C_{16} (Groco 56R Rubber Grade Stearic Acid)

60–65% stearic acid (Harwick F-1500)

65% C_{18} (Groco 57, 58)

70% stearic acid (Hystrene 7018; Industrene 7018, 7018 FG)

70–75% stearic acid (Harwick F-1000)

75–80% stearic acid (Harwick F-300)

80% stearic acid (Hystrene 8018)

85% stearic acid (Industrene 8518)

90% stearic acid (Hystrene 9018; Industrene 9018)

92% stearic acid (Hystrene 9718 NF FG; Industrene 8718 FG)

92+% stearic acid (Hystrene 9718)

95% stearic acid (Hydrofol Acid 1895)

100% conc. (Industrene 5016, 5016 FG)

Solubility:

Sol. in aliphatic hydrocarbons (Dar-Chem 11, 12, 13, 14)

Sol. in esters (Dar-Chem 11, 12, 13, 14)

Sol. in glycol ethers (Dar-Chem 11, 12, 13, 14)

Sol. in glycols (Dar-Chem 11, 12, 13, 14)

Sol. in min. oil USP (Dar-Chem 11, 12, 13, 14)

Sol. in water @ pH > 9 (Dar-Chem 11, 12, 13, 14)

Sp.gr.:
  0.84 (generic—Hall, Housmex, Royal, Smith)
  0.840–0.850 (80 C) (Loxiol G20)
Visc.:
  8–11 mPa•s (80 C) (Loxiol G20)
M.P.:
  70 C (generic—Hall)
  126–133 F (generic—Humko, Royal, Smith)
Set Pt.:
  54–55 C (Loxiol G20)

Solidification Pt.:
  52 C (generic—Hall)
  52–56 C (Harwick F-1500)
  52–62 C (Groco 56R Rubber Grade Stearic Acid)
  54.5–55.6 C (Groco 55L)
  54.5–56.5 C (Hystrene 5016, 5016 NF FG)
  55–60 C (Harwick F-300)
  56–60 C (Harwick F-1000)
  57–60 C (Groco 57)
  58–61 C (Groco 58)
  58–62 C (Industrene 7018 FG)
  61.0–62.5 C (Hystrene 7018 FG)
  64–68 C (Industrene 9018)
  66.5–68.0 C (Hystrene 9718 NF FG; Industrene 8718 FG)

Flash Pt.:
  > 180 C (Loxiol G20)

Acid No.:
  193 min. (Groco 56R Rubber Grade Stearic Acid)
  195–200 (Hystrene 9718)
  196–201 (Hystrene 9018, 9718 NF FG; Industrene 8718 FG, 9018)
  196–205 (Harwick F-1000, F-1500)
  197–209 (generic—Hall)
  198–203 (Hystrene 8018)
  199–205 (Hystrene 7018)
  200–205 (Hystrene 7018 FG)
  200–207 (Industrene 7018 FG)
  202–205 (Harwick F-300)
  202–207 (Groco 57, 58)
  205 (Petrac 270)
  206–210 (Hystrene 5016 NF FG)
  207–210 (Hystrene 5016; Loxiol G20)
  207–211 (Groco 55L)

Stearic acid *(cont'd.)*

Iodine No.:
0–15 (generic—Royal, Smith)
0.5 max. (Hystrene 5016, 5016 NF FG, 7018 FG)
0.8 max. (Hystrene 9718 NF FG; Industrene 8718 FG)
1.0 max. (Groco 55L; Loxiol G20)
1.5 max. (Industrene 7018 FG)
2.0 (Petrac 270; Industrene 9018)
2.0 max. (Groco 58; Harwick F-300)
6.0 max. (Groco 57; Harwick F-1000)
9.5 (generic—Hall)
14 max. (Groco 56R Rubber Grade Stearic Acid)
15 max. (Harwick F-1500)
Saponification No.:
194–202 (Hydrofol Acid 1895)
196–201 (Hystrene 9718, 9718 NF FG)
196–202 (Industrene 8718 FG)
196–206 (Harwick F-1000, F-1500)
197–202 (Hystrene 9018; Industrene 9018)
198–205 (Industrene 8518)
199 (Petrac 250)
199–204 (Hystrene 8018)
200–206 (Hystrene 7018, 7018 FG)
200–208 (Industrene 7018 FG)
201–208 (Industrene 7018)
202–205 (Harwick F-300)
203–208 (Groco 57, 58)
205–213 (Hydrofol Acid 1855)
206 (Petrac 270)
206–211 (Hystrene 5016 NF FG)
206–214 (Hydrofol Acid 1655-CG-NF)
208–211 (Hystrene 5016; Industrene 5016)
Ref. Index:
1.436–1.437 (60 C) (Loxiol G20)

Stearic acid amide

SYNONYMS:
Octadecanamide
Stearamide (CTFA)
EMPIRICAL FORMULA:
$C_{18}H_{37}NO$

STRUCTURE:

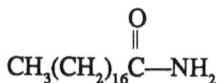

CAS No.:

124-26-5

TRADENAME EQUIVALENTS:

Armid 18 [Akzo/Armak]

Armoslip 18 [Akzo/Noury]

Crodamide S, SR [Croda Ltd.]

Kemamide S [Humko]

Petrac Vyn-Eze [Synthetic Products]

CATEGORY:

Lubricant, slip agent, release agent, antiblock agent, antitack agent, stabilizer, dispersant, raw material, water repellent, thickener

APPLICATIONS:

Food applications: food packaging (Armid 18; Petrac Vyn-Eze)

Household detergents: (Armid 18)

Industrial applications: coatings and films (Armid 18; Kemamide S; Petrac Vyn-Eze); dyes and pigments (Armid 18); lubricating/cutting oils (Kemamide S); metalworking (Kemamide S); paint mfg. (Armid 18); plastics (Armoslip 18; Crodamide S, SR; Kemamide S; Petrac Vyn-Eze); printing inks (Armid 18; Kemamide S); textile/leather processing (Armid 18; Kemamide S); waxes (Armid 18; Kemamide S)

PROPERTIES:

Form:

Solid (Kemamide S)

Flake (Armid 18; Armoslip 18; Petrac Vyn-Eze)

Pellets (Armoslip 18; Petrac Vyn-Eze)

Dry powder (Petrac Vyn-Eze)

Wax (Crodamide S, SR)

Color:

APHA 10 (Petrac Vyn-Eze)

Gardner 5 max. (Kemamide S)

Gardner 7 (Armid 18)

Odor:

Characteristic mild bland (Petrac Vyn-Eze)

Composition:

90% active (Armid 18)

99.5% total amide (Petrac Vyn-Eze)

100% active (Kemamide S)

Solubility:

Slightly sol. in organic solvents @ R.T. (Kemamide S)

Insol. in water (Armid 18; Kemamide S)

Stearic acid amide *(cont'd.)*

Sp.gr.:
0.52 (100 C) (Armid 18)
M.P.:
98–108 C (Kemamide S)
99–109 C (Armid 18)
104 C (Petrac Vyn-Eze)
Flash Pt.:
225 C (Armid 18)
Acid No.:
4.0 max. (Kemamide S)
Iodine No.:
3.0 max. (Kemamide S)
STD. PKGS.:
55-gal steel drums, cardboard cartons (Armid 18)

Stearoamphoacetate (CTFA)

SYNONYMS:
Stearoamphoglycinate
EMPIRICAL FORMULA:
$C_{24}H_{48}N_2O_4 \cdot Na$
STRUCTURE:

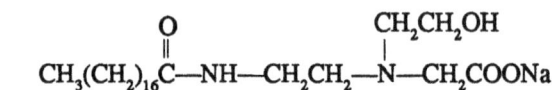

CAS No.:
68608-63-9
TRADENAME EQUIVALENTS:
Amphoterge S [Lonza]
Miranol DM, DM Conc. 45 [Miranol]
CATEGORY:
Lubricant, softener, conditioner, antistat
APPLICATIONS:
Cosmetic industry preparations: (Miranol DM Conc. 45); conditioners (Amphoterge S)
Household applications: fabric softener (Amphoterge S; Miranol DM, DM Conc. 45)
Industrial applications: textile/leather processing (Miranol DM, DM Conc. 45)
PROPERTIES:
Form:
Paste (Amphoterge S; Miranol DM Conc. 45)
Creamy paste; readily pourable at 60 C (Miranol DM)

Color:
   White (Miranol DM)
Composition:
   25% active (Amphoterge S)
   25–27% solids (Miranol DM)
Solubility:
   Disp. in water (Miranol DM, DM Conc. 45)
Ionic Nature:
   Amphoteric (Amphoterge S; Miranol DM)
pH:
   5.4–6.0 (65 C) (Miranol DM)
Biodegradable: (Amphoterge S; Miranol DM)

Stearyl betaine (CTFA)

SYNONYMS:
   N-(Carboxymethyl)-N,N-dimethyl-1-octadecanaminium hydroxide, inner salt
   1-Octadecanaminium, N-(carboxymethyl)-N,N-dimethyl-, hydroxide, inner salt
   Stearyl dimethyl glycine
EMPIRICAL FORMULA:
   $C_{22}H_{45}NO_2$
STRUCTURE:

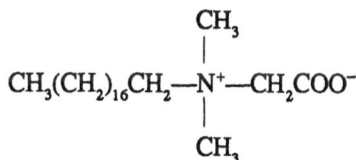

CAS No.:
   820-66-6
TRADENAME EQUIVALENTS:
   Alkateric STB [Alkaril]
   Chimin BX [Tessilchimica]
   Varion SDG [Sherex]
CATEGORY:
   Detergent, softener, surfactant, dispersant, wetting agent, antistat
APPLICATIONS:
   Cosmetic industry preparations: (Alkateric STB); hair preparations (Varion SDG)
   Household products: (Alkateric STB); fabric softener (Varion SDG)
   Industrial cleaners: institutional cleaners (Alkateric STB)

Stearyl betaine (cont'd.)

PROPERTIES:
Form:
   Liquid (Alkateric STB; Varion SDG)
   Soft paste (Chimin BX)
Color:
   Pale yellow (Alkateric STB)
Composition:
   40% active (Chimin BX)
   50% active (Varion SDG)
   50% active in water/alcohol (Alkateric STB)
Ionic Nature:
   Amphoteric (Alkateric STB; Chimin BX; Varion SDG)
pH:
   Neutral (Alkateric STB)

Stearyl dimethyl amine oxide

SYNONYMS:
   N,N-Dimethyl-1-octadecanamine-N-oxide
   1-Octadecanamine, N,N-dimethyl-, N-oxide
   Octadecyl dimethylamine oxide
   Stearamine oxide (CTFA)
   Stearylamine oxide
EMPIRICAL FORMULA:
   $C_{20}H_{43}NO$
STRUCTURE:

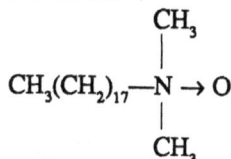

CAS No.:
   2571-88-2
TRADENAME EQUIVALENTS:
   Ammonyx SO [Millmaster-Onyx]
   Aromox DM18D-W [Akzo/Armak; Akzo BV]
   Conco XA-S [Continental Chem.]
   Incromine Oxide S [Croda]
   Jordamox SDA [PPG-Mazer]
   Ninox S [Stepan Europe]

Rewominox S 300 [Rewo GmbH]

Schercamox DMS [Scher]

CATEGORY:

Emollient, conditioner, softener, lubricant, emulsifier, detergent, thickener, viscosity builder, foaming agent, foam stabilizer, stabilizer, antistat, wetting agent

APPLICATIONS:

Bath products: (Ammonyx SO)

Cosmetic industry preparations: (Aromox DM18D-W; Conco XA-S; Incromine Oxide S; Jordamox SDA); conditioners (Schercamox DMS); creams and lotions (Ammonyx SO; Incromine Oxide S; Schercamox DMS); cream rinses (Ammonyx SO; Schercamox DMS); hair preparations (Schercamox DMS); shampoos (Schercamox DMS); shaving preparations (Schercamox DMS)

Household detergents: (Conco XA-S; Jordamox SDA)

Industrial applications: (Jordamox SDA); dyes and pigments (Conco XA-S); electroplating (Conco XA-S); photography (Conco XA-S); polymers/polymerization (Conco XA-S); textile/leather processing (Conco XA-S)

Industrial cleaners: (Conco XA-S)

PROPERTIES:

Form:

Liquid (Rewominox S 300)

Paste (Ammonyx SO; Aromox DM18D-W; Incromine Oxide S; Jordamox SDA; Ninox S; Schercamox DMS)

Color:

White (Incromine Oxide S)

White to off-white (Schercamox DMS)

Odor:

Typical (Schercamox DMS)

Composition:

24.5–26.5% active (Incromine Oxide S; Jordamox SDA)

25% active (Ammonyx SO; Rewominox S 300; Schercamox DMS)

25% active min. in water (Aromox DM18D-W)

30% active (Conco XA-S)

Solubility:

Sol. in most alcohols (Schercamox DMS)

Sol. in cottonseed oil (Aromox DM18D-W)

Sol. in ethanol (Aromox DM18D-W; Conco XA-S)

Sol. in glycol ethers (Schercamox DMS)

Sol. in glycols (Schercamox DMS)

Sol. in hexylene glycol (Conco XA-S)

Sol. in isopropanol (Aromox DM18D-W; Conco XA-S)

Sol. in isopropyl esters (Aromox DM18D-W)

Sol. in min. oil (Aromox DM18D-W)

Sol. in polyols (Schercamox DMS)

Stearyl dimethyl amine oxide *(cont'd.)*

Sol. in triols (Schercamox DMS)
Sol. in water (Aromox DM18D-W); slightly sol. in water (Schercamox DMS); forms
  emulsions in water (Conco XA-S)
Ionic Nature:
Nonionic (Ammonyx SO; Incromine Oxide S; Jordamox SDA; Rewominox S 300)
Cationic/nonionic (Schercamox DMS)
Cationic/amphoteric (Conco XA-S)
M.W.:
326 (avg.) (Schercamox DMS)
Sp.gr.:
0.90 (Schercamox DMS)
0.95 (Aromox DM18D-W)
0.99 (25/20 C) (Ammonyx SO)
Visc.:
90 cps (Aromox DM18D-W)
Pour Pt.:
< 0 C (Aromox DM18D-W)
Flash Pt.:
> 200 F (Ammonyx SO)
Stability:
Good (Schercamox DMS)
Acid and alkali stable (Conco XA-S)
pH:
7.0–8.0 (Aromox DM18D-W)
7.0 ± 1.0 (1% sol'n.) (Schercamox DMS)
Biodegradable: (Conco XA-S)
TOXICITY/HANDLING:
Avoid prolonged contact with skin and eyes (Incromine Oxide S)
STORAGE/HANDLING:
Store in a cool, dry place (Incromine Oxide S)
STD. PKGS.:
55-gal (425 lb net) lined drums (Incromine Oxide S)
55-gal poly-lined drums (Schercamox DMS)

Stearyl dimethyl benzyl ammonium chloride

SYNONYMS:
Benzenemethanaminium, N,N-dimethyl-N-octadecyl-, chloride
Benzyl dimethyl stearyl ammonium chloride
N,N-Dimethyl-N-octadecylbenzenemethanaminium chloride
Stearalkonium chloride (CTFA)

Stearyl dimethyl benzyl ammonium chloride *(cont'd.)*

EMPIRICAL FORMULA:

$C_{27}H_{50}N \cdot Cl$

STRUCTURE:

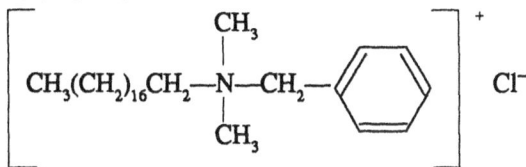

CAS No.:

122-19-0

TRADENAME EQUIVALENTS:

Ablumine 280 [Taiwan]

Ammonyx 4, 4B, 485, 490, 4002 [Millmaster-Onyx]

Amyx A-25S [Clough]

Arquad DM18B-90 [Akzo/Armak]

Carsoquat SDQ-25, SDQ-85 [Lonza]

Catinal OB-80E [Toho]

Cation S [Sanyo]

Crapol AU-23 [Pulcra SA]

Cycloton M270C/18, M270C/85, SCS [Alcolac]

Dehyquart STC-25 [Henkel]

Empigen BCP/25 [Albright & Wilson/Australia]

Incroquat S-85, SDQ-25 [Croda; Croda Universal]

Jordaquat JS-25 [PPG-Mazer]

Mackernium SDC-25, SDC-50, SDC-85 [McIntyre]

Maquat SC-18 [Mason]

Nissan Cation S2-100 [Nippon Oils & Fats]

Standamul STC-25 [Henkel]

Stedbac [Hexcel]

Triton CG-400, CG-500, X-400 [Rohm & Haas]

Varisoft SDC, SDC-65%, SDC-100P, SDC-W [Sherex]

CATEGORY:

Conditioner, softener, emollient, emulsifier, antistat, surfactant, base, dispersant, germicide, coagulant

APPLICATIONS:

Cosmetic industry preparations: (Arquad DM18B-90; Carsoquat SDQ-25); conditioners (Ablumine 280; Carsoquat SDQ-25, SDQ-85; Crapol AU-23; Cycloton M270C/18, M270C/85, SCS; Empigen BCP/25; Incroquat S-85, SDQ-25; Mackernium SDC-25, SDC-50, SDC-85; Stedbac; Varisoft SDC, SDC-65%, SDC-100P, SDC-W); creams and lotions (Ammonyx 4, 4B, 485, 490, 4002; Amyx A-25S; Crapol AU-23; Dehyquart STC-25); hair preparations (Ablumine 280; Ammonyx 4, 4B, 485, 490, 4002; Amyx A-25S; Carsoquat SDQ-25, SDQ-85; Catinal OB-80E; Cation S; Crapol AU-23; Dehyquart STC-25; Empigen BCP/25; Incroquat S-

Stearyl dimethyl benzyl ammonium chloride *(cont'd.)*

85; Jordaquat JS-25; Mackernium SDC-25, SDC-50, SDC-85; Standamul STC-25; Stedbac; Varisoft SDC, SDC-65%, SDC-100P, SDC-W); shampoos (Arquad DM18B-90); skin preparations (Ammonyx 4, 4B, 485, 490, 4002; Amyx A-25S; Incroquat SDQ-25)

Food applications: (Nissan Cation S2-100)

Industrial applications: dyes and pigments (Incroquat SDQ-25; Nissan Cation S2-100); paper mfg. (Incroquat SDQ-25; Nissan Cation S2-100); petroleum industry (Nissan Cation S2-100); plastics (Nissan Cation S2-100); textile/leather processing (Carsoquat SDQ-25; Dehyquart STC-25; Incroquat SDQ-25; Nissan Cation S2-100; Standamul STC-25); water treatment (Nissan Cation S2-100)

Pharmaceutical applications: (Arquad DM18B-90)

PROPERTIES:

Form:

Liquid (Triton CG-400, CG-500, X-400)

Paste (Ammonyx 4, 4B; Amyx A-25S; Crapol AU-23; Cycloton M270C/18; Incroquat SDQ-25; Jordaquat JS-25; Mackernium SDC-25, SDC-50, SDC-85; Maquat SC-18; Varisoft SDC, SDC-65%, SDC-W)

Mobile paste (Empigen BCP/25)

Thick creamy paste (Carsoquat SDQ-25)

Slurry (Dehyquart STC-25; Standamul STC-25)

Solid (Cation S; Cycloton M270C/85)

Flake (Maquat SC-18)

Powder (Ammonyx 485, 490, 4002; Arquad DM18B-90; Maquat SC-18; Stedbac; Varisoft SDC-100P)

Crystalline powder (Nissan Cation S2-100)

Waxy flake (Carsoquat SDQ-85; Incroquat S-85)

Waxy (Catinal OB-80E)

Color:

White (Arquad DM18B-90; Carsoquat SDQ-25, SDQ-85; Dehyquart STC-25; Incroquat S-85; Jordaquat JS-25; Mackernium SDC-25, SDC-85; Standamul STC-25; Triton CG-400, CG-500; Varisoft SDC-W)

Cream white (Triton X-400)

White or light yellow (Nissan Cation S2-100)

Gardner 2 max. (Varisoft SDC)

Odor:

Mild and sweet (Carsoquat SDQ-25)

Composition:

17–19% quat., 24–26% solids (Incroquat SDQ-25)

17–19% quat., 24–28% solids (Ammonyx 4)

16–18% quat., 24–26% solids (Ammonyx 4B)

20–21% active (Carsoquat SDQ-25)

24–26% solids (Varisoft SDC)

24–26% solids in water (Varisoft SDC-W)

Stearyl dimethyl benzyl ammonium chloride (cont' d.)

25% active (Cycloton M270C/180; Dehyquart STC-25; Empigen BCP/25; Jordaquat
JS-25; Maquat SC-18 paste; Standamul STC-25; Triton X-400)

25% active in water with 5% 2-propanol (Triton CG-400)

25% active min. (Amyx A-25S)

25 ± 1% active (Crapol AU-23)

50% active (Mackernium SDC-50)

50% active in water with 11% 2-propanol (Triton CG-500)

65–70% active (Varisoft SDC-65%)

85% active (Carsoquat SDQ-85; Cycloton M270C/85; Maquat SC-18 flake)

85% active min. (Incroquat S-85; Nissan Cation S2-100)

85% quat. min., 87% solids min. (Ammonyx 485)

90% active min. (Arquad DM18B-90)

90% quat. min., 92% solids min. (Ammonyx 490)

94% active (Maquat SC-18 powder)

94% quat. min., 96% solids min. (Ammonyx 4002)

100% active (Stedbac; Varisoft SDC-100P)

Solubility:

Sol. in water (Arquad DM18B-90); disp. in water (Ablumine 280; Carsoquat SDQ-85;
Crapol AU-23)

Ionic Nature:

Cationic (Ammonyx 4, 4B, 485, 490, 4002; Amyx A-25S; Carsoquat SDQ-25, SDQ-
85; Crapol AU-23; Cycloton M270C/18, M270C/85; Incroquat S-85, SDQ-25;
Standamul STC-25; Stedbac; Triton CG-400, CG-500, X-400; Varisoft SDC-65%,
SDC-100P, SDC-W)

Sp.gr.:

0.45 (25/20 C) (Ammonyx 485)

0.47 (25/20 C) (Ammonyx 490)

0.52 (25/20 C) (Ammonyx 4002)

0.99 (25/20 C) (Ammonyx 4, 4B)

Density:

7.7 lb/gal (50 C) (Triton CG-500)

8.0 lb/gal (50 C) (Triton CG-400, X-400)

M.P.:

140 F (Carsoquat SDQ-85; Incroquat S-85)

Pour Pt.:

110 F (Triton CG-400, CG-500)

Flash Pt.:

90 F (Seta CC) (Triton CG-500)

111 F (Seta CC) (Triton CG-400)

116 F (PM) (Varisoft SDC)

122 F (Ammonyx 485)

130 F (Ammonyx 490)

170 F (Ammonyx 4002)

Stearyl dimethyl benzyl ammonium chloride *(cont'd.)*

> 200 F (Ammonyx 4, 4B); (PM) (Varisoft SDC-2)
> 212 F (TOC) (Triton X-400)

Stability:

Stable under acid and alkaline conditions (Incroquat SDQ-25)

pH:

3.0–4.0 (1% disp.) (Carsoquat SDQ-25); (1% sol'n.) (Incroquat SDQ-25)

3.0–5.0 (Varisoft SDC-65%)

3.5–4.0 (Varisoft SDC-W)

5.0–7.0 (0.5% sol'n.) (Carsoquat SDQ-85; Incroquat S-85)

5.0–8.0 (1% sol'n.) (Arquad DM18B-90)

Surface Tension:

34 dynes/cm (1%) (Triton X-400)

TOXICITY/HANDLING:

Skin irritant, severe eye irritant (Arquad DM18B-90)

Avoid prolonged contact with skin and eyes (Incroquat S-85)

Avoid prolonged contact with skin (Incroquat SDQ-25)

STORAGE/HANDLING:

Avoid contact with strong oxidizing agents, anionics; product will kill bacteria in biological disposal systems (Arquad DM18B-90)

Store in a cool, dry place (Incroquat S-85)

Store in a dry area at temps. ≤ 120 F (Incroquat SDQ-25)

STD. PKGS.:

10-kg paper box (Nissan Cation S2-100)

50-kg fiberboard kegs (Arquad DM18B-90)

44-gal (150 lb net) Leverpak with polyliner (Incroquat S-85)

55-gal (425 lb net) Liquipak containers (Incroquat SDQ-25)

Stearyl erucamide *(CTFA)*

SYNONYMS:

13-Docosenamide, N-octadecyl-

N-Octadecyl-13-docosenamide

EMPIRICAL FORMULA:

$C_{40}H_{79}NO$

STRUCTURE:

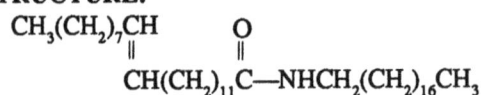

CAS No.:

10094-45-8

TRADENAME EQUIVALENTS:
HTSA #3 [Hexcel]
Kemamide E-180 [Humko/Witco]

CATEGORY:
Lubricant, slip agent, antiblock agent, release agent, defoamer, water repellent, corrosion inhibitor, grinding aid, dispersant, intermediate

APPLICATIONS:
Household applications: (Kemamide E-180); detergents (Kemamide E-180)
Industrial applications: (Kemamide E-180); asphalts (Kemamide E-180); dyes and pigments (Kemamide E-180); lubricating/cutting oils (Kemamide E-180); metalworking (Kemamide E-180); paint mfg. (Kemamide E-180); petroleum products (Kemamide E-180); plastics (HTSA #3; Kemamide E-180); polishes (Kemamide E-180); printing inks (Kemamide E-180); textile/leather processing (Kemamide E-180); waxy products (Kemamide E-180)

PROPERTIES:
Form:
Free-flowing powder (Kemamide E-180)
Color:
White (Kemamide E-180)
Solubility:
Sol. in alcohols (HTSA #3)
Very sol. in benzene (HTSA #3)
Sol. < 50 g/100 g @ 50 C in chloroform (Kemamide E-180)
Sol. 50 g/100 g @ 50 C in dichloroethane (Kemamide E-180)
Sol. in ethyl acetate (HTSA #3)
Sol. 30 g/100 g @ 50 C in isopropanol (Kemamide E-180)
Sol. 25 g/100 g @ 50 C in MEK (Kemamide E-180)
Sol. 40 g/100 g @ 50 C in VM&P naphtha (Kemamide E-180)
Sol. < 50 g/100 g @ 50 C in toluene (Kemamide E-180)
Density:
0.8074 (110 C) (Kemamide E-180)
B.P.:
824 F (HTSA #3)
M.P.:
69–77 C (HTSA #3)
72–75 C (Kemamide E-180)
Flash Pt.:
258 C (CC) (Kemamide E-180)
Acid No.:
10 max. (Kemamide E-180)
Iodine No.:
44 (Kemamide E-180)

Stearyl erucamide (cont'd.)

Amine No.:
10 max. (Kemamide E-180)
Sp. Heat:
0.34 Cal/g (55 C) (Kemamide E-180)
TOXICITY/HANDLING:
Nontoxic; handle powders to prevent eye irritation; wear protective goggles and gloves when handlig (Kemamide E-180)
STORAGE/HANDLING:
Handle powders to prevent dust explosions; not classified as flammable, but will burn if ignited; keep away from strong oxidizing agents; should not remain at temps. higher than 175–185 C for long periods (Kemamide E-180)

Stearyl stearate (CTFA)

SYNONYMS:
Octadecanoic acid, octadecyl ester
EMPIRICAL FORMULA:
$C_{36}H_{72}O_2$
STRUCTURE:

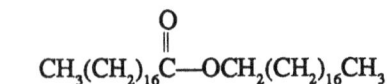

$$CH_3(CH_2)_{16}C\!\!-\!\!OCH_2(CH_2)_{16}CH_3$$

CAS No.:
2778-96-3
TRADENAME EQUIVALENTS:
Cyclochem SS [Alcolac]
Falco S-18S [Fallek]
Lexol SS [Inolex]
Liponate SS [Lipo]
Rilanit STS-T [Henkel KGaA]
CATEGORY:
Emollient, thickener, viscosity control agent, bodying agent, gloss enhancer, lubricant, fatting agent, emulsifier, dispersant
APPLICATIONS:
Bath products: (Liponate SS); bath oils (Lexol SS)
Cosmetic industry preparations: (Falco S-18S); creams and lotions (Lexol SS; Liponate SS); hair preparations (Cyclochem SS); skin preparations (Cyclochem SS)
Industrial applications: ceramics (Rilanit STS-T); lubricating/cutting oils (Rilanit STS-T); metalworking (Rilanit STS-T); textile/leather processing (Rilanit STS-T)

PROPERTIES:
Form:
 Solid (Rilanit STS-T)
 Flake (Cyclochem SS; Liponate SS)
Color:
 Off-white (Cyclochem SS; Liponate SS)
Solubility:
 Sol. in oils (Lexol SS); miscible (Liponate SS)
 Insol. in water (Cyclochem SS; Liponate SS); insol. to slightly sol. (Falco S-18S)
M.P.:
 53 C (Cyclochem SS)
Acid No.:
 5.0 max. (Cyclochem SS; Liponate SS)
Saponification No.:
 103–117 (Liponate SS)
 110 (Cyclochem SS)

Stearyl trimethyl ammonium chloride

SYNONYMS:
 1-Octadecanaminium, N,N,N-trimethyl-, chloride
 Octadecyl trimethyl ammonium chloride
 Steartrimonium chloride (CTFA)
 N,N,N-Trimethyl-1-octadecanaminium chloride
EMPIRICAL FORMULA:
 $C_{21}H_{46}N \cdot Cl$
STRUCTURE:

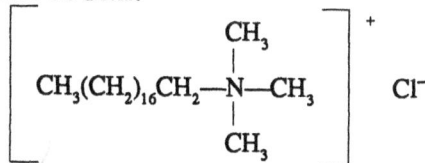

CAS No.:
 112-03-8
TRADENAME EQUIVALENTS:
 Arquad 18-50 [Akzo/Armak]
 Quartamin 86W [Kao]
 Nissan Cation AB [Nippon Oils & Fats]
CATEGORY:
 Conditioner, softener, antistat, dyeing aid, leveling agent, corrosion inhibitor, stabilizer, germicide, dispersant, coagulant

Stearyl trimethyl ammonium chloride (cont'd.)

APPLICATIONS:
   Cosmetic industry preparations: conditioners (Quartamin 86W); hair preparations (Quartamin 86W)
   Farm products: herbicides (Nissan Cation AB)
   Food applications: (Nissan Cation AB)
   Industrial applications: dyes and pigments (Arquad 18-50; Nissan Cation AB); lubricating/cutting oils (Arquad 18-50); paper mfg. (Arquad 18-50; Nissan Cation AB); petroleum industry (Arquad 18-50; Nissan Cation AB); plastics (Arquad 18-50; Nissan Cation AB); textile/leather processing (Arquad 18-50; Nissan Cation AB); water treatment (Nissan Cation AB)

PROPERTIES:
Form:
   Liquid (Arquad 18-50; Quartamin 86W; Nissan Cation AB)
Color:
   Light yellow (Nissan Cation AB)
   Gardner 7 max. (Arquad 18-50)
Composition:
   23% active min. (Nissan Cation AB)
   30% active (Quartamin 86W)
   49–52% quat in aq. isopropanol (Arquad 18-50)
Solubility:
   Sol. < 1% @ 20 C in water (Arquad 18-50)
Ionic Nature:
   Cationic (Arquad 18-50; Quartamin 86W; Nissan Cation AB)
M.W.:
   347 (of active) (Arquad 18-50)
Sp.gr.:
   0.878 (Arquad 18-50)
Flash Pt.:
   65 F (Seta) (Arquad 18-50)
HLB:
   15.7 (Arquad 18-50)
pH:
   5.0–8.0 (10% aq.) (Arquad 18-50)
Surface Tension:
   34 dynes/cm (0.1%) (Arquad 18-50)
STD. PKGS.:
   16-kg Atron can, 170-kg drum (Nissan Cation AB)

Synthetic beeswax (CTFA)

SYNONYMS:
   Beeswax, synthetic
CAS No.:
   977026-94-0
TRADENAME EQUIVALENTS:
   Cyclochem 326A [Alcolac]
   Lipobee 102 [Lipo]
   Syncrowax BB4 [Croda]
   Waxenol 821 S.B. [CasChem]
CATEGORY:
   Lubricant, conditioner, emollient
APPLICATIONS:
   Cosmetic industry preparations: skin preparations (Syncrowax BB4)
   Industrial applications: lubricating/cutting oils (Cyclochem 326A)
PROPERTIES:
Form:
   Very fine flakes/beads (Waxenol 821 S.B.)
   Wax (Syncrowax BB4)
Solubility:
   Sol. in oils (Syncrowax BB4); miscible in oils (Lipobee 102)
   Insol. in water (Lipobee 102)

Synthetic spermaceti

SYNONYMS:
   Cetyl esters (CTFA)
   Synthetic spermaceti wax
CAS No.:
   8002-23-1
   RD No.: 977067-67-6
TRADENAME EQUIVALENTS:
   Crodamol SS [Croda; Croda Ltd.] (NF)
   Cyclochem SPS [Alcolac]
   Kessco Synthetic Spermaceti NF [Stepan]
   Liponate SPS [Lipo]
   Spermwax [Robeco] (NF)
   Starfol Wax CG [Sherex]
CATEGORY:
   Emollient, thickener, bodying agent, viscosity control agent, slip agent

323

Synthetic spermaceti *(cont'd.)*

APPLICATIONS:
  Bath products: (Liponate SPS)
  Cosmetic industry preparations: (Crodamol SS; Spermwax; Starfol Wax CG); aerosol talc (Cyclochem SPS); cold creams (Kessco Synthetic Spermaceti NF); creams and lotions (Cyclochem SPS; Kessco Synthetic Spermaceti NF; Liponate SPS); makeup (Cyclochem SPS; Kessco Synthetic Spermaceti NF); toiletries (Spermwax)
  Pharmaceutical applications: (Crodamol SS); dermatologicals (Spermwax)
PROPERTIES:
Form:
  Crystalline solid (Crodamol SS)
  Solid (Spermwax)
  Flakes (Cyclochem SPS; Kessco Synthetic Spermaceti NF; Liponate SPS; Starfol Wax CG)
Color:
  White (Kessco Synthetic Spermaceti NF)
  Nearly white (Crodamol SS)
  White to cream (Liponate SPS)
  Off-white (Cyclochem SPS)
  APHA 100 max. (Starfol Wax CG)
Composition:
  100% active (Cyclochem SPS; Starfol Wax CG)
Solubility:
  Sol. in boiling alcohol (Kessco Synthetic Spermaceti NF)
  Sol. in chloroform (Kessco Synthetic Spermaceti NF)
  Sol. in ether (Kessco Synthetic Spermaceti NF)
  Sol. in min. oil (Starfol Wax CG)
  Sol. in oils (Crodamol SS); miscible with oils (Liponate SPS)
  Insol. in water (Crodamol SS; Kessco Synthetic Spermaceti NF; Liponate SPS)
  Sol. in other waxes, oils, hydrocarbons (Kessco Synthetic Spermaceti NF)
Ionic Nature:
  Nonionic (Cyclochem SPS; Starfol Wax CG)
Sp.gr.:
  0.82–0.84 (50 C) (Crodamol SS)
M.P.:
  43–47 C (Crodamol SS; Kessco Synthetic Spermaceti NF)
  46 C (Cyclochem SPS)
  46–49 C (Starfol Wax CG)
Acid No.:
  2.0 max. (Kessco Synthetic Spermaceti NF; Starfol Wax CG)
  3.0 max. (Cyclochem SPS)
  5.0 max. (Crodamol SS; Liponate SPS)
Iodine No.:
  1.0 max. (Starfol Wax CG)

Saponification No.:

109–117 (Kessco Synthetic Spermaceti NF; Starfol Wax CG)
109–120 (Crodamol SS; Liponate SPS)
115 (Cyclochem SPS)

Tallow dimethyl amine oxide

SYNONYMS:
Amines, tallow alkyl dimethyl, oxides
Tallow amine oxide (CTFA)

STRUCTURE:

$$CH_3$$
$$|$$
$$R-N \rightarrow O$$
$$|$$
$$CH_3$$

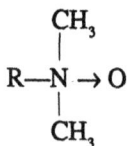

where R represents tallow fatty radicals

CAS No.:
RD No.: 977058-39-1

TRADENAME EQUIVALENTS:
Conco XA-T [Continental Chem.]

CATEGORY:
Detergent, foaming agent, emollient, foam stabilizer, stabilizer, dispersant, wetting agent

APPLICATIONS:
Cosmetic industry preparations: (Conco XA-T)
Household detergents: (Conco XA-T)
Industrial applications: dyes and pigments (Conco XA-T); electroplating (Conco XA-T); photography (Conco XA-T); polymers/polymerization (Conco XA-T); textile/leather processing (Conco XA-T)
Industrial cleaners: (Conco XA-T)

PROPERTIES:
Form:
Liquid (Conco XA-T)
Composition:
30% active (Conco XA-T)
Solubility:
Sol. in ethanol (Conco XA-T)
Sol. in hexylene glycol (Conco XA-T)
Sol. in isopropanol (Conco XA-T)
Forms emulsions in water (Conco XA-T)
Ionic Nature:
Cationic/amphoteric (Conco XA-T)
Biodegradable: (Conco XA-T)

326

Tetraglyceryl monooleate

SYNONYMS:
Polyglyceryl-4 oleate (CTFA)
CAS No.:
9007-48-1 (generic)
RD No.: 977057-95-6
TRADENAME EQUIVALENTS:
Nikkol Tetraglyn 1-O [Nikko]
Witconol 14 [Witco]
CATEGORY:
Lubricant, emulsifier, antifoaming agent, anticrystallizing agent
APPLICATIONS:
Industrial applications: (Witconol 14); aerosols (Witconol 14)
PROPERTIES:
Form:
Liquid (Nikkol Tetraglyn 1-O; Witconol 14)
Composition:
100% active (Nikkol Tetraglyn 1-O; Witconol 14)
Ionic Nature:
Nonionic (Nikkol Tetraglyn 1-O; Witconol 14)

Tridecyl stearate (CTFA)

SYNONYMS:
Octadecanoic acid, tridecyl ester
EMPIRICAL FORMULA:
$C_{31}H_{62}O_2$
STRUCTURE:

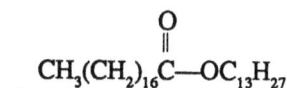

$$CH_3(CH_2)_{16}\overset{\displaystyle O}{\overset{\|}{C}}-OC_{13}H_{27}$$

CAS No.:
31556-45-3
TRADENAME EQUIVALENTS:
Emerest 2308 [Henkel/Emery]
Grocor 5721 [A. Gross]
Kemester 5721 [Humko/Witco]
Liponate TDS [Lipo]
CATEGORY:
Lubricant, wetting agent, emulsifier, carrier

Tridecyl stearate *(cont'd.)*

APPLICATIONS:
    Cosmetic industry applications: creams and lotions (Liponate TDS)
    Industrial applications: dyes and pigments (Grocor 5721; Kemester 5721); textile/
        leather processing (Emerest 2308; Grocor 5721; Kemester 5721)
PROPERTIES:
Form:
    Liquid (Emerest 2308; Grocor 5721; Liponate TDS)
Color:
    Water-white (Grocor 5721)
    Gardner 1 (Emerest 2308)
Composition:
    100% active (Emerest 2308; Grocor 5721; Liponate TDS)
Solubility:
    Sol. in isopropanol @ 5% (Emerest 2308)
    Sol. in min. oil @5% (Emerest 2308)
    Sol. in toluene @ 5% (Emerest 2308)
    Insol. in water @ 5% (Emerest 2308)
    Sol. in xylene @ 5% (Emerest 2308)
Ionic Nature:
    Nonionic (Emerest 2308)
Sp.gr.:
    0.86 (Grocor 5721)
Density:
    7.1 lb/gal (Emerest 2308)
Visc.:
    18 cSt (100 F) (Emerest 2308)
F.P.:
    6 C (Grocor 5721)
Pour Pt.:
    3 C (Emerest 2308)
Flash Pt.:
    440 F (Emerest 2308)

Triisocetyl citrate *(CTFA)*

EMPIRICAL FORMULA:
    $C_{54}H_{104}O_7$

STRUCTURE:

$$
\begin{array}{c}
\quad\quad\quad O \\
\quad\quad\quad \| \\
CH_2C\!-\!OC_{16}H_{33} \\
| \\
\quad\quad O \\
\quad\quad \| \\
HO\!-\!C\!-\!C\!-\!OC_{16}H_{33} \\
| \\
\quad\quad O \\
\quad\quad \| \\
CH_2C\!-\!OC_{16}H_{33}
\end{array}
$$

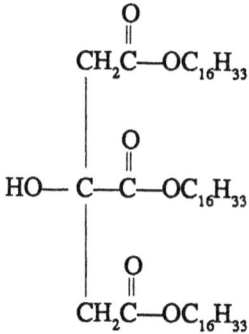

TRADENAME EQUIVALENTS:

Hetester TICC [Heterene]

CATEGORY:

Emollient, dispersant

APPLICATIONS:

Cleansers: cleansing creams (Hetester TICC)

Cosmetic industry preparations: emulsions (Hetester TICC); makeup (Hetester TICC)

PROPERTIES:

Form:

Clear liquid (Hetester TICC)

Color:

Yellow (Hetester TICC)

Composition:

99.5% active (Hetester TICC)

Solubility:

Sol. in castor oil @ 5% (Hetester TICC)

Sol. in 95% SD-40 ethanol @ 5% (Hetester TICC)

Sol. in min. oil @ 5% (Hetester TICC)

Sol. in octyl palmitate @ 5% (Hetester TICC)

Sol. in oleyl alcohol @ 5% (Hetester TICC)

Insol. in propylene glycol @ 5% (Hetester TICC)

Sol. in safflower oil @ 5% (Hetester TICC)

Insol. in water @ 5% (Hetester TICC)

Acid No.:

4.0 max. (Hetester TICC)

Saponification No.:

165–180 (Hetester TICC)

Wheat germ oil (CTFA)

SYNONYMS:
Oil of wheat germ
CAS No.:
8006-95-9
TRADENAME EQUIVALENTS:
EmCon W [Fanning]
Lipovol WGO [Lipo]
Wheat Germ Oil CLR 0.26% (αβ) Tocopherol [Henkel Canada]
CATEGORY:
Emollient, flavoring agent, glosser, vitamin E source
APPLICATIONS:
Bath products: bath/body oils (Lipovol WGO)
Cosmetic industry preparations: conditioners (Lipovol WGO); creams and lotions (EmCon W; Lipovol WGO); emulsions (Wheat Germ Oil CLR 0.26% (αβ) Tocopherol); hair preparations (EmCon W; Lipovol WGO; Wheat Germ Oil CLR 0.26% (αβ) Tocopherol); makeup (Lipovol WGO); shaving preparations (Lipovol WGO); skin preparations (EmCon W; Lipovol WGO; Wheat Germ Oil CLR 0.26% (αβ) Tocopherol)
Food applications: food additives (EmCon W)
PROPERTIES:
Form:
Oil (Wheat Germ Oil CLR 0.26% (αβ) Tocopherol)
Oil—may precipitate if chilled (Lipovol WGO)
Color:
Brown (Lipovol WGO; Wheat Germ Oil CLR 0.26% (αβ) Tocopherol)
Odor:
Characteristic, fatty (Lipovol WGO)
Faint characteristic (Wheat Germ Oil CLR 0.26% (αβ) Tocopherol)
Solubility:
Miscible with isopropyl esters (Lipovol WGO)
Miscible with min. oil (Lipovol WGO)
Sol. in water (EmCon W); insol. in water (Lipovol WGO)
Acid No.:
5 max. (Lipovol WGO)
Iodine No.:
120–140 (Lipovol WGO)

Saponification No.:
175–195 (Lipovol WGO)
Stability:
Stable in neutral and mildly acidic or alkaline systems (Lipovol WGO)

Zinc stearate (CTFA)

SYNONYMS:
Octadecanoic acid, zinc salt
Zinc octadecanoate

EMPIRICAL FORMULA:
$C_{18}H_{36}O_2 \cdot {}^1/_2Zn$

STRUCTURE:
$[CH_3(CH_2)_{16}COO]_2Zn$

CAS No.:
557-05-1

TRADENAME EQUIVALENTS:
Akrochem Zinc Stearate [Akron]
CS-250, -251 [Crystal Soap]
Dymsol RL7 [Harwick]
Hallcote ZS-50/50 [C.P. Hall]
Harwick Zinc Stearate H, USP [Harwick]
Hy Dense Zinc Stearate XM Powd., XM Ultra Fine [Mallinckrodt]
Hydro Zinc [Morton Int'l./Ventron]
Interstab ZN-18-1 [Akzo/Interstab]
Lubrazinc W, Superfine [Witco]
Nopco 1097-B, 1097-M [Henkel/Process]
Norfox ZNS [Norman, Fox]
Nuodex Micronized Zinc Stearate 1058 Densified [Tenneco]
Petrac ZN-41 [Synthetic Products] (USP)
Petrac ZN-42, ZW-45 [Synthetic Products]
Petrac ZN-44HS [Synthetic Products] (heat-stable)
Quikote [Morton Int'l./Ventron]
Stave Zinc Stearate [Stave]
Wet Zinc [Morton Int'l./Ventron]
Witco Zinc Stearate 11, 44, Disperso, Heat-Stable, LV, NW, Polymer Grade, Regular, USP [Witco]
Zincloid Synpro [Synthetic Products]
Zincote [Morton Int'l./Ventron]
Zinc Stearate Synpro [Synthetic Products]
Zinc Stearate (Wettable) [C.P. Hall]
Zinc Stearate (Wettable, Synpro) [Synthetic Products]

Generically sold by:
[C.P. Hall; Harwick; Henkel; Housmex; Mallinckrodt)

CATEGORY:
Lubricant, release agent, antitack agent, antistick agent, processing aid, dispersant, water repellent, dusting agent, flatting agent, suspending agent, anticaking agent, stabilizer, emulsifier

APPLICATIONS:
Cleansers: soaps (Witco Zinc Stearate USP)
Cosmetic industry preparations: (Witco Zinc Stearate USP); creams and lotions (Witco Zinc Stearate USP); toiletries (Witco Zinc Stearate USP)
Food applications: food packaging (Interstab ZN-18-1)
Industrial applications: (Hy Dense Zinc Stearate XM Powd., XM Ultra Fine); cement (Petrac ZW-45); fire fighting (Petrac ZN-41); metallurgy (Lubrazinc W, Superfine; Petrac ZN-41); paint mfg. (Petrac ZN-41, ZW-45; Witco Zinc Stearate Regular); paper mfg. (Petrac ZW-45); plastics (Interstab ZN-18-1; Lubrazinc W, Superfine; Nuodex Micronized Zinc Stearate 1058 Densified; Petrac ZN-41, ZN-42, ZN-44HS, ZW-45; Witco Zinc Stearate 11, 44, Heat-Stable, LV, NW, Polymer Grade, Regular); printing inks (Petrac ZW-45); rubber (generic—Hall, Harwick, Henkel; CS-250, -251; Hallcote ZS-50/50; Harwick Zinc Stearate H, USP; Hydro Zinc; Nopco 1097-B, 1097-M; Nuodex Micronized Zinc Stearate 1058 Densified; Petrac ZN-41, ZN-42; Quikote; Stave Zinc Stearate; Wet Zinc; Witco Zinc Stearate 11, 44, Disperso, NW, Regular; Zincloid Synpro; Zincote; Zinc Stearate Synpro; Zinc Stearate (Wettable); Zinc Stearate (Wettable, Synpro))
Pharmaceutical applications: (Witco Zinc Stearate USP)

PROPERTIES:
Form:
Liquid-paste (Nopco 1097-B, 1097-M)
Semiliquid (Hydro Zinc)
Soft paste (CS-250, -251; Quikote)
Paste (Wet Zinc; Zincloid Synpro; Zincote)
Particulate (Petrac ZN-41, ZN-42)
Fine particle size (Interstab ZN-18-1)
Impalpable powder (Norfox ZNS)
Fine powder (Akrochem Zinc Stearate; Stave Zinc Stearate)
Powder (generic—Hall, Harwick, Henkel; Harwick Zinc Stearate H, USP; Hy Dense Zinc Stearate XM Powd., XM Ultra Fine; Lubrazinc W, Superfine; Petrac ZN-44HS; Witco Zinc Stearate 11, 44, Heat-Stable, LV, NW, Polymer Grade, Powder, Regular, USP; Zinc Stearate Synpro; Zinc Stearate (Wettable); Zinc Stearate (Wettable, Synpro))
Free-flowing powder (generic—Housmex)
Aq. dispersion (Hallcote ZS-50/50; Petrac ZW-45)
Dispersion (Dymsol RL7; Witco Zinc Stearate Disperso)

Zinc stearate *(cont'd.)*

Fineness:
99% thru 200 mesh (Petrac ZN-42)
99% thru 325 mesh sieve (Lubrazinc W, Superfine)
99.5% thru 325 mesh (Petrac ZN-41, ZN-44HS)
99.7% thru 325 mesh sieve (Witco Zinc Stearate LV)
99.9% thru 325 mesh sieve (Witco Zinc Stearate 11, 44, Disperso, Heat-Stable, NW, Polymer Grade, Regular, USP)
100% thru 325 mesh (Akrochem Zinc Stearate)

Color:
White (generic—Harwick, Henkel; Housmex; Akrochem Zinc Stearate; CS-250, -251; Hallcote ZS-50/50; Harwick Zinc Stearate H, USP; Hydro Zinc; Lubrazinc W, Superfine; Nopco 1097-B, 1097-M; Nuodex Micronized Zinc Stearate 1058 Densified; Petrac ZW-45; Quikote; Stave Zinc Stearate; Wet Zinc; Witco Zinc Stearate 11, 44, Disperso, Heat-Stable, LV, NW, Polymer Grade, Regular, USP; Zincloid Synpro; Zincote; Zinc Stearate Synpro; Zinc Stearate (Wettable, Synpro))

Composition:
13.5% ZnO (Petrac ZN-44HS)
14% ZnO (Petrac ZN-42)
20% active in water (Zincloid Synpro)
23% active (Dymsol RL7)
24% NV solids in water (CS-251)
26% NV solids in water (CS-250)
45% active (Nopco 1097-B, 1097-M)
45% solids (Petrac ZW-45)
48–50% solids in water (generic—Mallinckrodt)
99.3% active (Witco Zinc Stearate Disperso)
99.5% active (Witco Zinc Stearate 44, Heat-Stable, LV, NW, Polymer Grade, Regular, USP)
99.7% active (Witco Zinc Stearate 11)
99.8% active (Lubrazinc W, Superfine)
100% active (Zinc Stearate Synpro)

Solubility:
Insol. in alcohols (Lubrazinc W, Superfine; Witco Zinc Stearate Polymer Grade)
Sol. in hot aliphatic hydrocarbons (Lubrazinc W, Superfine; Witco Zinc Stearate Polymer Grade)
Sol. in hot aromatic hydrocarbons (Lubrazinc W, Superfine; Witco Zinc Stearate Polymer Grade)
Sol. in benzene (Lubrazinc W, Superfine; Witco Zinc Stearate 11, 44, Disperso, Heat-Stable, LV, NW, Polymer Grade, Regular, USP)
Sol. in carbon tetrachloride (Lubrazinc W, Superfine; Witco Zinc Stearate 11, 44, Disperso, Heat-Stable, LV, NW, Polymer Grade, Regular, USP)
Insol. in esters (Lubrazinc W, Superfine; Witco Zinc Stearate 11, 44, Disperso, Heat-Stable, LV, NW, Polymer Grade, Regular, USP)

Insol. in ethanol (Lubrazinc W, Superfine; Witco Zinc Stearate 11, 44, Disperso, Heat-Stable, LV, NW, Polymer Grade, Regular, USP)

Insol. in ketones (Lubrazinc W, Superfine; Witco Zinc Stearate 11, 44, Disperso, Heat-Stable, LV, NW, Polymer Grade, Regular, USP)

Insol. in methanol (Lubrazinc W, Superfine; Witco Zinc Stearate 11, 44, Disperso, Heat-Stable, LV, NW, Polymer Grade, Regular, USP)

Sol. in min. oil (Lubrazinc W, Superfine; Witco Zinc Stearate 11, 44, Disperso, Heat-Stable, LV, NW, Polymer Grade, Regular, USP)

Sol. in toluene (Lubrazinc W, Superfine; Witco Zinc Stearate 11, 44, Disperso, Heat-Stable, LV, NW, Polymer Grade, Regular, USP)

Sol. in hot turpentine (Lubrazinc W, Superfine; Witco Zinc Stearate 11, 44, Disperso, Heat-Stable, LV, NW, Polymer Grade, Regular, USP)

Sol. in veg. oil (Lubrazinc W, Superfine; Witco Zinc Stearate 11, 44, Disperso, Heat-Stable, LV, NW, Polymer Grade, Regular, USP)

Disp. in water (CS-250, -251; Zinc Stearate (Wettable); Zinc Stearate (Wettable, Synpro)); water dilutable (Hallcote ZS-50/50); insol. (Lubrazinc W, Superfine; Petrac ZN-41; Witco Zinc Stearate 11, 44, Disperso, Heat-Stable, LV, NW, Polymer Grade, Regular, USP)

Sol. in waxes (Lubrazinc W, Superfine; Witco Zinc Stearate 11, 44, Disperso, Heat-Stable, LV, NW, Polymer Grade, Regular, USP)

Sol. in xylene (Lubrazinc W, Superfine; Witco Zinc Stearate 11, 44, Disperso, Heat-Stable, LV, NW, Polymer Grade, Regular, USP)

Sp.gr.:

0.95 (Quikote)

0.96 (CS-250, -251; Hydro Zinc; Wet Zinc; Zincote)

1.01 (Nopco 1097-B, 1097-M)

1.03 (Hallcote ZS-50/50)

1.09 (generic—Hall, Housmex; Witco Zinc Stearate 11, 44, Disperso, Heat-Stable, LV, NW, Polymer Grade, Regular, USP; Zinc Stearate Synpro; Zinc Stearate (Wettable))

1.095 (generic—Harwick, Henkel; Harwick Zinc Stearate H, USP)

1.10 (Lubrazinc W, Superfine; Nuodex Micronized Zinc Stearate 1058 Densified; Stave Zinc Stearate)

Density:

1.7 lb/gal (Akrochem Zinc Stearate)

19 lb/ft^3 (bulk, tapped) (Petrac ZN-41, ZN-44HS)

24 lb/ft^3 (bulk) (Petrac ZN-42)

Visc.:

400 cps (Petrac ZW-45)

M.P.:

118 C (generic—Hall, Harwick, Henkel; Harwick Zinc Stearate H, USP)

120 C (Akrochem Zinc Stearate)

121 C (Petrac ZN-41, ZN-42, ZN-44HS)

Zinc stearate *(cont'd.)*

123 C (Nuodex Micronized Zinc Stearate 1058 Densified)

124 C (Zinc Stearate (Wettable))

Softening Pt.:

119 C (Witco Zinc Stearate 11)

120 C (Lubrazinc W, Superfine; Witco Zinc Stearate 44, Disperso, LV, NW, Polymer Grade, Regular, USP)

122 C (Witco Zinc Stearate Heat-Stable)

pH:

Neutral (Petrac ZN-41)

8.7 (Petrac ZW-45)

STD. PKGS.:

Drums (Hallcote ZS-50/50; Nopco 1097-B, 1097-M)

Bag, carton (generic—Harwick, Henkel; Nuodex Micronized Zinc Stearate 1058 Densified)

Drum, bag, and carton (Harwick Zinc Stearate H, USP)

50-lb cartons (Zinc Stearate (Wettable))

50-lb cartons and bags (generic—Hall)

50-lb cartons or 25-lb bags (Zinc Stearate Synpro)

55-gal drums (Zincloid Synpro)

TRADENAME PRODUCTS AND GENERIC EQUIVALENTS

Ablumine 280 [Taiwan]—Stearyl dimethyl benzyl ammonium chloride
Ablunol S-80 [Taiwan]—Sorbitan monooleate
Accobetaine CL [Capital City]—Coco-betaine
Accomeen S2 [Capital City]—POE (2) soya amine
Acconon ETG [Capital City]—POE (26) glyceryl ether
Acconon GTO [Capital City]—Glyceryl trioleate
Accosoft 707 [Capital City]—Dihydrogenated tallow dimethyl ammonium chloride
Acetoquat CTAB [Aceto]—Cetyl trimethyl ammonium bromide
Acrawax C, C DF #1, C DF #2, C SG [Lonza]—N,N´-Ethylene bisstearamide
Acylan [Croda]—Acetylated lanolin
Adogen 442, 442H [Sherex]—Dihydrogenated tallow dimethyl ammonium chloride
Adogen 444 [Sherex]—Cetyl trimethyl ammonium chloride
Adogen 470 [Sherex]—Ditallow dimethyl ammonium chloride
Adol 52, 52 NF, 54, 520, 520 NF [Sherex]—Cetyl alcohol
Adol 63, 630, 640 [Sherex]—Cetearyl alcohol
Adol 66 [Sherex]—Isostearyl alcohol
Adol 80, 85 NF, 320, 330, 340 [Sherex]—Oleyl alcohol
Adol 90 NF [Sherex]—Oleyl alcohol (cosmetic/pharmaceutical grade)
Advawax 240 [Morton Int'l./Carstab]—Ethylene bisoleamide
Advawax 275, 280, 290 [Morton Int'l./Carstab]—N,N´-Ethylene bisstearamide
Aerosol 30 [Amer. Cyanamid]—Coco amidopropyl betaine
Ahco 832, 944 [ICI Americas]—Sorbitan monooleate
Akrochem Calcium Stearate [Akron]—Calcium stearate
Akrochem Zinc Stearate [Akron]—Zinc stearate
Akypoquat 131, 132 [Chem-Y GmbH]—Behenoyl-PG-trimonium chloride
Alacsan 7LUF [Alcolac]—Oleyl dimethyl benzyl ammonium chloride
Albalan [Westbrook Lanolin]—Lanolin wax
Aldo TO [Lonza]—Glyceryl trioleate
Alfol 16 [Vista]—Cetyl alcohol (NF)
Alkamide RDO [Alkaril; Alkaril Canada]—Ricinoleic diethanolamide
Alkamide STEDA [Alkaril]—N,N´-Ethylene bisstearamide
Alkamox C2, C2-0 [Alkaril]—N,N-Bis (2-hydroxyethyl) cocamine oxide
Alkamox LO [Alkaril; Alkaril UK; Alkaril Canada]—Lauryl dimethyl amine oxide

Alkamox ODM [Alkaril]—Oleyl dimethyl amine oxide

Alkamuls GTO [Alkaril]—Glyceryl trioleate

Alkamuls SMO [Alkaril]—Sorbitan monooleate

Alkaquat DMB-451, DMB-451 50%, DMB-451 80% [Alkaril]—Benzalkonium chloride

Alkasurf CO-20 [Alkaril]—POE (20) castor oil

Alkasurf CO-200 [Alkaril]—POE (200) castor oil

Alkateric BC [Alkaril; Alkaril UK; Alkaril Canada]—Coco-betaine

Alkateric CAB [Alkaril]—Coco amidopropyl betaine

Alkateric OB [Alkaril]—Oleyl betaine

Alkateric STB [Alkaril]—Stearyl betaine

Amerlate P, W [Amerchol; Amerchol Europe]—Isopropyl lanolate

Amine 12D [Berol Nobel]—Lauryl amine

Aminol CA-2 [Finetex]—Ricinoleic diethanolamide

Ammonyx 4, 4B, 485, 490, 4002 [Millmaster-Onyx]—Stearyl dimethyl benzyl ammonium chloride

Ammonyx CETAC, CETAC-30 [Millmaster-Onyx]—Cetyl trimethyl ammonium chloride

Ammonyx DMCD-40, LO [Millmaster-Onyx]—Lauryl dimethyl amine oxide

Ammonyx KP [Millmaster-Onyx]—Oleyl dimethyl benzyl ammonium chloride

Ammonyx SO [Millmaster-Onyx]—Stearyl dimethyl amine oxide

Amonyl 265BA [Seppic]—Coco-betaine

Amonyl 380 BA [Seppic]—Coco amidopropyl betaine

Ampho B11-34 [Capital City]—Coco-betaine

Amphocerin K [Henkel; Henkel KGaA]—Cetearyl alcohol

Amphoram CB A30 [Ceca SA]—Coco-betaine

Amphosol CA, CG [Stepan; Stepan Europe]—Coco amidopropyl betaine

Amphoterge S [Lonza]—Stearoamphoacetate

Amyx A-25S [Clough]—Stearyl dimethyl benzyl ammonium chloride

Anfomul S4, S50 [Croda Ltd.]—Sorbitan monooleate

Anfomul S43 [Croda Ltd.]—Sorbitan sesquioleate

Aqualose L30 [Westbrook Lanolin]—POE (30) lanolin

Aqualose L75, L75/50 [Westbrook Lanolin]—POE (75) lanolin (USP)

Aqualose SLT, SLW [Westbrook Lanolin]—Lanolin

Aqualose W20, W20/50 [Westbrook Lanolin]—POE (20) lanolin ether

Aremsol A [Ronsheim & Moore]—Coco amidopropyl betaine

Argobase 125 [Westbrook Lanolin]—Lanolin alcohol

Argonol 50 Pharmaceutical, 50 Super, 60 [Westbrook Lanolin]—Lanolin oil

Argowax Distilled, Standard [Westbrook Lanolin]—Lanolin alcohol

Arlacel 80 [ICI Americas]—Sorbitan monooleate

Arlacel 83 [ICI Americas; ICI Specialty]—Sorbitan sesquioleate (cosmetic/pharmaceutical grade)

Arlacel C [ICI Americas; ICI Specialty]—Sorbitan sesquioleate

TRADENAME PRODUCTS AND GENERIC EQUIVALENTS

Arlamol E [ICI Americas]—POP (15) stearyl ether

Armeen 12, 12D [Akzo/Armak; Akzo BV]—Lauryl amine

Armid 18 [Akzo/Armak]—Stearic acid amide

Armid E [Akzo]—Erucamide

Armid O [Akzo/Armak]—Oleyl amide

Armoslip 18 [Akzo/Noury]—Stearic acid amide

Armoslip CPM [Akzo/Noury]—Oleyl amide

Armoslip EXP [Akzo/Noury]—Erucamide

Armosoft L [Akzo/Armak]—Ditallow dimethyl ammonium chloride

Armotan MO [Akzo BV; Akzo Italia]—Sorbitan monooleate

Aromox C/12, C/12-W, CD/12 [Akzo/Armak]—N,N-Bis (2-hydroxyethyl) cocamine oxide

Aromox DM18D-W [Akzo/Armak; Akzo BV]—Stearyl dimethyl amine oxide

Aromox DMC, DMCD, DMC-W [Akzo/Armak; Akzo BV]—Coco amine oxide

Aromox DMMC-W [Akzo/Armak; Akzo BV]—Lauryl dimethyl amine oxide

Aromox T/12 [Akzo/Armak]—Bis (2-hydroxyethyl) tallow amine oxide

Arosurf PT50 [Sherex]—Benzalkonium chloride

Arosurf TA-100, TA-101 [Sherex]—Distearyl dimethyl ammonium chloride

Arquad 2HT, 2HT-75, 88 [Akzo/Armak; Akzo BV]—Dihydrogenated tallow dimethyl ammonium chloride

Arquad 16-29, 16-50 [Akzo/Armak; Akzo BV]—Cetyl trimethyl ammonium chloride

Arquad 18-50 [Akzo/Armak]—Stearyl trimethyl ammonium chloride

Arquad B-50 USP, B-90 USP, B-100, DMCB, DMMCB-50, DMMCB-75 [Akzo Chemie]—Benzalkonium chloride

Arquad DM18B-90 [Akzo/Armak]—Stearyl dimethyl benzyl ammonium chloride

Atlas G-3780A [Atlas Refinery]—POE (20) tallow amine

Atlas G-4884 [Atlas Refinery]—Sorbitan monooleate (tech.)

Atlas G-7596-J [ICI Americas]—POE (10) sorbitan monolaurate

Atlas KO [Guelph Soap]—Potassium oleate

Atpet 80, 100, 200 [ICI Americas; ICI Specialty]—Sorbitan monooleate

Barlox 12 [Lonza]—Coco amine oxide

Barquat MB-50 (USP), MB-80 (USP), MX-50, MX-80, OJ-50 [Lonza]—Benzalkonium chloride

Benol [Witco/Sonneborn]—Mineral oil (white NF)

Be Square 175, 185, 195 [Petrolite]—Microcrystalline wax

Bio-Quat 50-24 (USP), 50-25, 50-28, 50-30, 50-40 (USP), 50-42, 50-60, 50-65, 80-24 (USP), 80-28, 80-40 (USP), 80-42 [Bio-Lab; Lonza]—Benzalkonium chloride

Bio-Quat ASH-29 [Bio-Lab; Lonza]—Cetyl trimethyl ammonium chloride

Blandol [Witco/Sonneborn]—Mineral oil (white NF)

Bromat [Hexcel]—Cetyl trimethyl ammonium bromide

BTC 50, 50 USP, 65, 65 USP, 100, 835, 2565, 8248, 8249, E-8358 [Millmaster-Onyx]—Benzalkonium chloride

Bumyr [Amerchol]—Butyl myristate
Cachalot C-50, C-51 [M. Michel]—Cetyl alcohol
Cachalot O-15 [M. Michel]—Oleyl alcohol
Capmul O [Capital City]—Sorbitan monooleate
Carnation [Witco/Sonneborn]—Mineral oil (white NF)
Carsamine CB [Lonza]—Cocamidopropyl dimethylamine
Carsonam 3, 32, 33-S, 3147 [Lonza]—Coco amidopropyl betaine
Carsonam BCW [Lonza]—Coco-betaine
Carsoquat 621, 621 80% [Lonza]—Benzalkonium chloride
Carsoquat CT-429, CTM-29, CTM-429 [Lonza]—Cetyl trimethyl ammonium chloride
Carsoquat SDQ-25, SDQ-85 [Lonza]—Stearyl dimethyl benzyl ammonium chloride
Carsosoft S-75, S-90, S-90M [Lonza]—Quaternium-27
Catigene CS40, T 50, T 80, T 80 F [Stepan Europe]—Benzalkonium chloride
Catinal OB-80E [Toho]—Stearyl dimethyl benzyl ammonium chloride
Cation DS [Sanyo]—Distearyl dimethyl ammonium chloride
Cation G-40 [Sanyo]—Benzalkonium chloride
Cation S [Sanyo]—Stearyl dimethyl benzyl ammonium chloride
Cegesoft C17 [Henkel KGaA]—Myristyl lactate
Cegesoft C19 [Henkel KGaA]—Cetyl lactate
Cegesoft C25 [Henkel KGaA]—Octyl palmitate
CEP-33 [Hefti Ltd.]—Cetyl palmitate
Ceralan [Amerchol]—Lanolin alcohol
Ceraphyl 28 [Van Dyk]—Cetyl lactate
Ceraphyl 31 [Van Dyk]—Lauryl lactate
Ceraphyl 50S [Van Dyk]—Myristyl lactate
Ceraphyl 140 [Van Dyk]—Decyl oleate
Ceraphyl 368 [Van Dyk]—Octyl palmitate
Ceraphyl 375 [Van Dyk]—Isostearyl neopentanoate
Ceraphyl 424 [Van Dyk]—Myristyl myristate
Ceraphyl 494 [Van Dyk]—Isocetyl stearate
Ceraphyl ICA [Van Dyk]—Isocetyl alcohol
Cetal [Amerchol]—Cetyl alcohol (NF)
Cetax 16 [Aquatec Quimica]—Cetyl alcohol
Cetax 50 [Aquatec Quimica]—Cetearyl alcohol
Cetiol [Henkel Canada; Henkel KGaA]—Oleyl monooleate
Cetiol 868 [Henkel Canada]—Octyl stearate
Cetiol 1414E [Henkel Canada]—POE (3) myristyl ether myristate
Cetiol A [Henkel; Henkel Canada; Henkel KGaA]—Hexyl laurate
Cetiol B [Henkel Canada; Henkel KGaA]—Dibutyl adipate
Cetiol G16S [Henkel Canada]—Isocetyl stearate
Cetiol HE [Henkel; Henkel Canada; Henkel KGaA]—POE (7) glyceryl monococoate
  (self-emulsifying)

340

Cetiol J 600 [Henkel Canada; Henkel KGaA]—Oleyl erucate
Cetiol MM [Henkel Canada; Henkel KGaA]—Myristyl myristate
Cetiol V [Henkel; Henkel Canada; Henkel KGaA]—Decyl oleate
Cetostearyl Alcohol BP, NF [Croda]—Cetearyl alcohol
Cetyl-Stearyl Alcohol, NF [Alcolac]—Cetearyl alcohol
Chem I Sorb CSD 1035 [Polychem Dispersions]—Petrolatum
Chemadene NA-30 [Richardson]—Coco amidopropyl betaine
Chemax CO-200/50 [Chemax]—POE (200) castor oil
Chemax HCO-16 [Chemax]—POE (16) hydrogenated castor oil
Chemcol T-20 [Chemform]—POE (20) tallow amine
Chemeen S-2 [Chemax]—POE (2) soya amine
Chemeen T-20 [Chemax]—POE (20) tallow amine
Chemical 39 Base [Sandoz]—Stearamidoethyl ethanolamine
Chemical Base 6532 [Sandoz]—Stearamidoethyl diethylamine
Chemquat 16-50 [Chemax]—Cetyl trimethyl ammonium chloride
Chimin AX [Tessilchimica]—Coco amidopropyl betaine
Chimin BX [Tessilchimica]—Stearyl betaine
Clearlan 650, 1650, K50 [Henkel/Emery]—Lanolin (anhyd. USP)
CO-1695 [Procter & Gamble]—Cetyl alcohol (NF)
Cobee 76 [Stepan/PVO]—Coconut oil (refined, bleached, deodorized)
Collagen CLR [Henkel Canada]—Soluble collagen
Collagen Hydrolyzate Cosmetic 55, N-55, SD [Amerchol]—Hydrolyzed animal protein
Collasol [Croda; Croda Ltd.]—Soluble collagen
Comperlan LD9 [Henkel Canada]—Distearyl dimethyl ammonium chloride
Conco XA-L [Continental Chem.]—Lauryl dimethyl amine oxide
Conco XA-O [Continental Chem.]—Oleyl dimethyl amine oxide
Conco XA-S [Continental Chem.]—Stearyl dimethyl amine oxide
Conco XA-T [Continental Chem.]—Tallow dimethyl amine oxide
Conco XA-Y [Continental Chem.]—Coco amine oxide
Corona Lanolin [Croda Ltd.]—Lanolin (anhyd.)
Coronet Lanolin [Croda Ltd.]—Lanolin (super refined, cosmetic grade, BP)
Cosmetic Lanolin Anhydrous USP [Croda]—Lanolin (anhyd. USP)
Cralane LR-10 [Pulcra SA]—POE (75) lanolin
Crapol AU-20, AU-21, AU-24 [Pulcra SA]—Benzalkonium chloride
Crapol AU-23 [Pulcra SA]—Stearyl dimethyl benzyl ammonium chloride
Crapol FU-25 [Pulcra SA]—Dihydrogenated tallow dimethyl ammonium chloride
Crill 4, 50 [Croda]—Sorbitan monooleate (tech.)
Crill 43 [Croda; Croda Ltd.]—Sorbitan sesquioleate
Crodacol C [Croda]—Cetyl alcohol
Crodacol CS50 [Croda]—Cetearyl alcohol
Crodactil SISL [Croda]—Sodium isostearoyl lactylate
Crodafos N3 Acid [Croda; Croda Ltd.]—POE (3) oleyl ether phosphate

Crodafos N10 Acid [Croda; Croda Ltd.]—POE (10) oleyl ether phosphate
Crodalan IPL [Croda; Croda Ltd.]—Isopropyl lanolate
Crodamide E, ER [Croda Ltd.]—Erucamide
Crodamide O, OR [Croda Ltd.]—Oleyl amide
Crodamide S, SR [Croda Ltd.]—Stearic acid amide
Crodamol BS [Croda]—Butyl stearate
Crodamol CAP [Croda Ltd.]—Cetearyl octanoate
Crodamol CL [Croda]—Cetyl lactate
Crodamol DO [Croda Ltd.]—Decyl oleate
Crodamol ICS [Croda]—Isocetyl stearate
Crodamol IPM [Croda]—Isopropyl myristate
Crodamol IPP [Croda; Croda Ltd.]—Isopropyl palmitate
Crodamol ISNP [Croda Ltd.]—Isostearyl neopentanoate
Crodamol LL [Croda; Croda Ltd.]—Lauryl lactate
Crodamol ML [Croda; Croda Ltd.]—Myristyl lactate
Crodamol MM [Croda; Croda Ltd.]—Myristyl myristate
Crodamol OHS [Croda Ltd.]—Octyl hydroxystearate
Crodamol OP [Croda; Croda Ltd.]—Octyl palmitate
Crodamol PMP [Croda; Croda Ltd.]—POP (2) myristyl ether propionate
Crodamol SS [Croda; Croda Ltd.]—Synthetic spermaceti (NF)
Crodapur [Croda Ltd.]—Lanolin (tech.)
Crodasinic LS30, LS35 [Croda; Croda Ltd.]—Sodium lauroyl sarcosinate
Cropepsol [Croda Ltd.]—Hydrolyzed animal protein
Cropeptone [Croda Ltd.]—Hydrolyzed animal protein
Crosterene [Croda Universal]—Stearic acid
Crotein A, C, O, SPA, SPA 55, SPC, SPO [Croda; Croda Ltd.]—Hydrolyzed animal
    protein
Crotein AD, AD Anhyd., ADX [Croda]—AMP isostearic hydrolyzed animal protein
CS-250, -251 [Crystal Soap]—Zinc stearate
Cutina CP, CP-A [Henkel; Henkel KGaA]—Cetyl palmitate
Cycetyl Alcohol NF [Alcolac]—Cetyl alcohol
Cyclochem 326A [Alcolac]—Synthetic beeswax
Cyclochem CL [Alcolac]—Cetyl lactate
Cyclochem CP [Alcolac]—Cetyl palmitate
Cyclochem GTIS [Alcolac]—Glyceryl triisostearate
Cyclochem GTL [Alcolac]—Glyceryl trilaurate
Cyclochem GTO [Alcolac]—Glyceryl trioleate
Cyclochem INEO [Alcolac]—Isostearyl neopentanoate
Cyclochem LVL [Alcolac]—Lauryl lactate
Cyclochem ML [Alcolac]—Myristyl lactate
Cyclochem MM/M [Alcolac]—Myristyl myristate
Cyclochem MST [Alcolac]—Myristyl stearate

Cyclochem PETO [Alcolac]—Pentaerythritol tetraoleate
Cyclochem PETS [Alcolac]—Pentaerythritol tetrastearate
Cyclochem SPS [Alcolac]—Synthetic spermaceti
Cyclochem SS [Alcolac]—Stearyl stearate
Cyclogol Cetyl Alcohol NF [Alcolac]—Cetyl alcohol
Cyclogol Cetyl-Stearyl Alcohol [Alcolac]—Cetearyl alcohol
Cyclomide CODI [Alcolac]—Cocamidopropyl dimethylamine
Cyclomide IODI [Alcolac]—Isostearamidopropyl dimethylamine
Cyclomide SODI [Alcolac]—Stearamidopropyl dimethylamine
Cyclomox C [Alcolac]—Coco amine oxide
Cyclomox L [Alcolac]—Lauryl dimethyl amine oxide
Cyclophos PO3 [Alcolac]—POE (3) oleyl ether phosphate
Cyclopol SB-5 [Alcolac]—Disodium laneth-5 sulfosuccinate
Cycloteric BET-C30 [Alcolac]—Coco amidopropyl betaine (cosmetic grade)
Cycloteric BET-C41 [Alcolac]—Coco-betaine
Cycloteric BET-CB [Alcolac]—Coco amidopropyl betaine (cosmetic, glycerin-free)
Cycloteric BET-I-30 [Alcolac]—Isostearamidopropyl betaine
Cycloteric BET-W [Alcolac]—Coco amidopropyl betaine (tech.)
Cycloton D261C/70, D261C/75 [Alcolac]—Dihydrogenated tallow dimethyl ammonium chloride
Cycloton M214B/99, M214C/45, M214C/99 [Alcolac]—Myristyl trimethyl ammonium bromide
Cycloton M242B/99 [Alcolac]—Cetyl trimethyl ammonium bromide
Cycloton M242C/29 [Alcolac]—Cetyl trimethyl ammonium chloride
Cycloton M270C/18, M270C/85, SCS [Alcolac]—Stearyl dimethyl benzyl ammonium chloride
Cyncal 80% [Hilton-Davis]—Benzalkonium chloride
Dar-Chem 11 [Unichema]—Stearic acid (single-pressed)
Dar-Chem 12 [Unichema]—Stearic acid (double-pressed)
Dar-Chem 13, 14 [Unichema]—Stearic acid (triple-pressed)
Dark Green No. 2 [Witco]—Petrolatum (tech.)
Dastar [Croda Ltd.]—Cholesterol
Dehydag Wax 16 [Henkel KGaA]—Cetyl alcohol
Dehydag Wax 22 (Lanette) [Henkel KGaA]—Behenyl alcohol
Dehydag Wax O [Henkel; Henkel KGaA]—Cetearyl alcohol
Dehymuls SMO [Henkel KGaA]—Sorbitan monooleate
Dehymuls SSO [Henkel KGaA]—Sorbitan sesquioleate
Dehyquart A [Henkel]—Cetyl trimethyl ammonium chloride
Dehyquart DAM [Henkel KGaA]—Distearyl dimethyl ammonium chloride
Dehyquart SP [Henkel]—Quaternium-52
Dehyquart STC-25 [Henkel]—Stearyl dimethyl benzyl ammonium chloride
Dehyton AB-30 [Henkel Canada; Henkel KGaA]—Coco-betaine

Dehyton K [Henkel KGaA]—Coco amidopropyl betaine
Deltyl Extra [Givaudan]—Isopropyl myristate
Deriphat BA, BAW [Henkel]—Coco amidopropyl betaine
Dermalcare HL [Alcolac]—Hexyl laurate
DeSomeen TA-20 [DeSoto]—POE (20) tallow amine
DeSonic SMO [DeSoto]—Sorbitan monooleate
DeSotan SMO [DeSoto]—Sorbitan monooleate
Dimodan P, PM, PM 300 [Grindsted]—Lard monoglyceride (dist. sat.)
Dimodan S [Grindsted]—Lard monoglyceride (dist. unsat.)
DPPG [Gattefosse Ets.]—Propylene glycol dipelargonate
Drakeol 5, 6, 7, 8, 9, 10, 13, 15 [Penreco]—Mineral oil (light USP)
Drakeol 19, 21, 32, 34, 35 [Penreco]—Mineral oil (USP)
Draketex 50 [Penreco]—Mineral oil (light USP)
Drewmulse SMO [Stepan/PVO]—Sorbitan monooleate
Duron Micro Wax 170/180 (0423) [Dura]—Microcrystalline wax
Durtan 80 [Durkee]—Sorbitan monooleate
Dymsol RL7 [Harwick]—Zinc stearate
Dynacerin 660 [Huls AG]—Oleyl erucate
Dynasan 118 [Huls]—Glyceryl tristearate
EBS Wax [Borg-Warner]—N,N′-Ethylene bisstearamide
Emasol O-10 [Kao]—Sorbitan monooleate
Emasol O-15 R [Kao]—Sorbitan sesquioleate
Emcol 1655 [Witco]—Cocamidopropyl dimethylamine propionate
Emcol 3780 [Witco]—Stearamidopropyl dimethylamine lactate
Emcol 6613 [Witco]—Isostearamidopropyl dimethylamine lactate
Emcol CC-37-18 [Witco]—Coco-betaine
Emcol DG, NA-30 [Witco]—Coco amidopropyl betaine
Emcol ISML [Witco]—Isostearamidopropyl morpholine lactate
Emcol L [Witco]—Lauryl dimethyl amine oxide
EmCon W [Fanning]—Wheat germ oil
Emerest 1723 [Henkel/Emery]—Isopropyl lanolate
Emerest 2308 [Henkel/Emery]—Tridecyl stearate
Emerest 2310 [Henkel/Emery]—Isopropyl isostearate
Emerest 2314 [Henkel/Emery]—Isopropyl myristate
Emerest 2316 [Henkel/Emery]—Isopropyl palmitate
Emerest 2321, 2325, 2326 [Henkel/Emery]—Butyl stearate
Emerest 2324 [Henkel/Emery]—Isobutyl stearate
Emerest 2388 [Henkel/Emery]—Propylene glycol dipelargonate
Emerest 2410 [Henkel/Emery]—Glyceryl isostearate
Emerest 2423 [Henkel/Emery]—Glyceryl trioleate
Emerest 2485, 2486 [Henkel/Emery]—Pentaerythritol tetrapelargonate
Emersol 110, 120 [Henkel/Emery]—Stearic acid

344

Emersol 132 [Henkel/Emery]—Stearic acid (triple-pressed)
Emersol 143 [Henkel/Emery]—Palmitic acid
Emersol Palmitic Acid [Henkel/Emery]—Palmitic acid
Emersol Stearic Acid [Henkel/Emery]—Stearic acid
Emery 400 [Henkel/Emery]—Stearic acid
Emery 1650, 1656 [Henkel/Emery]—Lanolin (anhyd. USP)
Emery 1660 [Henkel/Emery]—Lanolin (ultra anhyd. USP, cosmetic grade)
Emery 1787 [Henkel/Emery]—Cetyl alcohol (NF)
Emery 5430, 6744, 6748 [Henkel/Emery]—Coco amidopropyl betaine
Emery Methyl Stearate [Henkel/Emery]—Methyl stearate
Emkapol PO-18 [Emkay]—Potassium oleate
Empigen BAC, BAC 90, BCB 50, BCF 80 [Albright & Wilson/Marchon]—Benzalkonium chloride
Empigen BCJ-50 [Albright & Wilson/Australia]—Oleyl dimethyl benzyl ammonium chloride
Empigen BCP/25 [Albright & Wilson/Australia]—Stearyl dimethyl benzyl ammonium chloride
Empigen OB [Albright & Wilson/Marchon]—Lauryl dimethyl amine oxide
Emsorb 2500 [Henkel/Emery]—Sorbitan monooleate
Emsorb 2502 [Henkel/Emery]—Sorbitan sesquioleate
Epal 16NF [Ethyl]—Cetyl alcohol
Ervol [Witco/Sonneborn]—Mineral oil (white NF)
Ethomeen S/12 [Akzo/Armak]—POE (2) soya amine
Ethosperse G-26 [Lonza]—POE (26) glyceryl ether
Ethoxylan 1685, 1686 [Henkel/Emery]—POE (75) lanolin
Ethylan GO-80 [Harcros UK]—Sorbitan monooleate
Etocas 20 [Croda Ltd.]—POE (20) castor oil
Etocas 60 [Croda Ltd.]—POE (60) castor oil
Etocas 200 [Croda Ltd.]—POE (200) castor oil
Exceparl IPM [Kao]—Isopropyl myristate
Exceparl IPP [Kao]—Isopropyl palmitate
Falco S-14S [Fallek]—Myristyl stearate
Falco S-18S [Fallek]—Stearyl stearate
Fanchem HL [Fanning]—Hydrogenated lanolin
Fancol HL [Fanning]—Hydrogenated lanolin
Fancol LA [Fanning]—Lanolin alcohol (Regular USP and Superfine USP grades)
Fancol OA, OA 50, OA 70, OA 80, OA 90, OA 95 [Fanning]—Oleyl alcohol
Fancor Lanolin [Fanning]—Lanolin
Fancorp Lanolin [Fanning]—Lanolin
Flexowax C, C Light [Lonza]—Microcrystalline wax
Flowco [Mallinckrodt]—Calcium stearate
Fluilan [Croda; Croda Ltd.]—Lanolin oil

FMB 65-15 Quat, 65-28 Quat, 451-5 Quat, 451-8 Quat, 500-15 Quat USP, 4500-5 Quat, 4500-8 Quat [Huntington Lab]—Benzalkonium chloride

Fonoline White, Yellow [Witco]—Petrolatum (USP)

Forlan [RITA]—Lanolin

Fortex [Petrolite]—Microcrystalline wax

Gafac MC-470F [GAF]—Sodium POE (4) lauryl ether phosphate

Gafac RM-510 [GAF]—POE (10) dinonyl phenyl ether phosphate

Gardiquat 12H, 1450 (USP), 1480 (USP), SV480 [Albright & Wilson (Australia)]—Benzalkonium chloride

Genamin CTAC [Hoechst-Celanese AG]—Cetyl trimethyl ammonium chloride

Genamin DSAC [Hoechst-Celanese AG]—Distearyl dimethyl ammonium chloride

Genamin KDM [Hoechst-Celanese]—Behenyl trimethyl ammonium chloride

Genamin KDM-F [Hoechst-Celanese; Hoechst-Celanese AG]—Behenyl trimethyl ammonium chloride

Genaminox CS, KC [Hoechst-Celanese AG]—Coco amine oxide

Generol 122 [Henkel]—Soy sterol

Generol 122E5 [Henkel]—POE (5) soya sterol

Generol 122E10 [Henkel]—POE (10) soya sterol

Generol 122E25 [Henkel]—POE (25) soya sterol

Gloria [Witco/Sonneborn]—Mineral oil (white USP)

Glucam E-10 [Amerchol]—Methyl gluceth-10

Glucam E-20 [Amerchol]—Methyl gluceth-20

Glucam P-10 [Amerchol]—POP (10) methyl glucose ether

Glucam P-20 [Amerchol]—POP (20) methyl glucose ether

Glycomul O [Lonza]—Sorbitan monooleate

Glycomul SOC [Lonza]—Sorbitan sesquioleate

GMM-33 [Hefti Ltd.]—Glyceryl myristate

Golden Dawn Grade 1, 2, Superfine [Westbrook Lanolin]—Lanolin (anhyd.)

Golden Dawn Lanolin [Westbrook Lanolin]—Lanolin (pharmaceutical/cosmetic grade)

Golden Fleece DF [Westbrook Lanolin]—Lanolin (extra refined anhyd.)

Golden Fleece Lanolin [Westbrook Lanolin]—Lanolin (pharmaceutical/cosmetic grade)

Golden Fleece P-80, P-95, RA [Westbrook Lanolin]—Lanolin (anhyd.)

Graden Butyl Stearate [Graden]—Butyl stearate

Grindtek AMOS 90 [Grindsted]—Acetylated lard glyceride

Grindtek MM 90 [Grindsted]—Glyceryl myristate

Grindtek MOP 90 [Grindsted]—Lard monoglyceride

Groco 55L, 56R Rubber Grade Stearic Acid, 57, 58 [A. Gross]—Stearic acid

Grocor 1000, 1200 [A. Gross]—Glyceryl trioleate

Grocor 5410, 5510 [A. Gross]—Butyl stearate

Grocor 5721 [A. Gross]—Tridecyl stearate

Grocor 8058 [A. Gross]—Methyl stearate

GTS-33 [Hefti Ltd.]—Glyceryl tristearate

Hallcote ZS-50/50 [C.P. Hall]—Zinc stearate
Hamposyl L-30, L-95 [W.R. Grace]—Sodium lauroyl sarcosinate
Hartaine CB-40 [Hart Prod.]—Coco-betaine
Hartolan, Super Hartolan [Croda; Croda Ltd.]—Lanolin alcohol
Hartolite [Croda; Croda Ltd.]—Lanolin alcohol
Hartox DMCD [Hart Prod.]—Coco amine oxide
Harwick Calcium Stearate [Harwick]—Calcium stearate
Harwick F-300, F-1000, F-1500 [Harwick]—Stearic acid
Harwick Zinc Stearate H, USP [Harwick]—Zinc stearate
HD Eutanol [Henkel]—Oleyl alcohol
HD-Ocenol [Henkel Canada]—Oleyl alcohol
HD-Ocenol 92/96 [Henkel; Henkel Canada]—Oleyl alcohol
Hetamine 5L25 [Heterene]—Stearamidopropyl dimethylamine lactate
Hetester MS [Heterene]—Myristyl stearate
Hetester PMA [Heterene]—Propylene glycol myristyl ether acetate
Hetester TICC [Heterene]—Triisocetyl citrate
Hetoxamine S-2 [Heterene]—POE (2) soya amine
Hetoxamine T20 [Heterene]—POE (20) tallow amine
Hetoxide C-2 [Heterene]—POE (2) castor oil
Hetoxide C-200, C-200-50% [Heterene]—POE (200) castor oil
Hetoxide HC-16 [Heterene]—POE (16) hydrogenated castor oil
Hetoxide G-7 [Heterene]—POE (7) glyceryl ether
Hetoxide G-26 [Heterene]—POE (26) glyceryl ether
Hetoxol OA-3 Special [Heterene]—POE (3) oleyl ether
Hetoxol PLA [Heterene]—POP (30) lanolin ether
Hetoxol SP-15 [Heterene]—POP (15) stearyl ether
Hetsorb L-10 [Heterene]—POE (10) sorbitan monolaurate
Hodag GTO [Hodag]—Glyceryl trioleate
Hodag SMO [Hodag]—Sorbitan monooleate
HTSA #1 [Hexcel]—Oleyl palmitamide
HTSA #3 [Hexcel]—Stearyl erucamide
Huber SM [J.M. Huber]—Mica (surface-treated)
Huber WG-1, WG-2 [J.M. Huber]—Mica (water-ground)
Hyamine 3500, 3500 NF [Lonza]—Benzalkonium chloride
Hy Dense Calcium Stearate HP Gran., RSN Powd. [Mallinckrodt]—Calcium stearate
Hy Dense Zinc Stearate XM Powd., XM Ultra Fine [Mallinckrodt]—Zinc stearate
Hydrenol DD [Henkel Canada]—Cetearyl alcohol
Hydrofol Acid 1655-CG-NF, 1855, 1895 [Sherex]—Stearic acid
Hydrofol Acid 1690 [Sherex]—Palmitic acid
Hydroxylan [Fanning]—Hydroxylated lanolin
Hydro Zinc [Morton Int'l./Ventron]—Zinc stearate
Hystrene 4516, 8016, 9016 [Humko/Witco]—Palmitic acid

Hystrene 5016 [Humko/Witco]—Stearic acid (low IV triple-pressed)

Hystrene 5016 NF FG [Humko/Witco]—Stearic acid (triple-pressed food grade)

Hystrene 7018, 7018 FG, 8018, 9018, 9718, 9718 NF FG [Humko/Witco]—Stearic acid

Icomeen T-20 [BASF]—POE (20) tallow amine

Imwitor 780 [Hüls AG]—Glyceryl isostearate

Incrocas 60 [Croda]—POE (60) castor oil

Incromate CDL [Croda]—Cocamidopropyl dimethylamine lactate

Incromate CDP [Croda]—Cocamidopropyl dimethylamine propionate

Incromate IDL [Croda]—Isostearamidopropyl dimethylamine lactate

Incromate ISML [Croda]—Isostearamidopropyl morpholine lactate

Incromate SDL [Croda]—Stearamidopropyl dimethylamine lactate

Incromectant AMEA-70 [Croda]—Acetyl monoethanolamide

Incromine BB [Croda]—Behenamidopropyl dimethylamine

Incromine CB [Croda]—Cocamidopropyl dimethylamine

Incromine IB [Croda]—Isostearamidopropyl dimethylamine

Incromine ISM [Croda]—Isostearamidopropyl morpholine

Incromine L-40, Oxide L [Croda]—Lauryl dimethyl amine oxide

Incromine OPB, OPM [Croda]—Oleamidopropyl dimethylamine

Incromine Oxide OD-50 [Croda]—Oleyl dimethyl amine oxide

Incromine Oxide S [Croda]—Stearyl dimethyl amine oxide

Incromine PB [Croda]—Palmitamidopropyl dimethylamine

Incromine SB [Croda]—Stearamidopropyl dimethylamine

Incronam 30 [Croda; Croda Universal]—Coco amidopropyl betaine

Incronam I-30 [Croda; Croda Universal]—Isostearamidopropyl betaine

Incroquat B-85 [Croda]—Behenyl dimethyl benzyl ammonium chloride

Incroquat BDQ-25P, Behenyl BDQ/P [Croda Universal]—Behenyl dimethyl benzyl ammonium chloride

Incroquat Behenyl TMC/P [Croda Universal]—Behenyl trimethyl ammonium chloride

Incroquat CTC-25, CTC-30 [Croda]—Cetyl trimethyl ammonium chloride

Incroquat I-85 [Croda]—Isostearamidopropalkonium chloride

Incroquat S-75CG [Croda]—Quaternium-27

Incroquat S-85, SDQ-25 [Croda; Croda Universal]—Stearyl dimethyl benzyl ammonium chloride

Incrosul LAFS [Croda]—Disodium laneth-5 sulfosuccinate

Incrosul OMS [Croda]—Disodium oleamido MEA-sulfosuccinate

Industrene 225, 226 [Humko/Witco]—Soy acid (dist.)

Industrene 225 FG, 226 FG [Humko/Witco]—Soy acid

Industrene 4516 [Humko/Witco]—Palmitic acid

Industrene 5016 [Humko/Witco]—Stearic acid (double-pressed)

Industrene 5016 FG, 7018, 7018 FG, 8518, 8718 FG, 9018 [Humko/Witco]—Stearic acid

Industrol CO-200, CO-200-50% [BASF]—POE (200) castor oil

Interstab CA-18-1 [Akzo/Interstab]—Calcium stearate

TRADENAME PRODUCTS AND GENERIC EQUIVALENTS

Interstab GT-8257 [Akzo/Interstab]—N,N´-Ethylene bisstearamide
Interstab ZN-18-1 [Akzo/Interstab]—Zinc stearate
Ionet S-80 [Sanyo]—Sorbitan monooleate
I.P.M. [Akzo Chemie Italia]—Isopropyl myristate
IPP [Henkel KGaA]—Isopropyl palmitate
IPS [Henkel KGaA]—Isopropyl stearate
Isopropylpalmitat [Henkel Canada]—Isopropyl palmitate
Isopropylstearat [Henkel Canada]—Isopropyl stearate
JAQ Powdered Quat [Huntington Lab]—Benzalkonium chloride
Jet Quat 2HT-75 [Jetco]—Dihydrogenated tallow dimethyl ammonium chloride
Jordamine DAPI [PPG-Mazer]—Isostearamidopropyl dimethylamine
Jordamine DAPL, DMCAPA, SHCFA [PPG-Mazer]—Cocamidopropyl dimethylamine
Jordamine DAPSA, S-13 [PPG-Mazer]—Stearamidopropyl dimethylamine
Jordamine S-13 Lactate [PPG-Mazer]—Stearamidopropyl dimethylamine lactate
Jordamox LDA [PPG-Mazer]—Lauryl dimethyl amine oxide
Jordamox ODA [PPG-Mazer]—Oleyl dimethyl amine oxide
Jordamox SDA [PPG-Mazer]—Stearyl dimethyl amine oxide
Jordaquat 350, 350 80%, 358 [PPG-Mazer]—Benzalkonium chloride
Jordaquat 522 [PPG-Mazer]—Isostearamidopropyl ethyldimonium ethosulfate
Jordaquat JN [PPG-Mazer]—Ricinoleamidopropyl ethyldimonium ethosulfate
Jordaquat JO-50 [PPG-Mazer]—Oleyl dimethyl benzyl ammonium chloride
Jordaquat JS-25 [PPG-Mazer]—Stearyl dimethyl benzyl ammonium chloride
Jortaine C, CAB-35, CFA-35 [PPG-Mazer]—Coco amidopropyl betaine
Jortaine CB40 [PPG-Mazer]—Coco-betaine
Kalcohl 60 [Kao]—Cetyl alcohol
Katapol PN-810 [GAF]—POE (20) tallow amine
Kaydol [Witco/Sonneborn]—Mineral oil (white USP)
Kemamide E [Humko/Witco]—Erucamide
Kemamide E-180 [Humko/Witco]—Stearyl erucamide
Kemamide O, U [Humko/Witco]—Oleyl amide
Kemamide P-181 [Humko/Witco]—Oleyl palmitamide
Kemamide S [Humko/Witco]—Stearic acid amide
Kemamide W-20 [Humko/Witco]—Ethylene bisoleamide
Kemamide W-35, W-39, W-40, W-40/300, W-40DF, W-45 [Humko/Witco]—N,N´-
    Ethylene bisstearamide
Kemamine BQ-2802C [Humko/Witco]—Behenyl dimethyl benzyl ammonium chloride
Kemamine P-690, P-690D [Humko/Witco]—Lauryl amine (dist.)
Kemamine Q-9702C [Humko/Witco]—Dihydrogenated tallow dimethyl ammonium
    chloride
Kemester 1000 [Humko/Witco]—Glyceryl trioleate
Kemester 5415 [Humko/Witco]—Isobutyl stearate
Kemester 5510 [Humko/Witco]—Butyl stearate

Kemester 5654 [Humko/Witco]—Ditridecyl adipate
Kemester 5721 [Humko/Witco]—Tridecyl stearate
Kemester 5822 [Humko/Witco]—Isocetyl stearate
Kemester 7018, 9018 [Humko/Witco]—Methyl stearate
Kemester BE [Humko/Witco]—Behenyl erucate
Kemester CP [Humko/Witco]—Cetyl palmitate
Kemester E [Humko/Witco]—Erucamide
Kemester MM [Humko/Witco]—Myristyl myristate
Kessco 653 [Stepan]—Cetyl palmitate
Kessco 654 [Stepan]—Cetyl myristate
Kessco BSC [Stepan]—Butyl stearate
Kessco Butyl Stearate Cosmetic, Distilled [Stepan]—Butyl stearate
Kessco IBS, Isobutyl Stearate [Stepan]—Isobutyl stearate
Kessco ICS, Isocetyl Stearate [Stepan]—Isocetyl stearate
Kessco IPM, Isopropyl Myristate [Stepan]—Isopropyl myristate
Kessco IPP, Isopropyl Palmitate [Stepan]—Isopropyl palmitate
Kessco-Isopropyl-myristat [Akzo Chemie BV]—Isopropyl myristate
Kessco-Isopropyl-palmitat [Akzo Chemie BV]—Isopropyl palmitate
Kessco-Isopropyl-stearat [Akzo Chemie BV]—Isopropyl stearate
Kessco Myristyl Myristate [Stepan]—Myristyl myristate
Kessco Octyl Palmitate [Stepan]—Octyl palmitate
Kessco Synthetic Spermaceti NF [Stepan]—Synthetic spermaceti (NF)
Klearol [Witco/Sonneborn]—Mineral oil (white NF)
Kuplur SMO [BASF]—Sorbitan monooleate
Lamequat L [Henkel KGaA]—Hydrolyzed animal protein
Lanacet 1705 [Henkel/Emery]—Acetylated lanolin
Lan-Aqua-Sol xtra-Hydrophilic 50, 100 [Fanning]—POE (75) lanolin
Lanesta L, S, SA 30 [Westbrook Lanolin]—Isopropyl lanolate
Laneto-50, -100 [RITA]—POE (75) lanolin
Laneto 60 [RITA]—POE (60) lanolin
Lanette 16 [Henkel Canada]—Cetyl alcohol
Lanette 22 [Henkel Canada]—Behenyl alcohol
Lanette O [Henkel Canada; Henkel KGaA]—Cetearyl alcohol
Lanfrax, 1777 Deodorized [Henkel/Emery]—Lanolin wax
Lanocerin [Amerchol]—Lanolin wax
Lan-O-Derm [Alcolac]—Isopropyl isostearate
Lanogel 41 [Amerchol]—POE (75) lanolin
Lanogene [Amerchol]—Lanolin oil
Lanol C [Seppic]—Cetyl alcohol
Lanol CS [Seppic]—Cetearyl alcohol
Lanolin Alcohols LG, LO, THG, THO [Van Schuppen]—Lanolin alcohol
Lantrol 1673, 1674 [Henkel/Emery]—Lanolin oil (anhyd., cosmetic grade)

Lantrol, 1674 Deodorized [Henkel/Emery]—Lanolin oil
Laurex CS, CS/D [Albright & Wilson; Albright & Wilson/Marchon]—Cetearyl alcohol
Lebon 2000 [Sanyo]—Coco amidopropyl betaine
Lebon GM [Sanyo]—Benzalkonium chloride
Lexaine C, CG-30, CS [Inolex]—Coco amidopropyl betaine
Lexamine 22 [Inolex]—Stearamidoethyl diethylamine
Lexamine B-13 [Inolex]—Behenamidopropyl dimethylamine
Lexamine C-13 [Inolex]—Cocamidopropyl dimethylamine
Lexamine O-13 [Inolex]—Oleamidopropyl dimethylamine
Lexamine P-13 [Inolex]—Palmitamidopropyl dimethylamine
Lexamine S-13 [Inolex]—Stearamidopropyl dimethylamine
Lexamine S-13 Lactate [Inolex]—Stearamidopropyl dimethylamine lactate
Lexein X250, X300, X350, X400, X450 [Inolex]—Hydrolyzed animal protein
Lexol EHP [Inolex]—Octyl palmitate
Lexol IPM [Inolex]—Isopropyl myristate
Lexol IPP [Inolex]—Isopropyl palmitate
Lexol PG 900 [Inolex]—Propylene glycol dipelargonate
Lexol SS [Inolex]—Stearyl stearate
Lilamin 163, 163D [Berol Nobel]—Lauryl amine (dist.)
Lipal EB [Aquatec Quimica]—Butyl stearate
Lipal LC [Aquatec Quimica]—Cetyl lactate
Lipal ST [Aquatec Quimica]—Isopropyl stearate
Lipamide MEAA [Lipo]—Acetyl monoethanolamide
Lipobee 102 [Lipo]—Synthetic beeswax
Lipocol C [Lipo]—Cetyl alcohol
Lipocol O [Lipo]—Oleyl alcohol
Lipolan, S, Distilled Lipolan [Lipo]—Hydrogenated lanolin
Lipolan R [Lipo]—Lanolin oil
Liponate 143M [Lipo]—POE (3) myristyl ether myristate
Liponate CL [Lipo]—Cetyl lactate
Liponate IPM [Lipo]—Isopropyl myristate
Liponate IPP [Lipo]—Isopropyl palmitate
Liponate ML [Lipo]—Myristyl lactate
Liponate MM [Lipo]—Myristyl myristate
Liponate PB-4 [Lipo]—Pentaerythrityl tetrabehenate
Liponate PO-4 [Lipo]—Pentaerythritol tetraoleate
Liponate PS-4 [Lipo]—Pentaerythritol tetrastearate
Liponate SPS [Lipo]—Synthetic spermaceti
Liponate SS [Lipo]—Stearyl stearate
Liponate TDS [Lipo]—Tridecyl stearate
Liponic EG-1 [Lipo]—POE (26) glyceryl ether
Liponic EG-7 [Lipo]—POE (7) glyceryl ether

Lipoquat R [Lipo]—Ricinoleamidopropyl ethyldimonium ethosulfate
Liposorb L-10 [Lipo]—POE (10) sorbitan monolaurate
Liposorb O [Lipo]—Sorbitan monooleate
Liposorb SQO [Lipo]—Sorbitan sesquioleate
Lipovol J [Lipo]—Jojoba oil (refined)
Lipovol WGO [Lipo]—Wheat germ oil
Lipowax C [Lipo]—N,N′-Ethylene bisstearamide
Liquilan [Amerchol]—Lanolin oil
Lonzaine 12C [Lonza]—Coco-betaine
Lonzaine C, CO [Lonza]—Coco amidopropyl betaine
Lonzest 143S [Lonza]—Myristyl propionate
Lonzest SMO [Lonza]—Sorbitan monooleate
Loramine L408 [Rewo]—Lauryl dimethyl amine oxide
Loxiol G20 [Henkel KGaA]—Stearic acid
Loxiol G21 [Henkel KGaA]—Hydroxystearic acid
Lubracal 48, 53, 60 [Witco]—Calcium stearate
Lubrazinc W, Superfine [Witco]—Zinc stearate
Lutensit K-LC, K-LC80, K-OC [BASF AG]—Benzalkonium chloride
Luvitol EHO [BASF AG]—Cetearyl octanoate
Mackalene 116 [McIntyre]—Cocamidopropyl dimethylamine lactate
Mackalene 117 [McIntyre]—Cocamidopropyl dimethylamine propionate
Mackalene 316 [McIntyre]—Stearamidopropyl dimethylamine lactate
Mackalene 416 [McIntyre]—Isostearamidopropyl dimethylamine lactate
Mackalene 426 [McIntyre]—Isostearamidopropyl morpholine lactate
Mackam 35, 35 HP [McIntyre]—Coco amidopropyl betaine
Mackam CAP [McIntyre]—Cocamidopropyl dimethylamine propionate
Mackam CB, CB-35, CB-LS [McIntyre]—Coco-betaine
Mackam ISA [McIntyre]—Isostearamidopropyl betaine
Mackam OB, OB-30 [McIntyre]—Oleyl betaine
Mackam RA [McIntyre]—Ricinoleamidopropyl betaine
Mackamide R [McIntyre]—Ricinoleic diethanolamide
Mackamine CO [McIntyre]—Coco amine oxide
Mackamine LO [McIntyre]—Lauryl dimethyl amine oxide
Mackamine O2 [McIntyre]—Oleyl dimethyl amine oxide
Mackanate OM [McIntyre]—Disodium oleamido MEA-sulfosuccinate
Mackernium SDC-25, SDC-50, SDC-85 [McIntyre]—Stearyl dimethyl benzyl ammonium chloride
Mackine 101 [McIntyre]—Cocamidopropyl dimethylamine
Mackine 301 [McIntyre]—Stearamidopropyl dimethylamine
Mackine 401 [McIntyre]—Isostearamidopropyl dimethylamine
Mackine 421 [McIntyre]—Isostearamidopropyl morpholine
Mackine 501 [McIntyre]—Oleamidopropyl dimethylamine

Mackine 601 [McIntyre]—Behenamidopropyl dimethylamine
Mafo CAB [PPG-Mazer]—Coco amidopropyl betaine
Maprolyte C [Stepan]—Coco amidopropyl betaine
Maprosyl 30 [Millmaster-Onyx]—Sodium lauroyl sarcosinate
Maquat LC-12S 50%, 80%, MC-1412 50%, 80%, MC-1416 50%, 80%, MC-6025 50%
    [Mason]—Benzalkonium chloride
Maquat SC-18 [Mason]—Stearyl dimethyl benzyl ammonium chloride
Marcol [Exxon]—Mineral oil (white)
Mazeen S2 [PPG-Mazer]—POE (2) soya amine
Mazol 159 [PPG-Mazer]—POE (7) glyceryl monococoate
Mearlmica MMCF, MMSV [Mearl]—Mica
Mekon White [Petrolite]—Microcrystalline wax
Merpoxen RO200 [Kempen]—POE (20) castor oil
Michel XO-144, XO-144B [M. Michel]—Cetyl alcohol
Michel XO-146 [M. Michel]—Isostearyl alcohol
Micro-Mica C-1000, C-3000 [KMG Minerals]—Mica
Micromesh No. 3 [Mearl]—Mica (water-ground)
Mineral Jelly No. 5, No. 10, No. 15, No. 20, No. 25 [Penreco]—Petrolatum
Miramine DD [Miranol]—Stearamidopropyl dimethylamine
Miranol DM, DM Conc. 45 [Miranol]—Stearoamphoacetate
Mirataine BD, CB, CBC, CBR [Miranol]—Coco amidopropyl betaine
Mirataine CDMB [Miranol]—Coco-betaine
Mirataine ODMB-35 [Miranol]—Oleyl betaine
MO-33-F [Hefti Ltd.]—Sorbitan monooleate
Monafax L10 [Mona]—POE (10) dinonyl phenyl ether phosphate
Monateric ADA, ADFA, CAB, COAB, MCB [Mona]—Coco amidopropyl betaine
Monomuls 90-10 [Grünau]—Lard monoglyceride (dist.)
Montane 80 [Seppic]—Sorbitan monooleate
Montane 83 [Seppic]—Sorbitan sesquioleate
Multiwax 180-M, ML-445, W-445, X-145A (NF) [Witco]—Microcrystalline wax
Multiwax W-835 [Witco]—Microcrystalline wax
MYM-33 [Hefti Ltd.]—Myristyl myristate
Mytab [Hexcel]—Myristyl trimethyl ammonium bromide
Myverol 18-40 [Eastman]—Lard monoglyceride (dist.)
Naturechem OHS [CasChem]—Octyl hydroxystearate
Neobee 62 [Stepan/PVO]—Glyceryl tristearate
Newcol 3-80, 3-85, 80 [Nippon Nyukazai]—Sorbitan monooleate
Nikkol Behenyl Alcohol 65, 80 [Nikko Ltd.]—Behenyl alcohol
Nikkol CO-20TX [Nikko Ltd.]—POE (20) castor oil
Nikkol CO-60TX [Nikko Ltd.]—POE (60) castor oil
Nikkol TW-30 [Nikko Ltd.]—POE (30) lanolin
Nikkol GM-18S [Nikko Ltd.]—Butyl stearate

Nikkol ICS-R [Nikko Ltd.]—Isocetyl stearate
Nikkol Jojoba Oil N [Nikko Ltd.]—Jojoba oil
Nikkol Jojoba Oil S [Nikko Ltd.]—Jojoba oil (decolorized, deodorized)
Nikkol N-SP [Nikko Ltd.]—Cetyl palmitate
Nikkol Sarcosinate LN [Nikko Ltd.]—Sodium lauroyl sarcosinate
Nikkol SO-10 [Nikko Ltd.]—Sorbitan monooleate
Nikkol SO-15 [Nikko Ltd.]—Sorbitan sesquioleate
Nikkol Tetraglyn 1-O [Nikko Ltd.]—Tetraglyceryl monooleate
Nimco 1780 [Henkel/Emery]—Lanolin alcohol (distilled, deodorized)
Nimlesterol 1730, 1732 [Henkel/Emery]—Cholesterol
Ninox L [Stepan]—Lauryl dimethyl amine oxide
Ninox O [Stepan Europe]—Oleyl dimethyl amine oxide
Ninox S [Stepan Europe]—Stearyl dimethyl amine oxide
Niox EO-10 [Pulcra SA]—Cetyl alcohol
Niox EO-26 [Pulcra SA]—Cetearyl alcohol
Nissan Amine BB [Nippon Oils & Fats]—Lauryl amine
Nissan Cation AB [Nippon Oils & Fats]—Stearyl trimethyl ammonium chloride
Nissan Cation PB-40, PB-300 [Nippon Oils & Fats]—Cetyl trimethyl ammonium
    chloride
Nissan Cation S2-100 [Nippon Oils & Fats]—Stearyl dimethyl benzyl ammonium
    chloride
Nissan Nonion OP-80R [Nippon Oils & Fats]—Sorbitan monooleate
Nissan Nonion OP-83 RAT [Nippon Oils & Fats]—Sorbitan sesquioleate
Nopalcol 10-CO [Henkel/Process]—POE (20) castor oil
Nopco 1097-B, 1097-M [Henkel/Process]—Zinc stearate
Nopcote C-104 [Henkel/Process, Henkel-Nopco]—Calcium stearate
Nopcowax 22-DS [Henkel/Process]—N,N´-Ethylene bisstearamide
Noramium M2SH [Ceca SA]—Dihydrogenated tallow dimethyl ammonium chloride
Norfox B [Norman, Fox]—Sodium stearate
Norfox CS [Norman, Fox]—Calcium stearate
Norfox KO [Norman, Fox]—Potassium oleate
Norfox Sorbo S-80 [Norman, Fox]—Sorbitan monooleate
Norfox ZNS [Norman, Fox]—Zinc stearate
Novol [Croda]—Oleyl alcohol
Nuodex Micronized Zinc Stearate 1058 Densified [Tenneco]—Zinc stearate
Nutrilan H, I, L [Grunau GmbH]—Hydrolyzed animal protein
OHlan [Amerchol; Amerchol Europe]—Hydroxylated lanolin
Ointment Base No. 3, 4, 6 [Penreco]—Petrolatum (white USP)
Orzol [Witco/Sonneborn]—Mineral oil (white USP)
Oxamin LO [ICI Australia]—Lauryl dimethyl amine oxide
Pancogene S [Gattefosse Ets.]—Soluble collagen
Paralan [Croda Ltd.]—Lanolin (tech.)

TRADENAME PRODUCTS AND GENERIC EQUIVALENTS

Parol 70, 80, 100 [Penreco]—Mineral oil (tech.)

Pationic ISL [RITA]—Sodium isostearoyl lactylate

Patlac NAL [RITA]—Sodium lactate

Peneteck [Penreco]—Mineral oil (tech.)

Penreco Amber, Blond, Royal [Penreco]—Petrolatum (USP)

Penreco Cream, Frost, Lily, Regent, Snow, Super, Ultima [Penreco]—Petrolatum (white USP)

Penreco Green, Red [Penreco]—Petrolatum (tech.)

Perfecta [Witco]—Petrolatum

Perfecta USP [Witco]—Petrolatum (cosmetic/pharmaceutical grade, USP)

Petrac 250 [Synthetic Products]—Stearic acid (rubber grade)

Petrac 270 [Synthetic Products]—Stearic acid (tech.)

Petrac CP-11, CP-11LS, CP-11 LSG, CP-12, CP-22G [Synthetic Products]—Calcium stearate

Petrac Eramide [Synthetic Products]—Erucamide

Petrac MG-20, MG-20 NF [Synthetic Products]—Magnesium stearate

Petrac Slip-Eze [Synthetic Products]—Oleyl amide

Petrac Vyn-Eze [Synthetic Products]—Stearic acid amide

Petrac ZN-41 [Synthetic Products]—Zinc stearate (USP)

Petrac ZN-42, ZW-45 [Synthetic Products]—Zinc stearate

Petrac ZN-44HS [Synthetic Products]—Zinc stearate (heat-stable)

Petrolatum DLC [Harwick; Natrochem]—Petrolatum

Petrolatum Dry Liquid Conc. [Polymerics]—Petrolatum

Petrolatum RPB [Witco]—Petrolatum (tech.)

Petrolatum SR-172 [C.P. Hall]—Petrolatum

Petrolite C-700, C-1035 [Petrolite]—Microcrystalline wax

Pharmaceutical Lanolin Anhydrous USP [Croda]—Lanolin (anhyd. USP, pharmaceutical grade)

Plasthall DTDA [C.P. Hall]—Ditridecyl adipate

Polychol 20, 20-40 [Croda Ltd.]—POE (20) lanolin ether

Polyfac TA-20 [Stepan]—POE (20) tallow amine

Polymekon [Petrolite]—Microcrystalline wax

Polypeptide 37, LSN [Stepan]—Hydrolyzed animal protein

Polytrap 210 [Dow Corning/Wickhen]—Petrolatum

Promyr [Amerchol]—Isopropyl myristate

Propal [Amerchol]—Isopropyl palmitate

Prostearyl 15 [Croda; Croda Ltd.]—POP (15) stearyl ether

Protectol KLC 50, 80 [BASF AG]—Benzalkonium chloride

Protol [Witco/Sonneborn]—Mineral oil (white USP)

Protopet Alba, White 1S, White 2L, White 3L, Yellow 1E, Yellow 2A [Witco]—Petrolatum (USP)

QO-33-F [Hefti Ltd.]—Sorbitan sesquioleate

Quadrilan BC (BP grade) [Harcros UK]—Benzalkonium chloride
Quartamin 86W [Kao]—Stearyl trimethyl ammonium chloride
Quatrene CB, CB-50, CB-80, MB-50, MB-80 [Henkel]—Benzalkonium chloride
Querton 14Br, 14Br-40 [Berol Nobel]—Myristyl trimethyl ammonium bromide
Querton 16CL29, 16CL50, 24CL35 [Berol Nobel]—Cetyl trimethyl ammonium chloride
Querton 246 [Berol Nobel]—Benzalkonium chloride
Querton 442, 442-11, 442-82 [Berol Nobel]—Dihydrogenated tallow dimethyl ammonium chloride
Querton 442S, 442-Sx [Berol Nobel]—Distearyl dimethyl ammonium chloride
Querton 470 [Berol Nobel]—Ditallow dimethyl ammonium chloride
Quikote [Morton Int'l./Ventron]—Zinc stearate
Radia 7051 [Synfina-Oleofin]—Butyl stearate
Radia 7070 [Synfina-Oleofin]—Butyl myristate
Radia 7110 [Synfina-Oleofin]—Methyl stearate
Radia 7118 [Synfina-Oleofin]—Methyl laurate
Radia 7120 [Synfina-Oleofin]—Methyl palmitate
Radia 7129 [Synfina-Oleofin]—Octyl palmitate
Radia 7131 [Synfina-Oleofin]—Octyl stearate
Radia 7161, 7163, 7303, 7363 [Synfina-Oleofin]—Glyceryl trioleate
Radia 7171 [Synfina-Oleofin]—Pentaerythritol tetraoleate
Radia 7176 [Synfina-Oleofin]—Pentaerythritol tetrastearate
Radia 7190 [Synfina-Oleofin]—Isopropyl myristate
Radia 7195 [Synfina-Oleofin]—Isopropyl stearate
Radia 7200 [Synfina-Oleofin]—Isopropyl palmitate
Radia 7231 [Synfina-Oleofin]—Isopropyl oleate
Radia 7240 [Synfina-Oleofin]—Isobutyl stearate
Radia 7500 [Synfina-Oleofin]—Cetyl palmitate
Radia 7501 [Synfina-Oleofin]—Cetyl stearate
Radia 7514 [Synfina-Oleofin]—Pentaerythrityl tetrabehenate
Radiacid 200 [Synfina-Oleofin]—Hydroxystearic acid
Radiacid 408, 411, 420, 423 [Synfina-Oleofin]—Stearic acid
Radiacid 1060 [Synfina-Oleofin]—Calcium stearate
Radiamine 6163, 6164 [Synfina-Oleofin]—Lauryl amine (dist.)
Radiamuls 155, SORB 2155 [Synfina-Oleofin]—Sorbitan monooleate
Radiaquat 6442 [Synfina-Oleofin]—Dihydrogenated tallow dimethyl ammonium chloride
Radiaquat 6444 [Synfina-Oleofin]—Cetyl trimethyl ammonium chloride
Radiastar 1060 [Synfina-Oleofin]—Calcium stearate
Radiastar 1100 [Synfina-Oleofin]—Magnesium stearate
Radiasurf 7155 [Synfina-Oleofin]—Sorbitan monooleate
Retarder CA [Hart Chem. Ltd.]—Benzalkonium chloride
Rewolan LP [Rewo GmbH]—Isopropyl lanolate

TRADENAME PRODUCTS AND GENERIC EQUIVALENTS

Rewominox L408 [Rewo GmbH]—Lauryl dimethyl amine oxide
Rewominox S 300 [Rewo GmbH]—Stearyl dimethyl amine oxide
Rewoquat B50 [Rewo GmbH]—Benzalkonium chloride
Rewoquat W-7500, W-7500/H [Rewo; Rewo GmbH]—Quaternium-27
Rewoteric AM-B13 [Rewo GmbH]—Coco amidopropyl betaine
Rewoteric AM-R40 [Rewo GmbH]—Ricinoleamidopropyl betaine
Rheodol AO-10, SP-O10 [Kao]—Sorbitan monooleate
Rheodol AO-15 [Kao]—Sorbitan sesquioleate
Richamate 1655, MBF [Richardson]—Cocamidopropyl dimethylamine propionate
Richamate 3780 [Richardson]—Stearamidopropyl dimethylamine lactate
Richamate 6613 [Richardson]—Isostearamidopropyl dimethylamine lactate
Richamate ISML [Richardson]—Isostearamidopropyl morpholine lactate
Rilanit BS [Henkel KGaA]—Butyl stearate
Rilanit DBA [Henkel KGaA]—Dibutyl adipate
Rilanit EHS [Henkel KGaA]—Octyl stearate
Rilanit G16 [Henkel KGaA]—Cetyl alcohol
Rilanit GTO [Henkel KGaA]—Glyceryl trioleate
Rilanit GTS [Henkel KGaA]—Glyceryl tristearate
Rilanit HL [Henkel KGaA]—Hexyl laurate
Rilanit IBS [Henkel KGaA]—Isobutyl stearate
Rilanit IPM [Henkel KGaA]—Isopropyl myristate
Rilanit IPP [Henkel KGaA]—Isopropyl palmitate
Rilanit IPS [Henkel KGaA]—Isopropyl stearate
Rilanit OLO [Henkel KGaA]—Oleyl monooleate
Rilanit STS-T [Henkel KGaA]—Stearyl stearate
Ritacetyl [RITA]—Acetylated lanolin
Ritahydrox [RITA]—Hydroxylated lanolin
Ritalan [RITA]—Lanolin oil
Ritasol [RITA]—Isopropyl lanolate
Ritawax [RITA]—Lanolin alcohol
Roccal 50% Technical, II 50% (USP), MC-14 (USP) [Hilton-Davis]—Benzalkonium chloride
Rodea [Alcolac]—Ricinoleic diethanolamide (2:1)
Sandoz Amine Oxide XA-L [Sandoz]—Lauryl dimethyl amine oxide
Sanisol C, CPR, CR, CR 80%, HTPR, OPR, TPR [Kao]—Benzalkonium chloride
Sarkosyl NL-30 [Ciba-Geigy]—Sodium lauroyl sarcosinate
Satulan [Croda; Croda Ltd.]—Hydrogenated lanolin
Schercamox CMA [Scher]—N,N-Bis (2-hydroxyethyl) cocamine oxide
Schercamox DML [Scher]—Lauryl dimethyl amine oxide
Schercamox DMS [Scher]—Stearyl dimethyl amine oxide
Schercamox T-12 [Scher]—Bis (2-hydroxyethyl) tallow amine oxide
Schercemol 185 [Scher]—Isostearyl neopentanoate

Schercemol 318 [Scher]—Isopropyl isostearate
Schercemol 1688 [Scher]—Cetearyl octanoate
Schercemol BE [Scher]—Behenyl erucate
Schercemol CL [Scher]—Cetyl lactate
Schercemol CM [Scher]—Cetyl myristate
Schercemol CP [Scher]—Cetyl palmitate
Schercemol CS [Scher]—Cetyl stearate
Schercemol DO [Scher]—Decyl oleate
Schercemol GMIS [Scher]—Glyceryl isostearate
Schercemol ICS [Scher]—Isocetyl stearate
Schercemol IPM [Scher]—Isopropyl myristate
Schercemol IPO [Scher]—Isopropyl oleate
Schercemol LL [Scher]—Lauryl lactate
Schercemol MEM-3 [Scher]—POE (3) myristyl ether myristate
Schercemol ML [Scher]—Myristyl lactate
Schercemol MM [Scher]—Myristyl myristate
Schercemol MP [Scher]—Myristyl propionate
Schercemol MS [Scher]—Myristyl stearate
Schercemol OLO [Scher]—Oleyl monooleate
Schercemol OP [Scher]—Octyl palmitate
Schercemol OPG [Scher]—Octyl pelargonate
Schercemol PGDP [Scher]—Propylene glycol dipelargonate
Schercodine B [Scher]—Behenamidopropyl dimethylamine
Schercodine C [Scher]—Cocamidopropyl dimethylamine
Schercodine I [Scher]—Isostearamidopropyl dimethylamine
Schercodine M [Scher]—Myristamidopropyl dimethylamine
Schercodine O [Scher]—Oleamidopropyl dimethylamine
Schercodine P [Scher]—Palmitamidopropyl dimethylamine
Schercodine S [Scher]—Stearamidopropyl dimethylamine
Schercomid AME, AME-70 [Scher]—Acetyl monoethanolamide
Schercopol OMS-Na [Scher]—Disodium oleamido MEA-sulfosuccinate
Schercoquat IAS, IAS-LC [Scher]—Isostearamidopropyl ethyldimonium ethosulfate
Schercoquat IB [Scher]—Isostearamidopropalkonium chloride
Schercotaine CAB, CAB-G [Scher]—Coco amidopropyl betaine
Schercotaine IAB [Scher]—Isostearamidopropyl betaine
Secosyl [Stepan Europe]—Sodium lauroyl sarcosinate
Serdox NJAD20 [Servo]—POE (20) tallow amine
Simchin [RITA]—Jojoba oil
Sipoteric CB, COB [Alcolac]—Coco amidopropyl betaine
S-Maz 80 [PPG-Mazer]—Sorbitan monooleate
S-Maz 83R [PPG-Mazer]—Sorbitan sesquioleate
Softisan 100 [Huls; Huls AG]—Glyceryl trilaurate

Solan, 50 [Croda; Croda Ltd.]—POE (60) lanolin
Solangel 401 [Croda]—POE (75) lanolin
Sollagen [Hormel]—Soluble collagen
Solulan 75, L-575 [Amerchol]—POE (75) lanolin
Sonojell No. 4, No. 9 [Witco]—Petrolatum
Sontex 19, 21, 35 [Marathon Morco]—Mineral oil (white USP)
Sontex 55, 70 [Marathon Morco]—Mineral oil
Sontex 75 [Marathon Morco]—Mineral oil (white NF)
Sontex 75T, 85T, 95T [Marathon Morco]—Mineral oil (white tech.)
Sontex 100, 150 [Marathon Morco]—Mineral oil
Soprofor S/80 [Rhone-Poulenc Geronazzo]—Sorbitan monooleate
Sorba [Croda Ltd.]—Lanolin
Sorbax SMO [Chemax]—Sorbitan monooleate
Sorbon S-80 [Toho]—Sorbitan monooleate
Sorgen 30, S-30-H [Dai-ichi Kogyo Seiyaku]—Sorbitan sesquioleate
Sorgen 40, S-40-H [Dai-ichi Kogyo Seiyaku]—Sorbitan monooleate
Span 80 [ICI Americas]—Sorbitan monooleate
Sparkelan 656 [Henkel/Emery]—Lanolin (anhyd. USP)
Sparkelan 1656 [Henkel/Emery]—Lanolin (anhyd. USP, pharmaceutical grade)
Spermwax [Robeco]—Synthetic spermaceti (NF)
Standamox O1 [Henkel; Henkel Canada]—Oleyl dimethyl amine oxide
Standamul 1414-E [Henkel]—POE (3) myristyl ether myristate
Standamul 1616 [Henkel]—Cetyl palmitate
Standamul 7061 [Henkel]—Isocetyl stearate
Standamul CTA [Henkel]—Hexyl laurate
Standamul CTV [Henkel]—Decyl oleate
Standamul G-16 [Henkel]—Isocetyl alcohol
Standamul HE [Henkel]—POE (7) glyceryl monococoate (self-emulsifying)
Standamul STC-25 [Henkel]—Stearyl dimethyl benzyl ammonium chloride
Standapol AB-45 [Henkel]—Coco-betaine
Standapol BAW, BC-35 [Henkel]—Coco amidopropyl betaine
Standapol OLB-30, OLB-50 [Henkel]—Oleyl betaine
Starfol CP [Sherex]—Cetyl palmitate
Starfol IPM [Sherex]—Isopropyl myristate
Starfol IPP [Sherex]—Isopropyl palmitate
Starfol OO [Sherex]—Oleyl monooleate
Starfol Wax CG [Sherex]—Synthetic spermaceti
Starwax 100 [Petrolite]—Microcrystalline wax
Stave Zinc Stearate [Stave]—Zinc stearate
Stearic Acid-RECCO [R.E. Carroll]—Stearic acid (single, double, triple-pressed)
Stedbac [Hexcel]—Stearyl dimethyl benzyl ammonium chloride
Stepan C40, C41, C42, C43 [Stepan]—Methyl laurate

Stepan C60 [Stepan]—Methyl palmitate
Stepan C66 [Stepan]—Methyl stearate
Stepan D-50 [Stepan]—Isopropyl myristate
Stepan D-70 [Stepan]—Isopropyl palmitate
Sterling CAB [Canada Packers Ltd.]—Coco amidopropyl betaine
Sterling CH-grade, PH-grade, SH-grade [Canada Packers]—Hydrolyzed animal protein
Super Corona [Croda; Croda Ltd.]—Lanolin (anhyd. USP)
Super Refined Coconut Oil [Croda]—Coconut oil
Super-Sat [RITA]—Hydrogenated lanolin
Superfine Lanolin Anhydrous USP [Croda]—Lanolin (anhyd. USP)
Surco Coco Betaine [Stepan]—Coco amidopropyl betaine
Swanol CA-101 [Nikko Ltd.]—Benzalkonium chloride
Syncrowax BB4 [Croda]—Synthetic beeswax
Syncrowax HR-C [Croda]—Glyceryl tribehenate
Synoquart P50 [Aquatec Quimica]—Cetyl trimethyl ammonium chloride
Synprolam 35DMBQC [ICI PLC]—Benzalkonium chloride
Tech Pet F, M [Witco]—Petrolatum (tech.)
Tegamine 18 [Goldschmidt]—Stearamidopropyl dimethylamine
Tegamine P-7 [Goldschmidt]—Cocamidopropyl dimethylamine
Tegin E 66 [Goldschmidt AG]—Acetylated lard glyceride
Tego-Amid D5040 [Goldschmidt]—Cocamidopropyl dimethylamine
Tego-Amid O 18 [Goldschmidt]—Oleamidopropyl dimethylamine
Tego-Amid S18 [Goldschmidt AG]—Stearamidopropyl dimethylamine
Tego-Betaine C, L-7, L-10, S, T [Goldschmidt; Goldschmidt AG]—Coco amidopropyl
    betaine
Tequat BC [Tessilchimica SpA]—Cetyl trimethyl ammonium chloride
Teric 16M2 [ICI Australia]—POE (2) soya amine
Tewax TC1, TC2 [Tessilchimica SpA]—Cetearyl alcohol
Textamine Oxide LMW [Henkel Canada]—Lauryl dimethyl amine oxide
Trisolan 1720 [Henkel/Emery]—Isopropyl lanolate
Triton CG-400, CG-500, X-400 [Rohm & Haas]—Stearyl dimethyl benzyl ammonium
    chloride
Trylox 5918, CO-200, CO-200/50 [Henkel/Emery]—POE (200) castor oil
Trylox 5921, HCO-16 [Henkel/Emery]—POE (16) hydrogenated castor oil
Trymeen 6607, TAM-20 [Henkel/Emery]—POE (20) tallow amine
Ultraflex [Petrolite]—Microcrystalline wax
Uniflex BYS CP, Tech. [Union Camp]—Butyl stearate
Uniflex IBYS [Union Camp]—Isobutyl stearate
Unimate BYS [Union Camp]—Butyl stearate
Unimate EHP [Union Camp]—Octyl palmitate
Unimate IPM [Union Camp]—Isopropyl myristate
Unimate IPP [Union Camp]—Isopropyl palmitate

Varion AM-R40 [Sherex]—Ricinoleamidopropyl betaine
Varion CADG, CADG-HS, CADG-LS, CADG-W [Sherex]—Coco amidopropyl betaine
Varion SDG [Sherex]—Stearyl betaine
Variquat 50AC, 50AE, 50MC, 60LC, 80AC, 80AE, 80MC, 80ME, B345, LC60, LC80
    [Sherex]—Benzalkonium chloride
Variquat E228, E290 [Sherex]—Cetyl trimethyl ammonium chloride
Varisoft 110, 3262, DHT [Sherex]—Dihydrogenated tallow dimethyl ammonium chloride
    ride
Varisoft 475 [Sherex]—Quaternium-27
Varisoft SDC, SDC-65%, SDC-100P, SDC-W [Sherex]—Stearyl dimethyl benzyl
    ammonium chloride
Varonic L202 [Sherex]—POE (2) soya amine
Varox 365, 375 [Vanderbilt]—Lauryl dimethyl amine oxide
Velvetex AB-45, BC [Henkel; Henkel Canada]—Coco-betaine
Velvetex BA, BA-35, BC-35, BK-35 [Henkel; Henkel Canada]—Coco amidopropyl
    betaine
Velvetex OLB-30, OLB-50 [Henkel; Henkel Canada]—Oleyl betaine
Victory [Petrolite]—Microcrystalline wax
Vigilan Regular and Superfine [Fanning]—Lanolin oil
Vikol RQ [Vikon]—Benzalkonium chloride
Volpo 3, O3 [Croda Ltd.]—POE (3) oleyl ether
Volpo N3 [Croda Ltd.]—POE (3) oleyl ether (dist.)
Waxenol 810 [CasChem]—Myristyl myristate
Waxenol 815, 816 [CasChem]—Cetyl palmitate
Waxenol 821 S.B. [CasChem]—Synthetic beeswax
Wet Zinc [Morton Int'l./Ventron]—Zinc stearate
Wheat Germ Oil CLR 0.26% (αβ) Tocopherol [Henkel Canada]—Wheat germ oil
White Swan [Croda Ltd.]—Lanolin (anhyd. BP)
Wickenol 101 [CasChem]—Isopropyl myristate
Wickenol 111 [CasChem]—Isopropyl palmitate
Wickenol 127 [CasChem]—Isopropyl stearate
Wickenol 131 [CasChem]—Isopropyl isostearate
Wickenol 139 [CasChem]—Jojoba oil (synthetic)
Wickenol 141 [CasChem]—Butyl myristate
Wickenol 143 [CasChem]—Oleyl monooleate
Wickenol 155 [CasChem]—Octyl palmitate
Wickenol 156 [CasChem]—Octyl stearate
Wickenol 160 [CasChem]—Octyl pelargonate
Wickenol 171 [CasChem]—Octyl hydroxystearate
Wickenol 506 [CasChem]—Myristyl lactate
Wickenol 727 [CasChem]—POP (30) lanolin ether
Witco Calcium Stearate A, EA, F, G, Kosher, LM, M Impalpable, Polymer Grade,

Polymer Grade S, Regular, Superfine [Witco]—Calcium stearate

Witco EA [Witco]—Sodium stearate (food grade)

Witco Magnesium Stearate D [Witco]—Magnesium stearate

Witconol 14 [Witco]—Tetraglyceryl monooleate

Witco Sodium Stearate C-1, C-7, Heat Stable, Polymer Grade, T-1 [Witco]—Sodium stearate

Witco Zinc Stearate 11, 44, Disperso, Heat-Stable, LV, NW, Polymer Grade, Regular, USP [Witco]—Zinc stearate

Yeoman [Croda Ltd.]—Lanolin (anhyd. BP)

Zinc Stearate (Wettable) [C.P. Hall]—Zinc stearate

Zinc Stearate (Wettable, Synpro) [Synthetic Products]—Zinc stearate

Zinc Stearate Synpro [Synthetic Products]—Zinc stearate

Zincloid Synpro [Synthetic Products]—Zinc stearate

Zincote [Morton Int'l./Ventron]—Zinc stearate

Zoharquat 50, 80 [Zohar]—Benzalkonium chloride

GENERIC CHEMICAL SYNONYMS
AND CROSS REFERENCES

Acetamide MEA (CTFA). See Acetyl monoethanolamide

Acetamide, N-(2-hydroxyethyl)-. See Acetyl monoethanolamide

N-Acetyl ethanolamine. See Acetyl monoethanolamide

Adipic acid/dimethylaminohydroxypropyl diethylenetriamine copolymer. See Book VI

Alcohol C-16. See Cetyl alcohol (CTFA)

Alcohols, lanolin. See Lanolin alcohol (CTFA)

Alkyl dimethyl benzyl ammonium chloride. See Benzalkonium chloride (CTFA)

Amides, coco, N-[3-(dimethylamino) propyl]-. See Cocamidopropyl dimethylamine (CTFA)

Amides, coco, N-[3-(dimethylamino) propyl], lactates. See Cocamidopropyl dimethyl-amine lactate (CTFA)

Amides, coco, N-[3-(dimethylamino) propyl], propionates. See Cocamidopropyl di-methylamine propionate (CTFA)

Amines, coco alkyl dihydroxyethyl, oxides. See N,N-Bis (2-hydroxyethyl) cocamine oxide

Amines, coco alkyl dimethyl, oxides. See Coco amine oxide

Amines, tallow alkyl dihydroxyethyl, oxides. See Bis (2-hydroxyethyl) tallow amine oxide

Amines, tallow alkyl dimethyl, oxides. See Tallow dimethyl amine oxide

Ammonium, benzyldocosyldimethyl-, chloride. See Behenyl dimethyl benzyl ammo-nium chloride

1,4-Anhydro-D-glucitol, 6-(9-octadecenoate). See Sorbitan monooleate

Anhydrohexitol sesquioleate. See Sorbitan sesquioleate (CTFA)

Anhydrosorbitol monooleate. See Sorbitan monoooleate

Anhydrosorbitol sesquioleate. See Sorbitan sesquioleate (CTFA)

Anhydrous lanolin. See Lanolin (CTFA)

Beeswax, synthetic. See Synthetic beeswax (CTFA)

Behenalkonium chloride (CTFA). See Behenyl dimethyl benzyl ammonium chloride

(3-Behenoyloxy-2-hydroxypropyl) trimethyl ammonium chloride. See Behenoyl-PG-trimonium chloride (CTFA)

Behentrimonium chloride (CTFA). See Behenyl trimethyl ammonium chloride

Benzenemethanaminium, N,N-dimethyl-N-9-octadecenyl-, chloride. See Oleyl di-methyl benzyl ammonium chloride

Benzenemethanaminium, N,N-dimethyl-N-octadecyl-, chloride. See Stearyl dimethyl benzyl ammonium chloride

Benzenemethanaminium, N,N-dimethyl-N-[3-[(1-oxoisooctadecyl) amino] propyl]-, chloride. See Isostearamidopropalkonium chloride (CTFA)

Benzyl dimethyl stearyl ammonium chloride. See Stearyl dimethyl benzyl ammonium chloride

Benzyldocosyldimethyl ammonium chloride. See Behenyl dimethyl benzyl ammonium chloride

N,N-Bis(2-hydroxyethyl) ricinoleamide. See Ricinoleic diethanolamide

2,2-Bis [[(1-oxodocosyl) oxy] methyl]-1,3-propanediyl docosanoate. See Pentaerythritol tetrabehenate (CTFA)

Butanedioic acid, sulfo-, mono [2-[(1-oxo-9-octadecenyl) amino] ethyl] ester, disodium salt. See Disodium oleamido MEA-sulfosuccinate (CTFA)

n-Butyl myristate. See Butyl myristate

n-Butyl octadecanoate. See Butyl stearate (CTFA)

n-Butyl stearate. See Butyl stearate (CTFA)

Butyl n-tetradecanoate. See Butyl myristate

CADG. See Coco amidopropyl betaine

Calcium octadecanoate. See Calcium stearate (CTFA)

N-(Carboxymethyl)-N,N-dimethyl-1-octadecanaminium hydroxide, inner salt. See Stearyl betaine (CTFA)

N-(Carboxymethyl)-N,N-dimethyl-9-octadecen-1-aminium hydroxide, inner salt. See Oleyl betaine (CTFA)

N-(Carboxymethyl)-N,N-dimethyl-3-[(1-oxococonut) amino]-1-propanaminium hydroxide, inner salt. See Coco amidopropyl betaine

N-Carboxymethyl-N,N-dimethyl-3-[(1-oxoisooctadecyl) amino]-1-propanaminium hydroxide, inner salt. See Isostearamidopropyl betaine (CTFA)

N-(Carboxymethyl)-N,N-dimethyl-3-[(1-oxoricinoleyl) amino]-1-propanaminium hydroxide, inner salt. See Ricinoleamidopropyl betaine (CTFA)

CETAB. See Cetyl trimethyl ammonium bromide

Cetostearyl alcohol. See Cetearyl alcohol (CTFA)

Cetrimonium bromide (CTFA). See Cetyl trimethyl ammonium bromide

Cetrimonium chloride (CTFA). See Cetyl trimethyl ammonium chloride

Cetyl esters (CTFA). See Synthetic spermaceti

Cetylic acid. See Palmitic acid (CTFA)

Cetylic alcohol. See Cetyl alcohol (CTFA)

Cetyl/stearyl alcohol. See Cetearyl alcohol (CTFA)

Cetyl/stearyl 2-ethylhexanoate. See Cetearyl octanoate (CTFA)

Cholest-5-en-3-ol (3β)-. See Cholesterol (CTFA)

Cocamidopropyl betaine (CTFA). See Coco amidopropyl betaine

Cocamidopropyl dimethyl glycine. See Coco amidopropyl betaine

Cocamine oxide (CTFA). See Coco amine oxide

Coco amido betaine. See Coco amidopropyl betaine
Coco di-(hydroxyethyl) amine oxide. See N,N-Bis (2-hydroxyethyl) cocamine oxide
Coco dimethyl amine oxide. See Coco amine oxide
Coco dimethyl glycine. See Coco-betaine (CTFA)
Coconut amido betaine. See Coco amidopropyl betaine
Coconut betaine. See Coco-betaine (CTFA)
Coconut dimethylamine oxide. See Coco amine oxide
Cocoyl amide propylbetaine. See Coco amidopropyl betaine
Collagen hydrolysates. See Hydrolyzed animal protein (CTFA)
Copra oil. See Coconut oil (CTFA)
Cosmetic lanolin. See Lanolin (CTFA)
Cyclomethicone. See Book VI
Decyl-9-octadecenoate. See Decyl oleate (CTFA)
De-oiled lanolin. See Lanolin wax (CTFA)
Dewaxed lanolin. See Lanolin oil (CTFA)
Diethanolamine ricinoleic acid amide. See Ricinoleic diethanolamide
N-[2-(Diethylamino) ethyl] octadecanamide. See Stearamidoethyl diethylamine (CTFA)
Diethylaminoethyl stearamide. See Stearamidoethyl diethylamine (CTFA)
Dihydroxyethyl cocamine oxide (CTFA). See N,N-Bis (2-hydroxyethyl) cocamine oxide
Dihydroxyethyl tallowamine oxide (CTFA). See Bis (2-hydroxyethyl) tallow amine
    oxide
Dimethicone copolyol. See Book VI
Dimethylaminopropyl behenamide. See Behenamidopropyl dimethylamine (CTFA)
N-[3-(Dimethylamino) propyl cocamide lactate. See Cocamidopropyl dimethylamine
    lactate (CTFA)
N-[3-(Dimethylamino) propyl] coco amides. See Cocamidopropyl dimethylamine
    (CTFA)
N-[3-(Dimethylamino) propyl] coco amides, propionates. See Cocamidopropyl dimeth-
    ylamine propionate (CTFA)
N-[3-(Dimethylamino) propyl] docosanamide. See Behenamidopropyl dimethylamine
    (CTFA)
N-[3-(Dimethylamino) propyl] hexadecanamide. See Palmitamidopropyl dimethyl-
    amine (CTFA)
N-[3-(Dimethylamino) propyl] isooctadecanamide. See Isostearamidopropyl dimethyl-
    amine (CTFA)
Dimethylaminopropyl myristamide. See Myristamidopropyl dimethylamine (CTFA)
N-[3-(Dimethylamino) propyl] octadecanamide. See Stearamidopropyl dimethylamine
    (CTFA)
N-[3-Dimethylamino) propyl]-9-octadecenamide. See Oleamidopropyl dimethylamine
    (CTFA)
Dimethylaminopropyl oleamide. See Oleamidopropyl dimethylamine (CTFA)
Dimethylaminopropyl palmitamide. See Palmitamidopropyl dimethylamine (CTFA)

Dimethylaminopropyl stearamide. See Stearamidopropyl dimethylamine (CTFA)

N-[3-(Dimethylamino) propyl] tetradecanamide. See Myristamidopropyl dimethylamine (CTFA)

Dimethyl cocamine oxide. See Coco amine oxide

Dimethyl di(hydrogenated tallow) ammonium chloride. See Dihydrogenated tallow
dimethyl ammonium chloride

Dimethyl dioctadecyl ammonium chloride. See Distearyl dimethyl ammonium chloride

Dimethyl distearyl ammonium chloride. See Distearyl dimethyl ammonium chloride

Dimethyl ditallow ammonium chloride. See Ditallow dimethyl ammonium chloride

N,N-Dimethyl-1-dodecanamine-N-oxide. See Lauryl dimethyl amine oxide

Dimethyl lauryl amine oxide. See Lauryl dimethyl amine oxide

N,N-Dimethyl-1-octadecanamine-N-oxide. See Stearyl dimethyl amine oxide

N,N-Dimethyl-9-octadecen-1-amine-N-oxide. See Oleyl dimethyl amine oxide

N,N-Dimethyl-N-9-octadecenylbenzenemethanaminium chloride. See Oleyl dimethyl
benzyl ammonium chloride

N,N-Dimethyl-N-octadecylbenzenemethanaminium chloride. See Stearyl dimethyl
benzyl ammonium chloride

N,N-Dimethyl-N-octadecyl-1-octadecanaminium chloride. See Distearyl dimethyl
ammonium chloride

N,N-Dimethyl-N-[3-[(oxoisooctadecyl) amino] propyl] benzenemethanaminium chloride. See Isostearamidopropalkonium chloride (CTFA)

2,2-Dimethylpropanoic acid, isooctadecyl ester. See Isostearyl neopentanoate (CTFA)

Dimethyl tallow ammonium chloride, dihydrogenated. See Dihydrogenated tallow
dimethyl ammonium chloride

Disodium monooleamido MEA-sulfosuccinate. See Disodium oleamido MEA-sulfosuccinate (CTFA)

Distearyldimonium chloride (CTFA). See Distearyl dimethyl ammonium chloride

Ditallowalkonium chloride. See Dihydrogenated tallow dimethyl ammonium chloride

Ditallowdimonium chloride (CTFA). See Ditallow dimethyl ammonium chloride

Docosanamide, N-[3-(dimethylamino) propyl]-. See Behenamidopropyl dimethylamine
(CTFA)

1-Docosanaminium, N,N,N-trimethyl-, chloride. See Behenyl trimethyl ammonium
chloride

Docosanoic acid, 2,2-bis [[(1-oxodocosyl) oxy] methyl]-1,3-propanediyl ester. See
Pentaerythritol tetrabehenate (CTFA)

Docosanoic acid, 1,2,3-propanetriyl ester. See Glyceryl tribehenate (CTFA)

1-Docosanol. See Behenyl alcohol (CTFA)

13-Docosenamide. See Erucic acid amide

13-Docosenamide, N-octadecyl-. See Stearyl erucamide (CTFA)

13-Docosenoic acid, docosyl ester. See Behenyl erucate (CTFA)

13-Docosenoic acid, 9-octadecenyl ester. See Oleyl erucate (CTFA)

Docosyl 13-docosenoate. See Behenyl erucate (CTFA)

1-Dodecanamine. See Lauryl amine
1-Dodecanamine, N,N-dimethyl, N-oxide. See Lauryl dimethyl amine oxide
Dodecanoic acid, 1,2,3-propanetriyl ester. See Glyceryl trilaurate
Dodecanoic acid, hexyl ester. See Hexyl laurate (CTFA)
Dodecanoic acid, methyl ester. See Methyl laurate (CTFA)
Dodecylamine. See Lauryl amine
Dodecyl 2-hydroxypropanoate. See Lauryl lactate (CTFA)
Erucamide (CTFA). See Erucic acid amide
Erucic acid, oleyl ester. See Oleyl erucate (CTFA)
Ester of isocetyl alcohol and stearic acid. See Isocetyl stearate (CTFA)
Ester of isopropyl alcohol and myristic acid. See Isopropyl myristate (CTFA)
Ester of isostearyl alcohol and neopentanoic acid. See Isostearyl neopentanoate (CTFA)
Ester of lauryl alcohol and lactic acid. See Lauryl lactate (CTFA)
Ester of myristyl alcohol and lactic acid. See Myristyl lactate (CTFA)
Ester of myristyl alcohol and myristic acid. See Myristyl myristate (CTFA)
Ester of myristyl alcohol and stearic acid. See Myristyl stearate (CTFA)
N,N´-1,2-Ethanediylbisoctadecanamide. See N,N´-Ethylene bisstearamide
N,N´-1,2-Ethanediylbis-9-octadecenamide. See Ethylene bisoleamide
Ethanolaminoethyl stearamide. See Stearamidoethyl ethanolamine (CTFA)
Ethanol, 2,2´-iminobis-, N-coco alkyl, N-oxide. See N,N-Bis (2-hydroxyethyl) cocamine oxide
Ethanol, 2,2´-iminobis-, N-tallow alkyl, N-oxide. See Bis (2-hydroxyethyl) tallow amine oxide
N-Ethyl-N,N-dimethyl-3-[(1-oxoisooctadecyl) amino]-1-propanaminium ethyl sulfate. See Isostearamidopropyl ethyldimonium ethosulfate (CTFA)
N-Ethyl-N-N-dimethyl-3-[(1-oxoricinoleyl) amino]-1-propanaminium ethosulfate. See Ricinoleamidopropyl ethyldimonium ethosulfate (CTFA)
Ethylene dioleamide (CTFA). See Ethylene bisoleamide
Ethylene distearamide (CTFA). See N,N´-Ethylene bisstearamide
2-Ethylhexanoic acid, cetyl/stearyl ester. See Cetearyl octanoate (CTFA)
2-Ethylhexyl hexadecanoate. See Octyl palmitate (CTFA)
2-Ethylhexyl octadecanoate. See Octyl stearate (CTFA)
2-Ethylhexyl oxystearate. See Octyl hydroxystearate (CTFA)
2-Ethylhexyl palmitate. See Octyl palmitate (CTFA)
2-Ethylhexyl pelargonate. See Octyl pelargonate (CTFA)
2-Ethylhexyl stearate. See Octyl stearate (CTFA)
Fatty acids, lanolin, isopropyl esters. See Isopropyl lanolate (CTFA)
Fatty acids, soya. See Soy acid (CTFA)
D-Glucitol, 1,4-anhydro-, 6-(9-octadecenoate). See Sorbitan monooleate
Glycereth-7 (CTFA). See POE (7) glyceryl ether
Glycereth-26 (CTFA). See POE (26) glyceryl ether
Glycerides, lard mono-. See Lard monoglyceride

Glycerides, lard mono-, acetates. See Acetylated lard glyceride (CTFA)

Glycerol trioleate. See Glyceryl trioleate

Glyceryl monoisostearate. See Glyceryl isostearate (CTFA)

Glyceryl monomyristate. See Glyceryl myristate (CTFA)

Glycine, N-methyl-N-(1-oxododecyl)-, sodium salt. See Sodium lauroyl sarcosinate (CTFA)

Heavy mineral oil. See Mineral oil (CTFA)

1-Heptadecanol, 16-methyl-. See Isostearyl alcohol

Hexadecanamide, N-9-octadecenyl-. See Oleyl palmitamide (CTFA)

Hexadecanamide, N-[3-(dimethylamino) propyl]-. See Palmitamidopropyl dimethyl-amine (CTFA)

1-Hexadecanaminium, N,N,N-trimethyl-, bromide. See Cetyl trimethyl ammonium bromide

1-Hexadecanaminium, N,N,N-trimethylchloride. See Cetyl trimethyl ammonium chloride

n-Hexadecanoic acid. See Palmitic acid (CTFA)

Hexadecanoic acid, 2-ethylhexyl ester. See Octyl palmitate (CTFA)

Hexadecanoic acid, hexadecyl ester. See Cetyl palmitate (CTFA)

Hexadecanoic acid, methyl ester. See Methyl palmitate (CTFA)

Hexadecanoic acid, 1-methylethyl ester. See Isopropyl palmitate (CTFA)

1-Hexadecanol. See Cetyl alcohol (CTFA)

n-Hexadecyl alcohol. See Cetyl alcohol (CTFA)

n-Hexadecyl hexadecanoate. See Cetyl palmitate (CTFA)

n-Hexadecyl-2-hydroxypropanoate. See Cetyl lactate (CTFA)

n-Hexadecyl lactate. See Cetyl lactate (CTFA)

1-Hexadecyloctadecanoate. See Cetyl stearate (CTFA)

n-Hexadecyl stearate. See Cetyl stearate (CTFA)

Hexadecyl trimethyl ammonium chloride. See Cetyl trimethyl ammonium chloride

Hexanedioic acid, dibutyl ester. See Dibutyl adipate (CTFA)

Hydrolyzed collagen. See Hydrolyzed animal protein (CTFA)

12-Hydroxy-N,N-bis (2-hydroxyethyl)-9-octadecenamide. See Ricinoleic dietha-nolamide

N-(2-Hydroxyethyl) acetamide. See Acetyl monoethanolamide

N-[2-[(2-Hydroxyethyl) amino] ethyl] octadecanamide. See Stearamidoethyl etha-nolamine (CTFA)

12-Hydroxyoctadecanoic acid. See Hydroxystearic acid (CTFA)

Hydroxyoctadecanoic acid, 2-ethylhexyl ester. See Octyl hydroxystearate (CTFA)

2-Hydroxypropanoic acid, dodecyl ester. See Lauryl lactate (CTFA)

2-Hydroxypropanoic acid, monosodium salt. See Sodium lactate (CTFA)

2-Hydroxypropanoic acid, tetradecyl ester. See Myristyl lactate (CTFA)

12-Hydroxystearic acid. See Hydroxystearic acid (CTFA)

2,2´-Iminobisethanol, N-coco alkyl, N-oxide. See N,N-Bis (2-hydroxyethyl) cocamine oxide

2,2´-Iminobisethanol, N-tallow alkyl, N-oxide. See Bis (2-hydroxyethyl) tallow amine oxide

IPM. See Isopropyl myristate (CTFA)

IPP. See Isopropyl palmitate (CTFA)

Isohexadecanol. See Isocetyl alcohol (CTFA)

Isohexadecyl alcohol. See Isocetyl alcohol (CTFA)

Isooctadecanamide, N-[3-(dimethylamino) propyl]-. See Isostearamidopropyl dimethyl-amine (CTFA)

Isooctadecanamide, N-[3-(4-morpholinyl) propyl]-. See Isostearamidopropyl morpholine (CTFA)

Isooctadecanoic acid, monoester with 1,2,3-propanetriol. See Glyceryl isostearate (CTFA)

Isooctadecanoic acid, 1,2,3-propanetriol ester. See Glyceryl triisostearate

Isooctadecanol. See Isostearyl alcohol

Isopropyl n-hexadecanoate. See Isopropyl palmitate (CTFA)

Isostearamidopropyl dimethylamine, lactic acid salt. See Isostearamidopropyl dimethyl-amine lactate (CTFA)

Isostearic hydrolyzed animal protein, aminomethyl propanol salt. See AMP isostearic hydrolyzed animal protein (CTFA)

N-3-Isostearyl amidopropyl) N,N-dimethyl, N-benzyl ammonium chloride. See Isostear-amidopropalkonium chloride (CTFA)

Isostearyl triglyceride. See Glyceryl triisostearate

Laneth-20 (CTFA). See POE (20) lanolin ether

Lanolin, acetates. See Acetylated lanolin (CTFA)

Lanolin, anhydrous. See Lanolin (CTFA)

Lanolin, anhydrous USP. See Lanolin (CTFA)

Lanolin fatty acids, isopropyl esters. See Isopropyl lanolate (CTFA)

Lanolin, hydrogenated. See Hydrogenated lanolin (CTFA)

Lanolin, hydroxylated. See Hydroxlyated lanolin (CTFA)

Lard glyceride (CTFA). See Lard monoglyceride

Lauramine (CTFA). See Lauryl amine

Lauramine oxide (CTFA). See Lauryl dimethyl amine oxide

Lauric acid, hexyl ester. See Hexyl laurate (CTFA)

Lauryl amine oxide. See Lauryl dimethyl amine oxide

Light mineral oil. See Mineral oil (CTFA)

Liquid paraffin. See Mineral oil (CTFA)

Magnesium octadecanoate. See Magnesium stearate (CTFA)

Meroxapol 108. See Book VI

Meroxapol 171. See Book VI

Meroxapol 172. See Book VI

Meroxapol 174. See Book VI

Meroxapol 178. See Book VI

Meroxapol 251. See Book VI

Meroxapol 252. See Book VI

Meroxapol 254. See Book VI

Meroxapol 255. See Book VI

Meroxapol 258. See Book VI

Meroxapol 311. See Book VI

Meroxapol 312. See Book VI

Meroxapol 314. See Book VI

Methyl dodecanoate. See Methyl laurate (CTFA)

1-Methylethyl hexadecanoate. See Isopropyl palmitate (CTFA)

1-Methylethyl octadecanoate. See Isopropyl stearate (CTFA)

1-Methylethyl-9-octadecenoate. See Isopropyl oleate (CTFA)

1-Methylethyl tetradecanoate. See Isopropyl myristate (CTFA)

16-Methyl-1-heptadecanol. See Isostearyl alcohol

Methyl hexadecanoate. See Methyl palmitate (CTFA)

Methyl octadecanoate. See Methyl stearate (CTFA)

N-Methyl-N-(1-oxododecyl) glycine, sodium salt. See Sodium lauroyl sarcosinate
  (CTFA)

2-Methylpropyl octadecanoate. See Isobutyl stearate (CTFA)

Methyl-1-tallow amido ethyl-2-tallow imidazolinium-methyl sulfate. See Quaternium-
  27 (CTFA)

Monomyristin. See Glyceryl myristate (CTFA)

N-[3-(4-Morpholinyl) propyl- isooctadecanamide. See Isostearamidopropyl morpholine
  (CTFA)

Muscovite mica. See Mica (CTFA)

Myreth-3 myristate (CTFA). See POE (3) myristyl ether myristate

Myrtrimonium bromide (CTFA). See Myristyl trimethyl ammonium bromide

Nonanoic acid, 2-ethylhexyl ester. See Octyl pelargonate (CTFA)

Nonanoic acid, 1-methyl-1,2-ethanediyl ester. See Propylene glycol dipelargonate
  (CTFA)

Nonyl nonoxynol-10 phosphate (CTFA). See POE (10) dinonyl phenyl ether phosphate

Octadecanamide. See Stearic acid amide

Octadecanamide, N-[2-(diethylamino) ethyl]-. See Stearamidoethyl diethylamine
  (CTFA)

Octadecanamide, N-[3-(dimethylamino) propyl]-. See Stearamidopropyl dimethylamine
  (CTFA)

Octadecanamide, N,N´-1,2-ethanediylbis-. See N,N´-Ethylene bisstearamide

Octadecanamide, N-[2-[(2-hydroxyethyl) amino] ethyl]-. See Stearamidoethyl ethanolamine (CTFA)

1-Octadecanamine, N,N-dimethyl-, N-oxide. See Stearyl dimethyl amine oxide

1-Octadecanaminium, N-(carboxymethyl)-N,N-dimethyl-, hydroxide, inner salt. See

Stearyl betaine (CTFA)

1-Octadecanaminium, N,N,N-trimethyl-, chloride. See Stearyl trimethyl ammonium chloride

1-Octadecanaminium, N,N-dimethyl-N-octadecyl-, chloride. See Distearyl dimethyl ammonium chloride

n-Octadecanoic acid. See Stearic acid (CTFA)

Octadecanoic acid, 2,2-bis [[(1-oxooctadecyl) oxy] methyl]-1,2-propanediyl ester. See Pentaerythritol tetrastearate (CTFA)

Octadecanoic acid, butyl ester. See Butyl stearate (CTFA)

Octadecanoic acid, calcium salt. See Calcium stearate (CTFA)

Octadecanoic acid, 2-ethylhexyl ester. See Octyl stearate (CTFA)

Octadecanoic acid, 1-hexadecyl ester. See Cetyl stearate (CTFA)

Octadecanoic acid, 12-hydroxy-. See Hydroxystearic acid (CTFA)

Octadecanoic acid, hydroxy-, 2-ethylhexyl ester. See Octyl hydroxystearate (CTFA)

Octadecanoic acid, isohexadecyl ester. See Isocetyl stearate (CTFA)

Octadecanoic acid, magnesium salt. See Magnesium stearate (CTFA)

Octadecanoic acid, methyl ester. See Methyl stearate (CTFA)

Octadecanoic acid, 1-methylethyl ester. See Isopropyl stearate (CTFA)

Octadecanoic acid, 2-methylpropyl ester. See Isobutyl stearate (CTFA)

Octadecanoic acid, octadecyl ester. See Stearyl stearate (CTFA)

Octadecanoic acid, 1,2,3-propanetriyl ester. See Glyceryl tristearate

Octadecanoic acid, sodium salt. See Sodium stearate (CTFA)

Octadecanoic acid, tetradecyl ester. See Myristyl stearate (CTFA)

Octadecanoic acid, tridecyl ester. See Tridecyl stearate (CTFA)

Octadecanoic acid, zinc salt. See Zinc stearate (CTFA)

9-Octadecenamide. See Oleyl amide

9-Octadecenamide, N-[3-(dimethylamino) propyl]-. See Oleamidopropyl dimethyl-amine (CTFA)

9-Octadecenamide, N,N'-1,2-ethanediylbis-. See Ethylene bisoleamide

9-Octadecenamide, 12-hydroxy-N,N-bis (2-hydroxyethyl)-. See Ricinoleic dietha-nolamide

9-Octadecen-1-amine, N,N-dimethyl-, N-oxide. See Oleyl dimethyl amine oxide

9-Octadecen-1-aminium, N-(carboxymethyl-N,N-dimethyl-, hydroxide, inner salt. See Oleyl betaine (CTFA)

9-Octadecenoate, decyl ester. See Decyl oleate (CTFA)

9-Octadecenoic acid, 2,2-bis [[(1-oxo-9-octadecenyl) oxy] methyl]-1,3-propanediyl es-ter. See Pentaerythritol tetraoleate (CTFA)

9-Octadecenoic acid, 1-methylethyl ester. See Isopropyl oleate (CTFA)

9-Octadecenoic acid, 9-octadecenyl ester. See Oleyl monooleate

9-Octadecenoic acid, potassium salt. See Potassium oleate (CTFA)

9-Octadecenoic acid, 1,2,3-propanetriyl ester. See Glyceryl trioleate

9-Octadecen-1-ol. See Oleyl alcohol (CTFA)

9-Octadecenyl 13-docosenoate. See Oleyl erucate (CTFA)
N-9-Octadecenyl hexadecanamide. See Oleyl palmitamide (CTFA)
Octadecyl dimethylamine oxide. See Stearyl dimethyl amine oxide
N-Octadecyl-13-docosenamide. See Stearyl erucamide (CTFA)
Octadecyl trimethyl ammonium chloride. See Stearyl trimethyl ammonium chloride
Oil of wheat germ. See Wheat germ oil (CTFA)
Oils, jojoba. See Jojoba oil (CTFA)
Oils, lanolin. See Lanolin oil (CTFA)
Olealkonium chloride (CTFA). See Oleyl dimethyl benzyl ammonium chloride
Oleamide (CTFA). See Oleyl amide
Oleamine oxide (CTFA). See Oleyl dimethyl amine oxide
Oleic acid, oleyl ester. See Oleyl monooleate
Olein. See Glyceryl trioleate
Oleth-3 (CTFA). See POE (3) oleyl ether
Oleth-3 phoshate (CTFA). See POE (3) oleyl ether phosphate
Oleth-10 phosphate (CTFA). See POE (10) oleyl ether phosphate
Oleylamine oxide. See Oleyl dimethyl amine oxide
Oleyl dimethyl glycine. See Oleyl betaine (CTFA)
Oleyl oleate (CTFA). See Oleyl monooleate
Oleyl triethoxy monodiphosphate. See POE (3) oleyl ether phosphate
Palmityl alcohol. See Cetyl alcohol (CTFA)
Palmityl trimethyl ammonium chloride. See Cetyl trimethyl ammonium chloride
Paraffin oil. See Mineral oil (CTFA)
PEG-4. See Book VI
PEG-6. See Book VI
PEG-6-32. See Book VI
PEG-12. See Book VI
PEG-20. See Book VI
PEG-32. See Book VI
PEG-75. See Book VI
PEG-150. See Book VI
PEG-2 castor oil (CTFA). See POE (2) castor oil
PEG-20 castor oil (CTFA). See POE (20) castor oil
PEG-60 castor oil (CTFA). See POE (60) castor oil
PEG (60) castor oil . See POE (60) castor oil
PEG 100 castor oil. See POE (2) castor oil
PEG-200 castor oil (CTFA). See POE (200) castor oil
PEG (200) castor oil. See POE (200) castor oil
PEG 1000 castor oil. See POE (20) castor oil
PEG-10 dinonyl phenyl ether phosphate. See POE (10) dinonyl phenyl ether phosphate
PEG 500 dinonyl phenyl ether phosphate. See POE (10) dinonyl phenyl ether phosphate
PEG-45/dodecyl glycol copolymer. See Book VI

PEG-7 glyceryl cocoate (CTFA). See POE (7) glyceryl monococoate
PEG-7 glyceryl ether. See POE (7) glyceryl ether
PEG (7) glyceryl ether. See POE (7) glyceryl ether
PEG (7) glyceryl monococoate. See POE (7) glyceryl monococoate
PEG-26 glyceryl ether. See POE (26) glyceryl ether
PEG (26) glyceryl ether. See POE (26) glyceryl ether
PEG-16 hydrogenated castor oil (CTFA). See POE (16) hydrogenated castor oil
PEG (16) hydrogenated castor oil. See POE (16) hydrogenated castor oil
PEG-30 lanolin (CTFA). See POE (30) lanolin
PEG (30) lanolin. See POE (30) lanolin
PEG-60 lanolin (CTFA). See POE (60) lanolin
PEG (60) lanolin. See POE (60) lanolin
PEG-75 lanolin (CTFA). See POE (75) lanolin
PEG 4000 lanolin. See POE (75) lanolin
PEG-20 lanolin ether. See POE (20) lanolin ether
PEG 1000 lanolin ether. See POE (20) lanolin ether
PEG-3 myristyl ether myristate. See POE (3) myristyl ether myristate
PEG (3) myristyl ether myristate. See POE (3) myristyl ether myristate
PEG-3 oleyl ether. See POE (3) oleyl ether
PEG (3) oleyl ether. See POE (3) oleyl ether
PEG-3 oleyl ether phosphate. See POE (3) oleyl ether phosphate
PEG (3) oleyl ether phosphate. See POE (3) oleyl ether phosphate
PEG-10 oleyl ether phosphate. See POE (10) oleyl ether phosphate
PEG 500 oleyl ether phosphate. See POE (10) oleyl ether phosphate
PEG-10 sorbitan laurate (CTFA). See POE (10) sorbitan monolaurate
PEG 500 sorbitan monolaurate. See POE (10) sorbitan monolaurate
PEG 100 soya amine. See POE (2) soya amine
PEG-2 soyamine (CTFA). See POE (2) soya amine
PEG-5 soya sterol (CTFA). See POE (5) soya sterol
PEG (5) soya sterol. See POE (5) soya sterol
PEG-10 soya sterol (CTFA). See POE (10) soya sterol
PEG-25 soya sterol (CTFA). See POE (25) soya sterol
PEG (25) soya sterol. See POE (25) soya sterol
PEG 500 soya sterol. See POE (10) soya sterol
PEG-20 tallow amine (CTFA). See POE (20) tallow amine
PEG 1000 tallow amine. See POE (20) tallow amine
Pentaerythrityl tetrabehenate. See Pentaerythritol tetrabehenate (CTFA)
Pentaerythrityl tetraoleate. See Pentaerythritol tetraoleate (CTFA)
Pentaerythrityl tetrastearate. See Pentaerythritol tetrastearate (CTFA)
Petrolatum amber. See Petrolatum (CTFA)
Petrolatum white. See Petrolatum (CTFA)
Petroleum jelly. See Petrolatum (CTFA)

Petroleum wax, microcrystalline. See Microcrystalline wax (CTFA)

Phenylmethylpolysiloxane. See Book VI

Polyacrylamide. See Book VI

Polyglyceryl-4 oleate (CTFA). See Tetraglyceryl monooleate

Polyisobutene. See Book VI

Polyquaternium-1. See Book VI

Polyquaternium-2. See Book VI

Polyquaternium-5. See Book VI

Polyquaternium-6. See Book VI

Polyquaternium-7. See Book VI

Polyquaternium-10. See Book VI

Polyquaternium-11. See Book VI

Polysiloxane polyalkylene copolymer. See Book VI

Polysiloxane-polyether copolymer. See Book VI

Potassium 9-octadecenoate. See Potassium oleate (CTFA)

PPG-9. See Book VI

PPG-12. See Book VI

PPG-17. See Book VI

PPG-26. See Book VI

PPG-30. See Book VI

PPG-30 lanolin ether (CTFA). See POP (30) lanolin ether

PPG (30) lanolin ether. See POP (30) lanolin ether

PPG-10 methyl glucose ether (CTFA). See POP (10) methyl glucose ether

PPG (10) methyl glucose ether. See POP (10) methyl glucose ether

PPG-20 methyl glucose ether (CTFA). See POP (20) methyl glucose ether

PPG (20) methyl glucose ether. See POP (20) methyl glucose ether

PPG-1 myristyl ether acetate. See Propylene glycol myristyl ether acetate (CTFA)

PPG-2 myristyl ether propionate (CTFA). See POP (2) myristyl ether propionate

PPG (2) myristyl ether propionate. See POP (2) myristyl ether propionate

PPG-15 stearyl ether (CTFA). See POP (15) stearyl ether

PPG (15) stearyl ether. See POP (15) stearyl ether

1-Propanaminium, N-(carboxymethyl)-N,N-dimethyl-3-[(1-oxococonut) amino]-, hydroxide, inner salt. See Coco amidopropyl betaine

1-Propanaminium, N-(carboxymethyl)-N,N-dimethyl-3-[(1-oxoisooctadecyl) amino], hydroxide, inner salt. See Isostearamidopropyl betaine (CTFA)

1-Propanaminium, N-(carboxymethyl)-N,N-dimethyl-3-[(1-oxoricinoleyl) amino]-, hydroxide, inner salt. See Ricinoleamidopropyl betaine (CTFA)

1-Propanaminium, N-ethyl-N,N-dimethyl-3-[(1-oxooctadecyl) amino]-ethyl sulfate. See Isostearamidopropyl ethyldimonium ethosulfate (CTFA)

1-Propanaminium, N-ethyl-N-N-dimethyl-3-[(1-oxoricinoleyl) amino]-ethosulfate. See Ricinoleamidopropyl ethyldimonium ethosulfate (CTFA)

1,2,3-Propanetriol tridodecanoate. See Glyceryl trilaurate

1,2,3-Propanetriol trioctadecanoate. See Glyceryl tristearate
Propanoic acid, 2,2-dimethyl-, isooctadecyl ester. See Isostearyl neopentanoate (CTFA)
Propanoic acid, 2-hydroxy-, compd. with N-[3-(dimethylamino) propyl]-16-methylhep-
    tadecanamide (1:1). See Isostearamidopropyl dimethylamine lactate (CTFA)
Propanoic acid, 2-hydroxy-, compd. with N-[3-(dimethylamino) propyl] octadecan-
    amide. See Stearamidopropyl dimethylamine lactate (CTFA)
Propanoic acid, 2-hydroxy-, compd. with N-[3-(4-morpholinyl) propyl] isooctadecan-
    amide. See Isostearamidopropyl morpholine lactate (CTFA)
Propanoic acid, 2-hydroxy-, dodecyl ester. See Lauryl lactate (CTFA)
Propanoic acid, 2-hydroxy-, hexadecyl ester. See Cetyl lactate (CTFA)
Propanoic acid, 2-hydroxy-, monosodium salt. See Sodium lactate (CTFA)
Propanoic acid, 2-hydroxy-, tetradecyl ester. See Myristyl lactate (CTFA)
Propylene glycol dinonanoate. See Propylene glycol dipelargonate (CTFA)
Protein hydrolysate from collagen. See Hydrolyzed animal protein (CTFA)
Proteins, collagen, hydrolysate. See Hydrolyzed animal protein (CTFA)
PVP/dimethylaminoethyl-methacrylate copolymer. See Book VI
Quaternary ammonium compounds, bis (hydrogenated tallow alkyl) dimethyl, chlorides.
    See Dihydrogenated tallow dimethyl ammonium chloride
Quaternary ammonium compounds, (carboxymethyl) (coco alkyl) dimethyl, hydroxides,
    inner salt. See Coco-betaine (CTFA)
Quaternary ammonium compounds, (carboxymethyl) (3-cocoamidopropyl) dimethyl,
    hydroxides, inner salts. See Coco amidopropyl betaine
Quaternary ammonium compounds, dimethyl ditallow alkyl, chlorides. See Ditallow
    dimethyl ammonium chloride
Quaternium-5. See Distearyl dimethyl ammonium chloride
Quaternium-18 (CTFA). See Dihydrogenated tallow dimethyl ammonium chloride
Quaternium-48. See Ditallow dimethyl ammonium chloride
Ricinoleamide DEA (CTFA). See Ricinoleic diethanolamide
Ricinoleamidopropyl dimethyl glycine. See Ricinoleamidopropyl betaine (CTFA)
Ricinoleoyl diethanolamide. See Ricinoleic diethanolamide
SMO. See Sorbitan monooleate
Sodium isostearoyl-2-lactylate. See Sodium isostearoyl lactylate (CTFA)
Sodium laureth-4 phosphate (CTFA). See Sodium POE (4) lauryl ether phosphate
Sodium octadecanoate. See Sodium stearate (CTFA)
Sodium PEG 200 lauryl ether phosphate. See Sodium POE (4) lauryl ether phosphate
Soluble animal collagen. See Soluble collagen (CTFA)
Sorbitan mono-9-octadecenoate. See Sorbitan monooleate
Sorbitan, 9-octadecenoate (2:3). See Sorbitan sesquioleate (CTFA)
Sorbitan oleate (CTFA). See Sorbitan monooleate
Stearalkonium chloride (CTFA). See Stearyl dimethyl benzyl ammonium chloride
Stearamide (CTFA). See Stearic acid amide
Stearamine oxide (CTFA). See Stearyl dimethyl amine oxide

Stearic acid, 2-methylpropyl ester. See Isobutyl stearate (CTFA)

Stearoamphoglycinate. See Stearoamphoacetate (CTFA)

Stearoxy dimethicone. See Book VI

Steartrimonium chloride (CTFA). See Stearyl trimethyl ammonium chloride

Stearylamine oxide. See Stearyl dimethyl amine oxide

Stearyl dimethyl glycine. See Stearyl betaine (CTFA)

Sulfobutanedioic acid, mono [2-[1-(oxo-9-octadecenyl) amino] ethyl] ester-, disodium
    salt. See Disodium oleamido MEA-sulfosuccinate (CTFA)

Synthetic spermaceti wax. See Synthetic spermaceti

Tallow amine oxide (CTFA). See Tallow dimethyl amine oxide

Tetradecanamide, N-[3-(dimethylamino) propyl]-. See Myristamidopropyl dimethyl-
    amine (CTFA)

1-Tetradecanaminium, N,N,N-trimethyl-, bromide. See Myristyl trimethyl ammonium
    bromide

Tetradecanoic acid, butyl ester. See Butyl myristate

Tetradecanoic acid, hexadecyl ester. See Cetyl myristate (CTFA)

Tetradecanoic acid, 1-methylethyl ester. See Isopropyl myristate (CTFA)

Tetradecanoic acid, monoester with 1,2,3-propanetriol. See Glyceryl myristate (CTFA)

Tetradecanoic acid, tetradecyl ester. See Myristyl myristate (CTFA)

1-Tetradecanol, propanoate. See Myristyl propionate (CTFA)

Tetradecyl 2-hydroxypropanoate. See Myristyl lactate (CTFA)

Tetradecyl tetradecanoate. See Myristyl myristate (CTFA)

Tribehenin. See Glyceryl tribehenate (CTFA)

Triester of glycerin and oleic acid. See Glyceryl trioleate

Triglyceride of oleic acid. See Glyceryl trioleate

Triisostearin (CTFA). See Glyceryl triisostearate

Trilaurin (CTFA). See Glyceryl trilaurate

N,N,N,-Trimethyl-1-docosanaminium chloride. See Behenyl trimethyl ammonium chlo-
    ride

N,N,N-Trimethyl-1-hexadecanaminium bromide. See Cetyl trimethyl ammonium bro-
    mide

N,N,N-Trimethyl-1-hexadecanaminium chloride. See Cetyl trimethyl ammonium chlo-
    ride

Trimethyl hexadecyl ammonium chloride. See Cetyl trimethyl ammonium chloride

N,N,N-Trimethyl-1-octadecanaminium chloride. See Stearyl trimethyl ammonium chlo-
    ride

N,N,N-Trimethyl-1-, tetradecanaminium bromide. See Myristyl trimethyl ammonium
    bromide

Triolein (CTFA). See Glyceryl trioleate

Tristearin (CTFA). See Glyceryl tristearate

Waxes, lanolin. See Lanolin wax (CTFA)

Waxes, microcrystalline. See Microcrystalline wax (CTFA)

GENERIC CHEMICAL SYNONYMS AND CROSS REFERENCES

White mineral oil. See Mineral oil (CTFA)
Wool fat. See Lanolin (CTFA)
Wool wax. See Lanolin (CTFA)
Wool wax alcohol. See Lanolin alcohol (CTFA)
Zinc octadecanoate. See Zinc stearate (CTFA)

TRADENAME PRODUCT MANUFACTURERS

Aceto Chemical Co., Inc.
126-02 Northern Blvd.
Flushing, NY 11368

Akron Chemical Co.
255 Fountain St.
Akron, OH 44304

Akzo Chemie America
Akzo Chemie America/Armak Chemical
300 S. Riverside Plaza
Chicago, IL 60606

Akzo Chemie America/Interstab Chem.
500 Jersey Ave., POB 638
New Brunswick, NJ 08903

Akzo Chemie America/Noury Chem.
2153 Lockport-Olcott Rd.
Burt, NY 14094

Akzo Chemie America/Pioneer Chem. Div.
PO Box 237, Route 73
Maple Shade, NJ 08052

Akzo Chemie UK Ltd.
1-5 Queens Rd., Hersham
Waltham-on-Thames
Surrey KT12 5NL UK

Akzo Chemie
PO Box 186
LS Arnhem Netherlands 6800

Akzo Chemie B.V.
POB 975
3800 AZ Amersfoort, Netherlands

Akzo Chemie Italia SpA.
Via Vismara, 20020 Arese
Milano, Italy

Albright & Wilson Inc.
PO Box 26229
Richmond, VA 23260

Albright & Wilson (Australia) Ltd.
610 St. Kilda Rd., PO Box 4544
Melbourne 3001, Australia

Albright & Wilson Ltd./
Marchon, Whitehaven Works
Whitehaven
Cumbria CA28 9QQ, UK

Albright & Wilson Ltd./Phosphates Div.
PO Box 3, Hagley Rd. W., Oldbury, Warley
W. Midlands B68 0NN, UK

Marchon Espanola SA
Carretera Montblanc Km 2, 4
Alcover (Tarragona), Spain

Marchon France SA
BP 19, F-55300
St. Mihiel, France

Marchon Italiana SpA
Casella Postale No. 30, 1-46043
Castiglione delle Stiviere, Italy

Alcolac Inc.
1099 Winterson Rd.
Linthicum, MD 21090

Alcolac Ltd.
490 Dufferin St.
Valleyfield, Quebec J6S 2B4 Canada

Alkaril Chemicals Inc.
Industrial Pkwy., PO Box 1010
Winder, GA 30680

TRADENAME PRODUCT MANUFACTURERS

Alkaril Chemicals Ltd.
3265 Wolfedale Road
Mississauga, Ontario L5C 1V8, Canada

Alkaril Chemicals (UK) Ltd.—See
GAF Europe

Amerchol Corp./
Unit of CPC International Inc.
PO Box 4051, 136 Talmadge Road
Edison, NJ 08818

Amerchol Europe
Havenstraat 84, B-1800
Vilvoorde, Belgium

Amerchol
D.F. Anstead, Ltd.
Victoria House, Radford Way
Billericay, Essex
England CM-12-ODE

Amerchol
Ikeda Corporation
New Tokyo Bldg., No. 3-1
Marunouchi 3-Chome
Chiyoda-Ku
Tokyo 100, Japan

American Cyanamid Co.
American Cyanamid Co./
Industrial Chem. Div.
Berdan Ave.
Wayne, NJ 07470

American Cyanamid Co./Agri. Div.
One Cyanamid Plaza
Wayne, NJ 07470

American Cyanamid Co./
Polymer & Chem. Dept.
Berdan Ave.
Wayne, NJ 07470

Cyanamid B.V.
Postbus 1523, 3000 BM
Rotterdam, The Netherlands

Cyanamid India Ltd.
Nyloc House, 254-D2 Dr. Annie Besant Rd.
Bombay 400 025 India

Cyanamid Quimica do Brasil Ltda.
Av. Imperatriz Leopoldina, 86
Sao Paulo, Brazil

Cyanamid Taiwan Corp.
8/F Union Commercial Bldg., 137, Nanking
E. Rd., Sec. 2
Taipei, Taiwan, R.O.C.

Aquatec Quimica S/A
Rua Sampaio Viana, 425, CX. Postal 4885
04004 Sao Paulo Sp., Brazil

Atlas Refinery, Inc.
142 Lockwood St.
Newark, NJ 07105

BASF Corp.
100 Cherry Hill Rd.
Parsippany, NJ 07054

BASF Canada Ltd.
PO Box 430
Montreal, Quebec H4L 4V8, Canada

BASF (UK) Ltd.
PO Box 4, Earl Rd., Cheadle Hulme
Cheadle, Cheshire 5K8 60QG, UK

BASF Belgium S.A.
avenue Hamoir-Iaan 14
B-1180 Bruxelles/Brussel
Belgium

BASF AG
ESA/WA-H 201
D-6700 Ludwigshafen, West Germany

BASF Espanola S.A.
Apartado 762
Barcelona 8, Spain

BASF S.A., Compagnie Francaise
MC-NT, 140, Rue Jules Guesde
92303 Levallois-Perret, France

BASF India, Ltd.
Maybaker House, S.K. Ahire Marg., PO Box
19108
Bombay 400 025 India

BASF Japan Ltd.
C.P.O. Box 1757
Toyko 100-91, Japan

Berol Nobel Inc.
274 Riverside Ave.
Westport, CT 06880

Berol Nobel
Rue Gachard 88, Bte 9
B-1050 Bruxelles, Belgium

Berol Nobel AB
S-444 01
Stenungsund, Sweden

Bio-Lab Inc.
627 E. College Ave.
Decatur, GA 30030

Borg-Warner Chemicals Inc.
International Center
Parkersburg, WV 26102

Borg-Warner Chemicals Inc./Sparmar Dispersants Div.
PO Box 2643
Spartanburg, SC 29304

Borg Warner Chemicals
20 Coventry Rd., Cubbington
Leamington Spa, Warks CV32 7JW, UK

Borg-Warner Chemical Europe B.V.
Cyprusweg 2, P.O. Box 8122
1005 AC Amsterdam, Havens West
The Netherlands

(Borg-Warner)
Ube Cycon, Limited
Daito Building 3F
7-1 Kasumigaseki 3-chome
Chiyoda-ku, Tokyo 100
Japan

Canada Packers Ltd./Chemical Div.
5100 Timberlea Blvd.
Mississauga, Ontario L4W 2S5, Canada

Canada Packers Inc./Edible Oils Div.
2200 St. Clair Ave. West
Toronto, Ontario M6N 1K4, Canada

Capital City Products Co.
PO Box 569
Columbus, OH 43216

R.E. Carroll/Branch of Asarco Inc.
1570 North Olden Ave., Box 139
Trenton, NJ 08601

CasChem Inc.
40 Avenue A
Bayonne, NJ 07002

Ceca SA
22 Place des Vosges, Cedex 54 92062
LaDefense 5 Paris, France

Chemax, Inc.
POB 6067, Highway 25 South
Greenville, SC 29606

Chemform Corp.
141 S.W. 8th St.
Pompano Beach, FL 33061

Chem-Y GmbH
Kupferstrasse 1
D4240 Emmerich, West Germany

Ciba-Geigy Corp.
PO Box 18300
Greensboro, NC 27419

Ciba-Geigy Corp./Plastics & Additives Div.
Three Skyline Dr.
Hawthorne, NY 10532

Ren Plastics/Ciba-Geigy Corp.
4917 Dawn Ave.
Lansing, MI 48823

Ciba-Geigy Corp.
CH-4002
Basle, Switzerland

Ciba-Geigy PLC
30 Buckingham Gate
London SW1E 6LH, UK

Ciba-Geigy Dyestuffs & Chemicals
Ashton New Road, Clayton
Manchester M11 4AR, UK

Clough Chemical Co., Ltd.
178 St. Pierre
St.-Jean, Quebec J3B 7B5, Canada

Continental Chemical Co.
270 Clifton Blvd.
Clifton, NJ 07015

Croda Inc.
51 Madison Ave.
New York, NY 10010

Croda Surfactants Inc.—See Croda Inc.

Croda Chemicals Ltd.
Cowick Hall, Snaith Goole
North Humberside DN14 9AA, UK

Croda Food Products Ltd.
Cowick Hall, Snaith, Goole
North Humberside, England DN14 9AA

Croda Surfactants Ltd.
Cowick Hall, Snaith, Goole
North Humberside, England DN14 9AA

Croda Universal Ltd.
Cowick Hall, Snaith, Goole
North Humberside, UK DN14 9AA

Croda Chemicals Group Pty. Ltd.
PO Box 1012
Richmond, North Victoria 3121
Australia

Croda Italiana Srl
Via Grocco, N917 27036
Mortara (PV), Italay

Croda do Brazil Ltda.
Rua Croda 230 Distrito Industrial
CEP 13.053 Campinas/SP-C.P. 1098
Brazil

Croda Japan KK
Aceman Building 5F 3 7
Tokuicho 1-Chome Highashi-ku, Osaka 540
Japan

Crystal Soap Inc.
601 West 8th St.
PO Box 950
Lansdale, PA 19446

Dai-ichi Kogyo Seiyaku Co., Ltd.
Miki Building, 3-12-1, Nihombashi, Chuo-ku
Tokyo, 103 Japan

DeSoto Inc.
Chemical Specialties Div.
PO Box 23523
Harahan, LA 70183

Dow Corning/Wickhen
Bracken Rd., PO Box 384
Montgomery, NY 12549

Dura Commodities Corp./ Durachem
111 Calvert St., POB 618
Harrison, NY 10528

Durkee Industrial Foods/ SCM Corp.
900 Union Commerce Bldg.
Cleveland, OH 44115

Eastman Chemical Products, Inc./ Subsid. of Eastman Kodak Co.
PO Box 431
Kingsport, TN 37662

Eastman Chemical International A.G.
Hemel Hempstead
P.O. Box 66
Kodak House, Station Road
Herts, HP1 1JU England

Eastman Japan Ltd.
Nishi-Shinbashi Mitsui Bldg.
1-24-14 Nishi-Shinbashi
Minato-Ku, Tokyo 105
Japan

Emkay Chemical Co.
319-325 Second St.
Elizabeth, NJ 07206

Ethyl Corp./Chem. Group
451 Florida Blvd.
Baton Rouge, LA 70801

Ethyl S.A.
Industrial Chemical Division
523 Avenue Louise Boite 19
B-1050 Bruxelles
Belgium

Exxon Company U.S.A.
PO Box 2180
Houston, TX 77001

Tomah Products, Inc./
Subsid. of Exxon Chemical Americas
1012 Terra Dr., PO Box 388
Milton, WI 53563

Exxon Chemical Ltd.
Arundel Towers, Portland Terrace
Southampton SO9 2GW, UK

Fallek Chemical Co.
2125 Central Ave.
Ft. Lee, NJ 07024

Fanning Corp., The
3117 North Clybourn Ave.
Chicago, IL 60618

Finetex Inc.
418 Falmouth Ave.
Elmwood Park, NJ 07407

Represented by:
Pennine Chemical Ltd.
Kent Works, Thomas St.
Conglton
Cheshire CW12 1QZ, UK

GAF Corp./Chemical Products
1361 Alps Rd.
Wayne, NJ 07470

GAF Europe
40 Alan Turing Rd., Surrey Research Park
Guildford, Surrey, UK

Gattefosse Corp.
3 Westchester Plaza
Elmsford, NY 10523

Gattefosse Etablissements
36 Chemin de Genas
69800 Saint Priest, France

Represented by:
Alfa Chemicals Ltd.
Broadway House, 7-9 Shute End
Workingham, Berkshire RG11 1BH, UK

Givaudan Corp.
100 Delawanna Ave.
Clifton, NJ 07014

L. Givaudan & Cie SA
CH-1214 Vernier/Geneva
Switzerland

Goldschmidt Chemical Corp.
914 Randolph Rd., Box 1299
Hopewell, VA 23860

Goldschmidt AG, Th.
Goldschmidtstr. 100, Postfach 101461
D4300 Essen-1, West Germany

W.R. Grace & Co,
W.R. Grace & Co./Organic Chem. Div.
55 Hayden Ave.
Lexington, MA 02173

Davison Chemical/Div. W.R. Grace & Co.
PO Box 2117
Baltimore, MD 21203

Emerson & Cuming/W.R. Grace & Co.
869 Washington St.
Canton, MA 02021

Evans Chemetics/W.R. Grace & Co.
90 Tokeneke Rd.
Darien, CT 06820

W.R. Grace Ltd.
Northdale House, North Circular Rd.
London NW10 7UH, UK

Graden Chemical Co., Inc.
426 Bryan St.
Havertown, PA 19083

Grindsted Products Inc.
201 Industrial Pkwy., POB 26
Industrial Airport, KS 66031

Grindsted do Brazil
Cs. Postal 20.999
01000, Sao Paulo, S.P., Brazil

Grindsted France S.A.R.L.
30 Ave. Amiral Lemonnier
F 78160 Marly Le Roi, France

TRADENAME PRODUCT MANUFACTURERS

Grindsted Products A/S
Edwin Rahrs Vej 38
DK-8220 Brabrand, Denmark

Grindsted Products Ltd.
Northern Way, Bury St. Edmunds
Suffolk, IP32 6NP, UK

Grindstedvaerket GmbH
Kellerbleek 3
D-2000 Hamburg 54, Germany

A. Gross & Co./
Div. of Millmaster Onyx Corp.
652 Doremus Ave., PO Box 818
Newark, NJ 07101

Chemische Fabrick Grünau GmbH
Robert-Hansen Str. 1, Postfach 1063, D-7918
Jllertissen
Bavaria, West Germany

Guelph Soap Co. Inc.
21 Surrey St. W.
Guelph, Ontario, Canada N1H 3R3

C.P. Hall Co.
7300 South Central Ave.
Chicago, IL 60638

Harcros Chemicals Inc.
5200 Speaker Rd., PO Box 2383
Kansas City, MO 66110

Harcros Chemicals UK Ltd.
Lankro House
PO Box 1, Eccles
Manchester, M3O 0BH, UK

Hart Chemicals Ltd.
256 Victoria Rd. South
Guelph, Ontario N1H 6K8 Canada

Hart Products Corp.
173 Sussex St.
Jersey City, NJ 07302

Harwick Chemical Corp./
Polymer Application
60 South Seiberling St.
Akron, OH 44305

Hefti Ltd. Chemical Products
PO Box 1623, CH-8048
Zurich, Switzerland

Henkel Corp.
480 Alfred Ave.
Teaneck, NJ 07666

Henkel Corp./Chemical Specialties Div.
255 W. Spring Valley Ave.
Maywood, NJ 07607

Henkel Corp./Process Chemicals
350 Mt. Kemble Ave.
Morristown, NJ 07960

Henkel/Emery Group
11501 Northlake Dr., PO Box 429557
Cincinnati, OH 45249

Henkel Chem. (Canada) Ltd.
9550 Ray Lawson Blvd.
Ville d'Anjou, Quebec H1J 1L3, Canada

Henkel Chemicals Ltd.
Organic Products Division
Merit House, The Hyde
Edgeware Rd., London NW9 5AB, UK

Henkel Argentina S.A.
Avda. E. Madero Piso 14
1106 Capital Federal
Argentina

Henkel KGaA
Postfach 1100 D-4000
Dusseldorf 1, West Germany

Henkel-Nopco SA, Process Chem. Div.
185 Ave. de Fontainebleau, 77310 St. Fargeau
Pontheirry, France

Heterene Chemical Co., Inc.
POB 247, 792 21 Ave.
Paterson, NJ 07513

Hexcel Corp./
Chemical Products Div.
215 N. Centennial St.
Zeeland, MI 49464

Hexcel Corp./Rezolin Div.
20701 Nordhoff St.
Chatsworth, CA 91311

Hilton-Davis Chemical Co./
Div. of Sterling Drug, Inc.
2235 Langdon Farm Rd.
Cincinnati, OH 45237

Represented by:
D.F. Censtead, Ltd.
Victoria House, Radford Way
Billericay Essex CM12 0DE, UK

Hodag Chemical Corp.
7247 N. Central Park Ave.
Skokie, IL 60076

Hoechst-Celanese Corp.
Route 202-206 North
Somerville, NJ 08876

Hoechst AG
Verhaufkanststoffe, D-6230
Frankfurt (M) 80, West Germany

Hoechst AG/Agricultural Div.
Postfach 80 03 20
D-6230 Frankfurt (80), West Germany

Hoechst Celanese Plastics Ltd.
78-80 St. Albans Rd.
Watford, Herts
England WD2-4AP

Hoechst Japan
10-33, 4-Chome-Akasaka, Minato-ku
Tokyo, Japan

Hormel
240 Cedar Knolls Av. POB GH
Cedar Knolls, NJ 07927

Represented by:
Marlow Chem. Co. Ltd.
17 College Ave., Maidenhead
Berkshire SL6 6BX, UK

J.M. Huber Corp.
Rte. 4
Macon, GA 31298

J.M. Huber Corp./Calcium Carbonate Div.
PO Box 4005
Quincy, IL 62305-4005

J.M. Huber Corp./Carbon Black Div.
3250 W. Market St., PO Box 5360
Akron, OH 44313

J.M. Huber Corp./Chemicals Div.
PO Box 310
Havre De Grace, MD 21078

Represented by:
Zeofinn Oy
SF-49460 Hillo
Hamina, Finland

Huls America Inc.
10 Link Dr.
Rockleigh, NJ 07647

Hüls AG, Chemische Werke
Postfach 1320 D-4370
Marl 1, West Germany

Huls (UK) Ltd.
Cedars House, Farnborough Common
Orpington, Kent BR6 7TE, UK

Humko Chemical Div.—*See Witco*

Huntington Laboratories, Inc.
670 East Tipton
Huntington, IN 46750

ICI Americas Inc.
New Murphy Rd. & Concord Pike
Wilmington, DE 19897

ICI Australia Operations Pty, Ltd.
ICI House, 1 Nicholson St.
Melbourne 300, Australia

ICI Europe Ltd.
Everslaan 45
B-3078 Kortenberg
Belgium

ICI Ltd./Organics Div., Chemical Auxiliaries
Business Group
Smith's Rd., Bolton
Lancs, BL3 2QJ, UK

ICI Plant Protection Div.
Fernhurst, Haslemere
Surrey GU27 3JE, UK

TRADENAME PRODUCT MANUFACTURERS

ICI PLC/
Petrochemicals & Plastics Div., England
PO Box 90, Wilton Middlesbrough
Cleveland TS6 8JE, UK

ICI Specialty Chemicals
Everslann 45 B-3078
Kortenberg, Belgium

ICI Japan Ltd.
Osaka Green Building
1, 3-Chome Kitahama
Higashi-Ku
J-Osaka 541
Japan

Inolex Chemical Co./
Div. of American Can Co.
Jackson & Swanson Sts.
Philadelphia, PA 19148

Inolex Corp./Personal Care Div.
4221 S. Western Blvd.
Chicago, IL 60609

Represented by:
Stanley Black, Ltd.
30/31 Islington Green
London N1 8DU, UK

Jetco Chemicals, Inc.
PO Box 1898
Corsicana, TX 75110

Kao Corp.,
International Chemicals Dept.
14-10 Nihonbashi, Kayabacho 1-Chome,
Chuo-ku
Tokyo 103, Japan

Kao Corp., Edible Fat and Oil Div.
14-10 Nihonbashi, Kayabacho 1-Chome,
Chuo-ku
Tokyo 103, Japan

Kempen, Elektrochemische Fabrik
Kempen GmbH
Postfach 100 260
D-4152 Kempen 1, West Germany

Kenobel S.A.—*See Berol Nobel*

KMG Minerals
P.O. Box 729
Kings Mountain, NC 28086

Lipo Chemicals, Inc.
207 19th Ave.
Paterson, NJ 07504

Represented by:
Blagden Campbell & Campbell Ltd.
A.M.P. House, Dingwall Rd.
Croydon CR9 3QU, UK

Lonza Inc.
22-10 Route 208
Fair Lawn, NJ 07410

Lonza Ltd.
Munchensteinerstrasse 38, POB CH-4002
Basle, Switzerland

Mallinckrodt, Inc.
PO Box 5439, Mallinckrodt & Second Sts.
St. Louis, MO 63147

Van Dyk & Co., Inc./Div. of Mallinckrodt
11 Williams St.
Belleville, NJ 07109

Van Dyk & Co.
Represented by:
Stanley Black Ltd.
The Colonnade, High St.
Chestnut, Hertfordshire EN8 0DJ, UK

Marathon Morco Co.
1502 Pine Ave.
Dickinson, TX 77539

Mason Chemical Co.
5352 West Belmont Ave.
Chicago, IL 60641

McIntyre Chemical Co., Ltd.
4851 S. St. Louis Ave.
Chicago, IL 60632

Mearl Corp.
41 E. 42nd St.
New York, NY 10017

Represented by:
Cornelius Chemical Co. Ltd.
St. James House
27-43 Eastern Rd., Romford
Essex RM1 3NN, London EE3, UK

M. Michel & Co., Inc.
90 Broad St.
New York, NY 10004

Millmaster-Onyx International
PO Box 1045
Fairfield, NJ 07007

Millmaster-Onyx UK
Marlborough House, 30-2 Yarm Rd.
Stockton-on-Tees, Cleveland, UK

Miranol Chemical Co., Inc.
68 Culver Rd., PO Box 411
Dayton, NJ 08810

Represented by:
Alfa Chemicals Ltd.
Broadway House, 7 Shute End
Wokingham Berkshire RG11 1BH, UK

Mona Industries, Inc.
PO Box 425, 76 E. 24th St.
Paterson, NJ 07544

Morton International
1275 Lake Ave.
Woodstock, IL 60098

Morton International/Carstab Div.
2000 West St.
Cincinnati, OH 45215

Morton International/Morton Chem. Div.
101 Carnegie Center
Princeton, NJ 08540, NJ 08540

Morton International/Ventron Div.
150 Andover St.
Danvers, MA 01923

Williams Div./Morton Thiokol Ltd.
Greville House, Hibernia Rd.
Hounslow, Middlesex TW3 3RX, UK

Morton Thiokol—*See Morton Int'l.*

Natrochem Inc.
Exley Ave.
P.O. Box 1205
Savannah, GA 31498

Nikko Chemicals Co., Ltd.
1-4-8 Nihonbashi-Bakurocho, Chuo-ku
Tokyo 103, Japan

Nippon Nyukazai Co., Ltd.
19-9 Ginza 3-Chome, Chuo-ku
Tokyo 104, Japan

Nippon Oils & Fats Co., Ltd.
10-1, Yaraku-Cho, 1-Chome, Chiyoda-Ku
Tokyo 100, Japan

Norman, Fox & Co.
5511 S. Boyle Ave., PO Box 58727
Vernon, CA 90058

Penreco/Div. Pennzoil Co.
106 S. Main St.
Butler, PA 16001

Petrolite Corp.
Specialty Polymers Group
6910 E. 14th St.
Tulsa, OK 74112

Petrolite Corp./Bareco Div.
6910 E. 14th St., PO Drawer K
Tulsa, OK 74112

Petrolite Ltd.
137 Finchley Rd.
London NW3 6JE, UK

Polychem Dispersions
16066 Industrial Parkway
Middlefield, OH 44062

Polymerics, Inc.
2828 Second St.
Cuyahoga Falls, OH 44221

PPG-Mazer
3938 Porett Dr.
Gurnee, IL 60031

Procter and Gamble Co./ Industrial Chemicals Div.
301 E. 6th St., PO Box 599
Cincinnati, OH 45201

Pulcra S.A.
Avda De Roma, 157,
Apartado De Correos 37011
Barcelona 08011, Spain

Rewo Chemicals Inc.
107B Allen Blvd.
E. Farmingdale, NY 11735

Rewo Chemicals Ltd.
Crown House, London Rd., Morden
Surrey SM4 5DU, UK

Rewo Chemische Werke GmbH
Postfach 1160
D-6497 Steinau, West Germany

Schering Quimica Rewo
Avenida San Julian s/Nr
Poligone Industrial conost.
Granollers
Barcelona, Spain

Rhone-Poulenc Inc.
PO Box 125, Black Horse Lane
Monmouth Junction, NJ 08852

Rhone-Poulenc Industries
22 Ave. Montaigne
75360 Paris Cedex 08, France

Rhone-Poulenc Chimie de Base
Division Specialites Chimiques
Dept. Biochimie
18, avenue d'Alsace - F - 92400 Courbevoie
Paris, France

Rhone-Poulenc Geronazzo SpA
Via Milano 78, 20021 Ospiate Di Bollate
Milano, Italy

The Richardson Co.
2400 Devon Ave.
Des Plaines, IL 60018

RITA Corp.
PO Box 556
Crystal Lake, IL 60014

Represented by:
Maprecos
4, Rue des Passe-Loups
7770 Fontaine
Le Port, France

Robeco Chemicals Inc.
99 Park Ave.
New York, NY 10016

Rohm & Haas Co.
Independence Mall West
Philadelphia, PA 19105

Rohm & Haas (Australia) Pty. Ltd.
969 Burke Rd., PO Box 115 Camberwell
Victoria 3124 Australia

Rohm and Haas (UK) Ltd.
Lennig House
2 Mason's Avenue
Croydon CR9-3NB
England

Rohm & Haas Co. European Operations
Chesterfield House, Bloomsbury Way
London WC1A 2TP, UK

Rohm and Haas Asia, Ltd.
Kaisei Building
8-10 Azabudai 1-Chome
Minato-ku, Tokyo 106 Japan

Ronsheim & Moore Ltd./ Div. of Hickson & Welch Ltd.
Ings Lane, Castleford
Yorkshire WF10 2JT, UK

Sandoz Chemicals Corp.
4000 Monroe Rd.
Charlotte, NC 28205

Sandoz Inc./Colors & Chemicals Div.
Route 10
East Hanover, NJ 07936

Sandoz Chemicals Corporation
Dorva, Quebec H9R 4P5
Canada

Sandoz Ltd./Chemicals Div.
Lichtstrasse 35
CH-4002 Basel, Switzerland

Sandoz Ltd.
Calverley Lane, Horsforth
Leeds LS18 4RP, UK

Sanyo Chemical Industries, Ltd.
11-1 Ikkyo Nomoto-cho Higashiyama-ku
Kyoto 605, Japan

Scher Chemicals Inc.
Industrial West & Styertowne Rd.
Clifton, NJ 07012

Represented by:
Chesham Chemicals Ltd.
Cunningham House, Bessborough Rd.
Harrow HA1 3DU, UK

Seppic
75 Quai d'Orsay
75321 Paris Cedex 07, France

Servo Chemische Fabriek B.V.
PO Box 1, 7490 AA
Delden, Holland

Sherex Chemical Co., Inc./
Subsid. of Schering A.G.
5777 Frantz Rd., P.O. Box. 646
Dublin, OH 43017

Rewo Chemische Werke GmbH
Postfach 1160
D-6497 Steinau, West Germany

Rewo Chemical Ltd.
Gorsey Lane, Widnes
Cheshire WA8 0HE, UK

Nihon Schering KK
6-64 Nishimiyahara
2-Chome
Yodogawa Ku, Osaka 532
Japan

Stave Chemical Co., Inc.
20 Marilyn St.
Basking Ridge, NJ 07920

Stepan Co.
Edens & Winnetka Roads
Northfield, IL 60093

Stepan Co./PVO Dept.
100 West Hunter Ave.
Maywood, NJ 07607

Stepan Europe
BP127
38340 Voreppe, France

Synfina-Olefin
4 Rue Jacques de Lalaing
1040 Bruxelles, Belgium

Synthetic Products Co./
A Dart & Kraft Co.
292 Longbrook Ave.
Stratford, CT 06497

Taiwan Surfactant Corp.
8-1 Floor, No. 106, Sec. 2, Changan East Road
Taipei, Taiwan, R.O.C.

Tenneco Chemicals Inc.
Turner Place, PO Box 365
Piscataway, NJ 08854

La Tessilchimica SpA
Via Baertsch 1
24100 Bergamo, Italy

Toho Chemical Industry Co., Ltd.
1-14-9 Kakigara-cho, Nihonbashi, Chuo-ku
Tokyo 103, Japan

Unichema Chemicals
4650 S. Racine Ave.
Chicago, IL 60609

Unichema International
Postfach 1280
D-4240 Emmerich, West Germany

Union Camp Corp.
1600 Valley Rd.
Wayne, NJ 07470

Van Dyk & Co., Inc.—*See Mallinck-rodt*

Van Schuppen Chemie
Nieuweweg
Veenendaal, Netherlands

R.T. Vanderbilt & Co., Inc.
30 Winfield St.
Norwalk, CT 06855

Vikon Chemical Co., Inc.
PO Box 1520
Burlington, NC 27215

Vista Chemical Co.
15990 N. Barkers Landing Rd., PO Box 19029
Houston, TX 77224

Vista Chemical Europe
Hilton Tower
Boulevard de Waterloo, #39
81000 Brussels, Belgium

Vista Chemical Far East, Incorporated
Kasumigaseki Building, 25th Floor
Post Office Box 110
Tokyo, Japan 100

Westbrook Lanolin Co./ Argonaut Works
Laisterdyke, Bradford
Yorks BD4 8AU, UK

Witco Chemical Corp./ Organics Div. & Sonnneborn Div.
277 Park Ave.
New York, NY 10017

Witco Chemical Corp./
Inorganic Specialties Div.
520 Madison Ave.
New York, NY 10022

Witco Chemical Corp./Argus Chem. Div.
633 Court St.
Brooklyn, NY 11231

Witco Chemical Corp./Humko Chemical Div.
PO Box 125, 755 Crossover Lane
Memphis, TN 38101

Witco Chemical Corp./Pearsall Chem. Div
PO Box 437, 2519 Fairway Park Drive
Houston, TX 77001

Witco Chemical Corp./Ultra Div.
2 Wood St.
Paterson, NJ 07524

Witco Chemical Corp./U.S. Peroxygen Div.
850 Morton Ave.
Richmond, CA 94804

Witco Canada Ltd.
2 Lansing Square/Suite 1200
Willowdale, Ontario M2J4Z4

Witco Chemical Ltd. (UK)
Union Lane, Droitwich
Worcester WR9 9BB, UK

Witco Chemical S.A./Cyclo Div.
10, Rue Cambaceres
75008 Paris, France

Witco Ltd.
PO Box 10245
26112 Haifa Bay, Israel

Witco S.A.
Rue Gravetel
Saint Pierre les Elbeuf 76320, France

Witco B.V.
P.O. Box 5
Koogaan de Zaan, the Netherlands

Zohar Detergent Factory
Kibbutz Dalia, Israel 18920

www.ingramcontent.com/pod-product-compliance
Lightning Source LLC
Chambersburg PA
CBHW060752220326
41598CB00022B/2411